CW01290540

States and Nature

Under what circumstances might climate change lead to negative security outcomes? Over the past fifteen years, a rapidly growing applied field and research community on climate security has emerged. While much progress has been made, we still don't have a clear understanding of why climate change might lead to violent conflict or humanitarian emergencies in some places and not others. Busby develops a novel argument – based on the combination of state capacity, political exclusion, and international assistance – to explain why climate change leads to especially bad security outcomes in some places but not others. This argument is then demonstrated through application to case studies from sub-Saharan Africa, the Middle East, and South Asia. This book will provide an informative resource for scholars and students of international relations and environmental studies, especially those working on security, conflict, and climate change, and on the emergent practice and study of this topic; it identifies where policy and research should be headed.

JOSHUA W. BUSBY is author of numerous studies on climate change, national security, and energy policy that have been published by peer-reviewed academic outlets and various think tanks. He was a leading researcher on two multi-million-dollar grants on climate and security from the US Department of Defense under its Minerva Initiative and has consulted on the topic for international organizations and think tanks.

The Politics of Climate Change

Climate change may be the most important political challenge of our time. This new series will address large questions about the politics of climate change and feature scholarship that is problem-driven, crosses traditional subfield boundaries, and meets the discipline's highest standards for innovation, clarity and empirical rigor. Although focused on political science, it will incorporate empirical work on climate politics from across the social sciences.

Series Editor
Michael Ross, UCLA

Editorial Board Members
Thomas Bernauer (ETH, Zurich)
Xun Cao (Penn State)
Navroz K. Dubash (Center for Policy Research, New Delhi)
Kathryn Hochstetler (London School of Economics)
Robert O. Keohane (Princeton University)
Matto Mildenberger (UC Santa Barbara)
Helen Milner (Princeton University)
Megan Mullin (Duke University)
Barry Rabe (University of Michigan)
Kenneth Scheve (Yale University)
Leah Stokes (UC Santa Barbara)
Dustin Tingley (Harvard University)

States and Nature

The Effects of Climate Change on Security

JOSHUA W. BUSBY

University of Texas at Austin LBJ School of Public Affairs

CAMBRIDGE UNIVERSITY PRESS

CAMBRIDGE
UNIVERSITY PRESS

University Printing House, Cambridge CB2 8BS, United Kingdom

One Liberty Plaza, 20th Floor, New York, NY 10006, USA

477 Williamstown Road, Port Melbourne, VIC 3207, Australia

314–321, 3rd Floor, Plot 3, Splendor Forum, Jasola District Centre,
New Delhi – 110025, India

103 Penang Road, #05-06/07, Visioncrest Commercial, Singapore 238467

Cambridge University Press is part of the University of Cambridge.

It furthers the University's mission by disseminating knowledge in the pursuit of education, learning, and research at the highest international levels of excellence.

www.cambridge.org
Information on this title: www.cambridge.org/9781108832465
DOI: 10.1017/9781108957922

© Joshua W. Busby 2022

This publication is in copyright. Subject to statutory exception and to the provisions of relevant collective licensing agreements, no reproduction of any part may take place without the written permission of Cambridge University Press.

First published 2022

Printed in the United Kingdom by TJ Books Limited, Padstow Cornwall

A catalogue record for this publication is available from the British Library.

ISBN 978-1-108-83246-5 Hardback
ISBN 978-1-108-95846-2 Paperback

Cambridge University Press has no responsibility for the persistence or accuracy of URLs for external or third-party internet websites referred to in this publication and does not guarantee that any content on such websites is, or will remain, accurate or appropriate.

To my wife Bethany and son Will who have endured my anxious efforts to finish this book in the middle of a global pandemic. Looking forward to a long holiday when we can safely celebrate.

Contents

List of Figures		*page* viii
List of Maps		x
List of Tables		xii
Acknowledgments		xiii
1	Introduction	1
2	Conceptualizing Climate and Security	19
3	The Argument, Method, and Mechanisms: State Capacity, Institutional Inclusion, and International Assistance	37
4	Droughts and Famine in Somalia and Ethiopia	71
5	Drought in the Middle East: Contrasting Fortunes in Syria and Lebanon	120
6	Cyclones in South Asia: The Experiences of Myanmar, Bangladesh, and India	176
7	Beyond Internal Conflict: The Practice of Climate Security	222
8	The Next Decade of Climate Security Research	244
9	Conclusion	266
Bibliography		272
Index		323

Figures

3.1	Simple pathway to loss of life	page 61
3.2	A complex emergency pathway to loss of life	63
3.3	Pathway to civil/communal conflict	64
3.4	Extended pathway to conflict through migration	67
4.1	Somalia *gu* rainfall deviations, 1980–2016	81
4.2	Somalia *dayr* rainfall deviations, 1980–2016	81
4.3	Ethiopia *belg* rainfall deviations, 1980–2016	82
4.4	Ethiopia *kiremt* rainfall deviations, 1980–2016	83
4.5	Somalia *gu* average temperature change, 1980–2016	83
4.6	Somalia *dayr* average temperature change, 1980–2016	84
4.7	Ethiopia *belg* average temperature change, 1980–2016	85
4.8	Ethiopia *kiremt* average temperature change, 1980–2016	85
4.9	Interannual variability in rainfall, Lower Shabelle, Somalia	88
4.10	Interannual variability in rainfall, Oromia region, Ethiopia	92
4.11	March to September rainfall in central/eastern Ethiopia, 1960–2015	93
4.12	Government effectiveness in Ethiopia and Somalia, 1996–2017	96
4.13	Bureaucratic quality in Ethiopia and Somalia, 1985–2017	97
4.14	Social exclusion in Ethiopia, Somalia, and Denmark, 1980–2018	105
4.15	Foreign assistance to Somalia, 1999–2013	108
4.16	Foreign assistance to Somalia, 2007–2017	110
4.17	Official development assistance, 1980–2016	110
4.18	Mortality, funding, and aid recipients, 2010–2012	112
5.1	Syria growing season rainfall deviations, 1973–2016	135
5.2	Syria average growing season temperature change, 1973–2016	135

5.3	Lebanon growing season rainfall deviations, 1973–2016	136
5.4	Lebanon average growing season temperature change, 1973–2016	136
5.5	Wheat production in Syria, 1972–2017	141
5.6	Wheat production in Lebanon, 1972–2017	141
5.7	Government effectiveness in Lebanon and Syria, 1996–2018	145
5.8	Bureaucratic quality in Lebanon and Syria, 1984–2017	146
5.9	Tax revenue as share of GDP, Lebanon and Syria	152
5.10	Social exclusion in Lebanon, Syria, and Denmark	154
5.11	Development assistance as share of government expenditure	160
5.12	Personal remittances as share of GDP, 2002–2017	160
6.1	Bureaucratic quality in Myanmar, Bangladesh, and India	198
6.2	Social exclusion in Myanmar, Bangladesh, India, and Denmark, 1970–2018	207
6.3	Net official development assistance in Myanmar, Bangladesh, and India	208
6.4	Aid per capita in Myanmar, Bangladesh, and India, 2000–2019 (constant 2016 US$)	209
6.5	Humanitarian funding, Myanmar, Bangladesh, and India, 2000–2018	210
6.6	Real GDP per capita, Myanmar, Bangladesh, and India, 1990–2017 (constant 2011 US$)	220
6.7	GDP growth (annual %), Myanmar, Bangladesh, and India, 1990–2019	221

Maps

4.1	Rainfall deviations in East Africa, October–December 2010	*page* 86
4.2	*Dayr* three-month anomaly (October, November, December 2010) (mm)	87
4.3	*Gu* three-month anomaly (April, May, June 2011) (mm)	87
4.4	Rainfall deviations, East Africa, February–September 2015	90
4.5	*Belg* three-month anomaly (February, March, April 2015) (mm)	91
4.6	*Kiremt* three-month anomaly (June, July, August 2015) (mm)	91
4.7	World Bank and African Development Bank projects 2009–2010 and climate security vulnerability	107
5.1	Agricultural stress in Syria, 2008	138
5.2	Agricultural stress in Lebanon, 2008	139
6.1	Storm tracks of 2008 Cyclone Nargis and 2010 Myanmar population density	189
6.2	Storm tracks of 1970 Bhola Cyclone and districts of East Pakistan	191
6.3	Storm tracks of 1991 Bangladeshi cyclone and Bangladesh population density	192
6.4	Storm tracks of 2007 Cyclone Sidr and 2000 low elevation coastal zone population density	193

6.5 Storm tracks of 1999 India cyclone and 2000 population density 194
6.6 Storm tracks of 2013 Cyclone Phailin and 2015 population density 195
6.7 Storms tracks of Cyclone Fani and projected 2020 population density 196

Tables

1.1	Somalia and Ethiopia	page 11
1.2	Syria and Lebanon	12
1.3	Myanmar, Bangladesh, and India	15
3.1	Weak capacity states – configurations 1–4	68
3.2	Strong capacity states – configurations 5–8	69
4.1	Comparing Somalia and Ethiopia	77
4.2	Governance indicators for Ethiopia and Somalia	96
5.1	Comparing Syria and Lebanon	129
5.2	Governance indicators for Syria and Lebanon	145
6.1	Comparing Myanmar, Bangladesh, and India	179
6.2	Notable cyclones in Myanmar, Bangladesh, and India	186
6.3	Comparison of major cyclones in Bangladesh	216

Acknowledgments

The research that forms the basis of this book was supported by the Skoll Foundation and the University of Texas. Draft versions were presented at the Resource Competition, Environmental Security, and Stability (RECESS) Group at the US Department of Defense, the Austin Forum on Diplomacy and Statecraft, the 2020 Southern Political Science Association Annual Conference, the Center for International Security and Cooperation at Stanford University, the London School of Economics, the Peace Research Institute of Oslo, and the Norwegian Institute of International Affairs.

I would like to thank Konstantin Ash, Halvard Buhaug, Joel Campbell, Tom Deligiannis, Lara Eid, Robert Falkner, Cullen Hendrix, Colin Kahl, Florian Krampe, Kevin Mazur, Emily Meierding, Jason Miklian, Malin Mobjörk, Jonathan Monten, Clionadh Raleigh, Sam Rowan, Ken Schultz, Peter Schwartzstein, Ole Jacob Sending, Ken Shadlen, Nina von Uexkull, Christian Webersik, and Emily Whalen for their comments and direction on the project. Special thanks to Shelby Bohannon for helping with child care during the pandemic.

1

Introduction

The atmospheric concentration of carbon dioxide, the main greenhouse gas, has reached levels not seen for 800,000 years.[1] Over the next 100 years, concentrations could reach levels not observed on earth for 200 million years.[2] While the future is uncertain, we appear to be on track for temperature increases of at least 2 degrees Celsius (3.6°F) above preindustrial levels. Eighteen of the nineteen warmest years on record have occurred since 2000.[3] While there are historical analogues from ancient history of the planet, we are entering into uncharted territory for our species in the modern era.

We have already begun to observe striking changes in temperature and rainfall patterns around the world that have profound implications for human societies. These general trends have been punctuated by extreme weather events, such as hurricanes Harvey and Maria that buffeted the United States in 2017, bushfires like those that burned extensive parts of Australia in 2020, extreme temperatures spikes like those regularly observed in the Middle East and South Asia, and prolonged and severe periods of drought that led cities like Chennai, Cape Town, and São Paulo to nearly run out of water.

What does climate change mean, then, for security? Answers to this question hinge on what we think constitute security threats. Are we simply worried about the risks of violent conflict or are we interested in broader threats to human well-being? We also need to have an appreciation of whose security we are talking about. Are we talking about violent threats to peace and security between or within countries or the broader well-being of individuals or communities within particular countries?

---

[1] CBS News 2014. [2] Foster, Royer, and Lunt 2017. [3] Samenow 2019.

This book takes a somewhat expansive understanding of security threats: They include but are not limited to violent conflict. Humanitarian emergencies that pose a risk to large-scale loss of life are also included. I seek to explain why climate change leads to negative security outcomes in some places and not others.

In brief, climate change is most likely to trigger conflict and humanitarian emergencies in countries that have: (1) *weak state capacity,* (2) *exclusive political institutions,* and (3) *foreign assistance that is blocked or delivered unevenly.* Where state capacity reflects a government's ability to prepare for climate shocks and help people in times of need, inclusive political institutions capture its willingness to help all or merely some of its citizens. International assistance can partially compensate for weak state capacity. Countries that have stronger state capacity, more political inclusion, and which can tap international assistance to help them are less likely to experience violence or humanitarian emergencies.

Not every climate hazard leads to equally bad outcomes. In other words, not every storm is a natural disaster. While some swift-onset climate hazards such as cyclones pose direct and immediate short-term threats because they occur quickly, other slow-onset hazards such as droughts develop over longer periods of time. Whether or not these events lead to large-scale loss of life is thus even more dependent on the preparedness of the country in question and the nature of the domestic response.

Populations buffeted by natural hazards are not passive. They respond to such hazards by drawing down savings, making use of emergency reserves, seeking family help and community resources, and even moving from inhospitable areas to ones that may offer better chances of survival. They also make claims upon institutions in a position to help them, including national, provincial, and local governments; aid agencies; faith groups; charities; and companies. If their pleas for assistance are not honored, this can lead to demands for redress through peaceful protest that can escalate into looting or more violent confrontation if initial requests are rebuffed. This timetable of escalation can be truncated if hazards arrive in the midst of ongoing conflict, where combatants are already organized, though these hazards may also affect their capacity to continue the fight. Thus, the range of security outcomes of concern range from large-scale loss of life due to exposure, famine, and thirst through to escalating violent conflicts from protests to small scale civil conflict to civil war.

The academic community has largely focused on a narrow set of questions related to whether climate change will lead to a variety of kinds of civil conflict. The reasons for this circumscribed vision are largely methodological. Social scientists ask questions they can answer with the available evidence. Many are uncomfortable talking about the future as it quickly lends itself to prognostication rather than evidence-driven analysis.[4]

What this means is that social scientists generally look to the past to try to understand the implications for the future. Many use past evidence of droughts, temperature change, rainfall volatility, and other physical phenomena to assess whether they historically contributed to negative security outcomes such as armed conflict. To some extent, this book is no exception. Rather than trying to imagine what a world of more climate extremes – some of which have no modern equivalent – may mean for security, I too look to how states have responded in the recent past to climate-related extreme weather. Future extremes may be beyond what we have observed, but this analysis of the recent past gives us a historically and theoretically grounded account of how countries respond to climate threats.

This book, however, does something a little different from much of the existing literature, which has narrowly defined security in terms of violent conflict. A handful of case studies have sought to surface the connections between climate change and specific conflicts, mostly notably the civil conflict in Darfur, Sudan,[5] and the Syrian civil war,[6] both of which have been offered as examples of conflicts where climate processes have been an important conflict accelerant or multiplier. While some of this work has been careful in its claims, the cases tend to suffer from the same problems of the earlier environmental security literature of the 1990s: single case studies of climate–conflict links in the absence of paired cases to identify the scope conditions for when climate processes lead to conflict – and when they do not. Moreover, as noted above, most of the climate security work has focused on different kinds of conflict as the primary security outcome of interest. I see humanitarian emergencies and the risk of large-scale loss of life as security outcomes of concern in their own right.

This book seeks to address key deficiencies in that earlier scholarship on environmental and climate security. By exploring paired cases of comparable physical exposure that have different social and political effects, I seek to identify the conditions under which climate hazards

---

[4] Gleditsch 1998, 394.   [5] Faris 2007, 2009; Ki-moon 2007.
[6] Werrell and Femia 2013; Kelley et al. 2015; Gleick 2014; Fountain 2015.

lead to negative security outcomes.[7] In the process, I try to provide more insight into the causal mechanisms that have been neglected or unsatisfyingly addressed in the newer, largely quantitative literature on climate security. I am also more inclusive in the scope of security outcomes of interest, moving beyond studies of violent conflict to include humanitarian emergencies.

Chapters 4, 5, and 6 take paired cases in different regions – Africa, the Middle East, and Asia – to demonstrate the promise and challenges of case-based analysis in this space. Chapter 4 explores why, in the 2010s, Somalia had a famine after drought but its neighbor Ethiopia did not. The chapter also takes advantage of within-case variation in Ethiopia to show why Ethiopia had a famine in the 1980s but not in the 2010s. Chapter 5 examines why, in the 2000s, Syria had a civil war in the wake of a serious drought but neighboring Lebanon did not. Chapter 6 investigates why a 2008 cyclone led to the deaths of 140,000 people in Myanmar while regional neighbors Bangladesh and India have experienced relatively few deaths following exposure to severe cyclones since 2000. This chapter also assesses how Bangladesh and India have reduced cyclone mortality over time.

To better understand what I do differently in this book, we need to understand something of the intellectual history of the field. Since wars between states are rare and became rarer in the latter half of the twentieth century, most of the academic discussion has focused on civil wars and other lower-level violence and social conflict within states, though there has been a fair amount of work dedicated to competition and cooperation over transnational river basins. An older literature from the 1990s began to assess the relationship between environmental change and security largely through case studies. While this was important scholarship, it was difficult to generalize from the limited set of cases to the wider world. Moreover, much of the foundational work in this space consisted of single case studies that traced the path from some environmental harm to violent conflict. As Marc Levy noted, this approach had its limits: "The more logical research strategy under the circumstances would be to compare societies facing similar environmental problems but exhibiting different levels of violent conflict. That would permit some precision in identifying the conditions under which environmental degradation generates violent conflict and when it does not."[8]

---

[7] This echoes a call in Koubi (2019) for more micro-level case studies.   [8] Levy 1995, 57.

In the 2000s, with the arrival of better, more fine-grained data on climate hazards and violence, a second generation of scholarship on environmental security emerged, much of it quantitative, to test the statistical relationships between climate factors and conflict outcomes. This newer literature makes more generalizable claims across many cases. However, it has also struggled to pin down precise causal mechanisms between diverse climate phenomena (too little rain, too much rain, unpredictable rain, high temperatures) and different forms of conflict. Moreover, the emphasis on conflict has come at a cost to other legitimate security concerns that worry policymakers, such as humanitarian emergencies.

While this book is intended to contribute to these two generations of scholarship on environmental security, I also, in Chapter 7, back out and ask the more general question why countries such as the United States and those in Europe should care about the security consequences of climate change in other countries. Chapter 7 provides a theoretically informed account and discusses how the insights of this book may inform foreign policy and security practice going forward. Chapter 8 then reviews where I think the academic field of climate and security should be headed. In Chapter 9, I close with some final observations.

In what follows, I summarize the main arguments and the contributions of each chapter.

In Chapters 2 and 3, I develop a more theoretically driven account of what constitute threats to security and under what conditions climate change will lead to negative security outcomes, including but not limited to conflict.

In brief, the argument is that, historically, national security threats meant armed attacks by foreign countries. Of course, that failed to encompass security threats emanating from within countries like rebel movements. With problems like terrorism, we can appreciate that non-state actors, not just state actors, can also cause security problems. What about harms that lack human agency like pandemics or climate change? Is there anything that makes them "security" problems and not simply very important problems?

The familiar (and still contested) claim is that some climate-related physical processes like droughts can, under certain circumstances, lead to conflict. Here, climate effects become causal factors in the breakdown of internal security within countries (with potential ramifications for other countries that might be affected by that situation). A more expansive way of thinking about climate change as a security threat is in terms of the gravity of harms and how this compares to an armed external attack. Are the level of damages posed by climate change equivalent to what could be imposed by an armed attack by a foreign adversary? Pandemics and climate change can rise to the level of security problems if they become disasters, that is, if they cause such grave harms (in terms of loss of life and damage to the economy) that. If an adversary were to threaten such damages, a state would be willing to wage war to stop them (though use of force would hardly be effective to combat climate change or pandemics).

While the extent of the damage is one reason to consider humanitarian emergencies as security challenges, they also frequently require military mobilization to deliver emergency supplies, conduct search and rescue, and restore order. The diversion of military assets for humanitarian relief thus imposes opportunity costs and means those assets, at least temporarily, cannot be used for other purposes. For this and other reasons, we can consider climate change a security challenge, even in the absence of escalation to violent conflict. I expand on this logic in Chapter 2.

Under what conditions might such security consequences occur? I answer this question in Chapter 3. The first dimension that matters is *state capacity*. While the field has a variety of definitions of state capacity, I use the term here to reflect bureaucratic and administrative capacity,[9] or what Fukuyama describes as the ability to execute policy.[10] States need capacity to deliver services.

At the most basic level, a state must have sufficient capacity to protect itself from armed attacks, both those that are external and those that come from internal threats. A state too weak to protect itself from invasion will cease to exist, and a state without sufficient capacity will be subject to constant coup attempts from within. While states may retain coercive power to repress violence, they may not possess bureaucratic or administrative capacity to provide services to their populace.

In the face of climate hazards, states need to have some infrastructural power to respond, as they are expected to provide for the needs of their citizens. Even if leaders possess limited preoccupation with the fate of their citizens, climate hazards may ultimately lead to more far-reaching consequences that threaten regime survival by making it impossible for the state to retain the loyalty of its citizens or to repress violence. Those with weak institutions lack the organizational capacity to respond to

---

[9] Hendrix 2010.   [10] Fukuyama 2013, 349.

climate-related hazards. While this is likely correlated with wealth, some polities may have capacity for functions related to emergency preparedness and response, despite being poor. Throughout this book, when I refer to state capacity, I am referencing capacity to deliver services and carry out policy rather than a state's repressive or coercive capacity, though these are related as a monopoly of force may be necessary to extend services over a wider swath of territory.

While state capacity determines whether states *can* respond to climate hazards, other factors shape whether they *will*. Following work by Colin Kahl, Daron Acemoglu and James Robinson, and Douglass North among others, the second dimension of relevance is *political inclusion*.[11] According to Acemoglu and Robinson, inclusive institutions are characterized by "power broadly distributed in society and [institutions that] constrain its arbitrary exercise."[12] If we think of state stability as being based on elite pacts between groups to share power and resolve differences through law and politics rather than by force, then political inclusion implies incorporation of all politically and militarily relevant subgroups in government decision-making and "fair" apportionment of resources and programs.[13]

As Cullen Hendrix notes, institutional inclusivity in practice includes federalism, efforts to devolve power regionally, an independent judiciary, and checks on executive power such as votes that require supermajorities and policies that give minorities voice opportunities.[14] These constraints on leaders and institutional practices to resolve conflicts are two reasons why inclusive institutions are less likely to suffer from top-down violent oppression or the emergence of conflicts that bubble up from dissatisfied groups.[15]

Polities with exclusive political institutions are likely to respond to climate-related hazards with measures largely limited to the political base of the regime such as coethnics or the leader's home region.[16] This leaves less favored regions with meager to no access to resources that would protect them from harm or enable them to respond to climate shocks, such as emergency provisions, food aid, water, shelter, medical attention, transport, and cash. On some level, this is consistent with

---

[11] Kahl 2006; Acemoglu and Robinson 2013; North, Wallis, and Weingast 2009. This approach also has some affinities with selectorate theory, where exclusive regimes have small selectorates that provide few public goods; Bueno de Mesquita et al. 2003.
[12] Acemoglu and Robinson 2013, 82.
[13] For similar thinking, see North, Wallis, and Weingast 2009. [14] Hendrix 2016, 3.
[15] Ibid. [16] Kahl 2006; Acemoglu and Robinson 2013.

Amartya Sen's observation, based on the Indian experience, that there has never been a famine in a functioning democracy.[17] However, as Chapter 3 details, while democratic governments are generally inclusive by design, other regimes may be inclusive without being democracies. All else being equal, my expectation is that states with more inclusive governments will be more willing to prepare for threats and to come to their citizens' aid in the wake of exposure to climate hazards.

While state capacity can limit the ability of governments to respond in the midst of a crisis, a third dimension – *international assistance* – can partially compensate for state weakness, both in the lead up to and in the aftermath of hazard exposure. While much of the literature on environment and security has sought to grapple with the role played by domestic institutions, the international connections have been understudied.[18] State responses are hugely important, but international assistance can help compensate for capacity constraints, by building capacity over time and by responding in emergency settings with food aid, humanitarian response, conflict mediation, peacekeepers, or other measures to address human suffering, restore order, or quell conflicts.[19] In studies of disaster risk reduction, there are fears that international aid might encourage moral hazard, where states rely on international aid rather than using their own resources to prepare for climate hazards. In work on sub-Saharan Africa, however, Bussell and her collaborators found these fears were overblown.[20]

In the study of aid and civil wars, moral hazard has been found to be more of a concern: Aid may extend civil wars as warring parties fear the end of conflict will see these resources dry up. Moreover, in that literature, aid flows can extend civil wars by providing lootable assets to one side of a conflict. Aid flows may pose a threat to rebel groups by providing the state with assets to strengthen its power. As a consequence, aid projects themselves may become targets if perceived as a threat to rebel authority.[21]

Here, I make a different argument. I emphasize, first, whether a country receives or permits external aid in the midst of an emergency. This matters more in countries with weak capacity, as they may lack the means to prepare for and independently respond to climate hazards. A second aspect is whether the external resources are distributed in a manner that

---

[17] Sen 1981. [18] Exceptions are work by Baechler 1999b, 1998.
[19] On the role of aid in conflict mitigation, see de Ree and Nillesen 2009; Findley 2018.
[20] Bussell 2014. [21] Findley 2018.

is broadly based on necessity or is captured by exclusive political institutions or sectarian forces that might be favored by the aid provider based on ideology, diaspora affinity groups, religion, formal alliances, etc.

To the extent that some groups receive aid in a crisis and not others, we should expect to see groups with limited to no access to aid in such circumstances suffer, potentially die, and nurture that memory of deprivation as a source of grievance. By aid or assistance, I do not simply mean overseas development assistance (ODA). Emergency relief or humanitarian aid is often not counted by donors as ODA. Moreover, though aid can come in the form of money, other forms of assistance are important including in-kind contributions of material support and even weapons. If those external resources are provided or distributed in a one-sided manner, this can mean large-scale loss of life in the event of exposure to a climate hazard, with favored groups receiving assistance while others suffer. Alternatively, one-sided external support can also provide one group of actors with the resources to fight or serve as a source of grievance for others largely excluded from access to those resources. Aid that is allowed in and distributed in a broad-based manner will likely diminish the risk of humanitarian suffering and follow-on security consequences.

We should expect the worst security consequences to occur in settings with weak state capacity, exclusive political institutions, and no or one-sided provision of international assistance. We should expect the best security outcomes (that is limited death from exposure to climate hazards and limited conflict) to occur in polities with high capacity, inclusive political institutions, and broad-based provision of aid. In between, there are a number of other possibilities, which are elaborated in Chapter 3.

In Chapter 4, I examine the paired cases of Ethiopia and Somalia, comparing them to each other and comparing Ethiopia to itself over time. In 2011, Somalia suffered a devastating famine, in which more than 250,000 people were estimated to have died above normal rates of expected mortality. By contrast, Ethiopia, which faced a similar exposure in 2015, did not face a comparable famine. This was quite a different outcome from the mid-1980s when northern Ethiopia faced a drought that took the lives of 400,000 to 600,000 people. The suffering in Ethiopia became a cause célèbre with the 1985 Live Aid concert, Band Aid, and led to an outpouring of public demands for assistance. However, by 2015, the country was able to avoid large-scale loss of life when again faced with a severe drought.

What set these two countries apart? Between 1992 and 2016, Somalia languished without a functioning government. Meanwhile, Ethiopia had developed considerable government capacity since the 1980s and in the decades after the Ethiopian People's Revolutionary Democratic Front seized power from the socialist Derg government in 1991.

However, these differences in state capacity do not, on their own, fully explain the differences between the two countries. Somalia has faced numerous severe droughts since the 1980s; only two – 1992 and 2011 – resulted in famine. What explains the difference in outcomes? In 1992 and 2011, international assistance was not permitted into the country until late stages of the famine and after large numbers of people had died. In the other two episodes, in the mid-1990s and the mid-2000s, aid was allowed into the country and local groups were in a position to assist in modest self-governance, even in the absence of a functioning state. Chapter 4 focuses on the drought of the 2011–12 period in Somalia.

Despite arguably having an authoritarian government, Ethiopia has increased state capacity and been amenable to foreign aid and humanitarian assistance, which has helped to alleviate famine risk. In 2015, Ethiopia faced another extreme drought, and the government was up to the challenge, avoiding large scale-loss of life. However, on the heels of this famine, a protest among the Oromo, a large but marginalized ethnic group, suggested that Ethiopia's government had strong state capacity but lacked a fully inclusive government, potentially putting at risk the long-term stability of the country. Subsequent events since the Oromo came to power have unfortunately made the situation worse. Chapter 4 explores the 2015 drought in Ethiopia and compares the case both to Somalia in 2011–12 and to Ethiopia itself in the mid-1980s (see Table 1.1 for a summary comparison).

Chapter 5 explores a different set of paired cases, examining the multi-year drought in Syria that began in 2006 and a similar drought that Lebanon faced around the same time. The drought in Syria has been implicated in the emergence of protests and the subsequent civil war. I argue that Syria possessed some state capacity but mismanaged water policy in the lead up to the drought, which made its effects worse. Moreover, because the government is characterized by exclusive political institutions that rewarded groups most loyal to the Assad family (namely, the minority Alawite group), other less favored groups suffered considerably in the wake of the drought. The drought was especially severe in the northeast of the country in the provinces of Hassakeh, Raqqa, and Deir al-Zor, but also affected other areas. The government failed to adequately respond to those affected.

TABLE 1.1 *Somalia and Ethiopia*

Country	Hazard events	Capacity	Institutions	International assistance	Outcomes
Somalia	Droughts 1992, 2000, 2004, 2005, 2008, 2011–12	Weak	Exclusive	Limited access 1992, 2011 Broad-based aid delivery	Two famines 1992, 2011–12 No famine 2000, 2004, 2005, 2008
Ethiopia	Droughts 1984, 2000, 2002, 2010, 2015	Weak 1984 → Improved state capacity 2000, 2002, 2010, 2015	Exclusive → Increasing Inclusivity but still somewhat exclusive	Limited access 1983–5 Broad-based aid delivery	Famine 1984 No famine 2000, 2002, 2010, 2015

What followed was accelerated migration to the cities, which had already been occurring before the drought. The pressure from migrants, in turn, put a strain on social services and created a cohort of people who were mobilizable when protests broke out in 2011. The government repressed that protest activity, which ultimately escalated to a civil war. Thus, the route to conflict in Syria was indirect, and one can ask the legitimate question whether protest activity would have escalated to civil war in any case. Here, the evidentiary chain requires us to connect drought and water mismanagement to migratory pressures to social conflict and violence.

By contrast, neighboring Lebanon, which also faced some of the same pressures from drought and is perhaps the most suitable comparison case, did not experience conflict after its own multiyear drought. Although drought led to a drop in wheat production and increased food prices, the Lebanese government shielded consumers from major price increases and blunted some of the impacts on producers. Lebanon had a slightly more capable government and its albeit creaky power-sharing agreement created avenues for different groups to have representation in government. Its more democratic system also provides voice opportunities for the people. Although Lebanon has faced public protests and periodic violence linked to the Hezbollah militia, it has not (yet) experienced another large-scale civil war (see Table 1.2 for a summary comparison).

Chapter 6 provides a different set of paired cases, taking several states in South Asia that face high cyclone risks – Myanmar, Bangladesh, and India. Cyclone risk coupled with low-elevation coastal zones radiates from Odisha and West Bengal in India through Bangladesh to Rakhine State in Myanmar. In May 2008, a major cyclone devastated the Ayeyarwady Delta in Myanmar and left 700,000 homeless. Three-quarters of the delta's livestock was killed, half of the fishing fleet

TABLE 1.2 *Syria and Lebanon*

Country	Hazard events	Capacity	Institutions	International assistance	Outcomes
Syria	Droughts 2006–10	Weak capacity	Exclusive	Limited access	Civil war 2011–
Lebanon	Droughts 2007–10	Intermediate capacity	Somewhat inclusive	Broad-based aid delivery	Protest activity but no civil war

sank, and a million acres of rice paddies were inundated with saltwater.[22] With a referendum pending on a new constitution, Myanmar's authoritarian regime was reluctant to allow much humanitarian aid into the country and rebuffed efforts by outside parties such as the United States to offer assistance. The US Navy, having made fifteen unsuccessful attempts to receive authorization to deliver aid, ultimately ordered its ships to depart in early June.[23] In the end, some 140,000 people died.[24] Myanmar had limited government capacity, exclusive political institutions, and resisted outside assistance.

Bangladesh provides a different portrait. Like Myanmar, it has suffered large-scale loss of life in the wake of cyclones, but the most deadly occurred in 1970, before it was an independent country and at a time when the international humanitarian aid architecture was limited in its geographic reach. As an independent country, Bangladesh has continued to struggle with natural disaster preparedness. In 1991, a Category 5 cyclone killed nearly 140,000 people, a reflection, in part, of limited government capacity. More would have died had not the American military assisted with large-scale mobilization of humanitarian aid.[25] By the 2000s, although Bangladesh continued to suffer from governance challenges, its disaster preparedness and response capacity had improved considerably, helped in no small part by its receptiveness to foreign assistance. Such assistance bolstered both the capacity of the state and civil society which provided broad-based support to vulnerable populations. In November 2007, Cyclone Sidr, a Category 5 cyclone, hit coastal Bangladesh, especially affecting Satkhira, Khulna, Barguna, Patuakhali, and Bagerhat among other districts in the southwest. The cyclone's effects were felt as far inland as Dhaka. More than 4,200 deaths were associated with the storm and as many as eight million people were affected. To some extent, this represented progress in adaptive capacity and resilience compared to earlier decades when large numbers of people died in the wake of comparable cyclones. Here, foreign aid and international disaster response helped support the government to build capacity over time and respond effectively to storms as they happened.

India provides another portrait in improvement in adaptive capacity over time, though arguably the country has been less reliant on donor assistance for disaster preparedness and response. In 1999, a devastating Category 5 cyclone smashed into Odisha state in eastern India on the Bay of Bengal.

---

[22] *The New York Times* 2009. [23] *The New York Times* 2008.
[24] Zarni and Taneja 2015. [25] Berke 1991.

Around 10,000 people were killed. In 2013, another Category 5 hurricane struck the same state. Only 50 people died, as the country evacuated more than 500,000 people from low-lying areas, the largest such evacuation in more than twenty-three years.[26] While donors such as the United States Agency for International Development (USAID) worked with India on early warning systems and disaster preparedness, India did not rely much on disaster aid for preparedness or recovery (see Table 1.3 for a summary comparison).[27] By 2019, when another powerful storm smashed into Odisha state, there were limited casualties, as in 2013, despite population growth.

While the previous chapters focus on internal security challenges within states, Chapter 7 steps back from cases like these and assesses the conditions under which climate security consequences in one part of the world constitute potential security challenges for other countries, particularly advanced industrialized countries such as the United States. Advanced industrialized countries might have the resources to support climate resilience, disaster risk reduction, early warning systems, and other measures to reduce the risks of large-scale negative security consequences. These countries will also likely be called upon in the event of large-scale humanitarian emergencies and, potentially, where conflicts emerge.

Since the mid-2000s, there has been a vigorous discussion in the policy community about the security consequences of climate change. In the United States, policymakers have examined this question largely through the lens of direct harms to that country and indirect consequences for its overseas interests. The emphasis is in a sense parochial, but the consequences that decision-makers worry about are rather broad, encompassing everything from direct threats to military bases and critical infrastructure, effects on the operational environment for war-fighting, increased demands for humanitarian intervention, and extend to concerns about state failure and the emergence of conflict in countries strategically important to the United States.[28] Policymakers from other countries such as the UK and Germany and those in the European Union have issued similar reports of their own,[29] and the issue has been elevated to become an item of conversation at the UN Security Council in a number of different forums.[30]

---

[26] *Press Trust of India* 2013.   [27] Konyndyk 2013.
[28] Busby 2008, 2007, 2016a; Department of Defense 2014.
[29] Morisetti 2014; WBGU 2007; Solana 2008.   [30] UN Security Council 2007.

TABLE 1.3 *Myanmar, Bangladesh, and India*

Country	Hazard events	Capacity	Institutions	International assistance	Outcomes
Myanmar	Cat. 4 cyclone 2008	Weak capacity	Exclusive	Limited aid delivery	Significant deaths
Bangladesh	Cat. 5 cyclones 1970 1991 2007	Weak capacity→ Improved capacity	Exclusive→ Becoming more inclusive	Broad-based aid delivery by 2000s	Mass deaths 1970, 1991 Limited deaths 2007
India	Cat. 5 cyclones 1999 2013 2019	Intermediate and increasing capacity	Somewhat inclusive	Limited reliance on aid	Significant deaths 1999 Limited deaths 2013, 2019

While countries will generally be most concerned about the direct effects of climate change on their homelands, many have international interests that could be affected by climate change, starting with effects on proximate neighbors and international waters near their own shores but extending to a range of other concerns. These include the effects of climate change on countries that are sources of important raw materials or are links in supply chains; countries that control access to vital sea lanes; or those where overseas military bases and diplomatic outposts are stationed. They extend to allies and potentially to those countries where there may be active military operations. In Chapter 7, I provide a more complete discussion of these concerns and how these have materialized in policy development, with particular emphasis on how the US government has understood climate security concerns.

The critical challenge external actors face is where states of concern have weak state capacity and/or exclusive political institutions. How do they build capacity and encourage more inclusion? In the wake of hazard exposure, regimes that otherwise might not be eligible or desirable locations for development or humanitarian assistance may require infusions of aid to forestall large-scale loss of life. Whether such assistance can facilitate more capable and just governance, or provides a lifeline for exclusionary regimes to survive, is a persistent dilemma for external actors.

In Chapter 8, I return to the questions raised in Chapter 2 about the current state of the literature on climate and security, where it should be heading, and how productive links between research and policy communities could be structured. I emphasize the need for the field to become more comfortable and develop techniques for talking about the future, when climate effects may be outside recent known experience of natural hazards. I also suggest that the field needs to bring climate mitigation – that is, reduction of greenhouse gases – more centrally into climate security research. Uncontrolled climate change will make the security consequences unmanageable. As climate change becomes elevated to high politics, decarbonization and a transition to cleaner energy sources will become more central strategic challenges in international relations between the great powers. Moreover, as the discussion in Chapter 7 on maladaptation notes, responses to climate change might be as – if not more – important in triggering conflict as the physical effects of climate change itself. To the extent that the academic field wishes to be useful to policy, I also discuss how academics need to be able to say more about what their findings mean for particular places that policymakers care about. Finally, I suggest that climate change poses wider theoretical

1 Introduction

challenges for the discipline of political science and should be elevated alongside anarchy as a structural parameter all countries must face.

In the short concluding Chapter 9, I make some final observations about the aims and ambitions for the book and address its relative state-centric focus.

As noted above, much of the climate–security field has focused on the narrow question of whether proxies for climate change will lead to conflict, with little discussion about how climate might affect conflict dynamics or other security outcomes of concern. As discussed in Chapter 2, the climate–conflict question has sometimes led to sharp debates between those scholars who find strong evidence to suggest proxies for climate change are associated with increased conflict risk, and those who do not.[31] Even among scholars who see climate change as contributing to conflict potential, a number see other drivers – like low socio-economic development, state capacity, intergroup inequality, and recent conflict history – as more important risk factors for conflict.[32] But these accounts speak to historic drivers of conflict and tell us little about the future and how worried we should be.

Some scholars see a large divide between pessimists and optimists, between those who see a world of emergent scarcity and those who believe in the robust adaptive capacity of the human species to avoid the worst effects. Where pessimists fear that humanity may not be up to the challenge of avoiding dangerous climate change, given how little has been done, optimists have more confidence that humanity can innovate and adapt, much as it has throughout its history – citing, for example, the technological developments that fed a rapidly growing population during the Green Revolution in the mid-twentieth century.[33]

Another set of critical scholars worry that "securitization" of the environment and climate change will reinforce militarization of the environment, contribute to us-versus-them nationalism rather than collective problem-solving, and the implementation of emergency measures that might contravene democratic policy processes. They see efforts to elevate climate change to a security threat as a political project and are thus more interested in the discourse of how and why the problem has come to be seen as a security threat and how this shapes policies.[34]

---

[31] Busby 2017b.   [32] Mach et al. 2019.   [33] Gleditsch 2021.
[34] Deudney 1990; Dalby 2009; Rothe 2017; Diez, von Lucke, and Wellmann 2016; Hardt 2017.

I do not fully align with any of these views. I accept that climate change constitutes a security threat, an even greater one than many optimists would admit and with far-reaching security consequences that go beyond armed conflict. I worry about excessive militarization of the climate security dimensions: the most effective instruments to deal with security dimensions of the problem do not involve the military but require investments in civilian agencies such as development agencies and diplomats. If the problem requires military involvement, so many other policy failures will have already taken place. However, *how* climate security becomes operationalized is a separate question from whether it *should* be.

In terms of the debate between pessimists and optimists, none of the worst outcomes associated with climate change are inevitable. Human beings have agency both in mitigating climate change as well as adapting to its effects and building resilience. That said, climate change, to some extent, is inevitable. Progress to date to reduce emissions has been limited, and there are far too many places on the planet that are ill-equipped to deal with the effects. That makes understanding the risks all the more urgent. My hope is that by identifying the structural risk factors, we can better understand where to concentrate efforts and what to do about the problem.

2

Conceptualizing Climate and Security

Weather anomalies happen all the time, often accentuated by natural processes like El Niño and La Niña.[1] However, the accumulation of record-breaking hot years and new developments like unprecedented temperatures at the poles suggest a broader change is well underway.[2] While scientists continue to learn more about how climate change will manifest in different parts of the world, the general contours are clear – more warming of the earth's surface, higher temperature extremes during the day with little respite at night, and more erratic rainfall.[3]

And, with changes in mean temperatures and rainfall, we are also witnessing what are called fatter tail effects, more temperature distributions lying in the extreme territory of very hot temperatures, more heavy rainfall events, and more extreme periods of prolonged drought.[4] Scientists have also become a lot better at the science of "attribution" and being able to say that individual events were made more likely by climate change.[5] The journalist David Wallace-Wells captured the dynamism of a climate system that is characterized by more extremes: "The truth is actually much scarier. That is, the end of normal; never normal again. We have already exited the state of environmental conditions that

---

[1] Real Climate 2017. [2] Northon 2017. [3] IPCC Working Group I 2014; IPCC 2011.
[4] Wagner and Weitzman 2015.
[5] Kelley et al. 2015; Fountain 2015; BMAS 2014; Climate Central 2014; Fountain 2017. The field of attribution, whereby scientists seek to connect individual weather events to climate change, is a young and somewhat controversial field, but increasingly scientists are able to tease out whether climate change enhanced the likelihood and severity of individual weather events.

allowed the human animal to evolve in the first place, in an unsure and unplanned bet on just what that animal can endure."[6]

What do these changes mean for security? In its Fifth Assessment Report published in 2014, the Intergovernmental Panel on Climate Change (IPCC) included a chapter on "human security" for the first time in the organization's nearly thirty-year history. In using the lens of "human security," the chapter is eclectic and covers everything from connections between climate change and armed conflict to other areas such as threats to livelihoods and cultural integrity.[7] The IPCC's inclusion of the chapter suggests the issue of climate and security has matured as a distinct area of inquiry.[8]

At the same time, the breadth of coverage also papers over some of the differences within the scholarly community between those who embrace the broad concept of "human security" and those who worry about conceptual stretching and are thus committed to a narrower focus on armed conflict.[9] My contribution is pitched somewhere between these two views, seeing human security as a bridge too far and overly vague while a focus on conflict is overly restrictive.

How then can we understand the links between climate change and security? I begin with my understanding of security before reviewing the virtues and limits of the research on environmental security to date. In Chapter 3, I make a synthetic argument about the intersection of state capacity, government inclusiveness, and external aid and how certain configurations make negative security outcomes more likely.

THE MEANING OF SECURITY

The traditional meaning of national security is "protection from organized violence caused by armed foreigners."[10] Although it historically referred to protecting the state from existential threats to its territorial integrity, it has a broader meaning than state survival. Countries have interests beyond their borders for which they may be willing to fight. These "vital interests" may be tied to the country's "way of life," its access

---

[6] Wallace-Wells 2019, 18. [7] Adger et al. 2014.
[8] This chapter builds on material developed in Busby 2008; Busby 2018b; Busby 2017b.
[9] For a sophisticated early treatment of human security and climate change, see Webersik 2010.
[10] Del Rosso 1995, 183.

to critical natural resources, and be considered so important that a challenge would threaten national security.[11]

Moves to link environmental issues to security date back to the mid-1980s when scholars and advocates sought to widen the concept to encompass environmental concerns, health, human rights, and development. In 1983, Richard Ullman wrote that defining security in military terms "causes states to concentrate on military threats and to ignore other and perhaps even more harmful dangers."[12] He called for a different approach based on harms that could (1) quickly and drastically cause a degradation in the quality of life of a people and (2) threaten to narrow the options available to governments and other actors in response. With this definition in hand, other issues like "natural" disasters such as droughts and floods or epidemics could rise to the level of concern long occupied by interstate war and internal violence.[13]

In the academic field of climate and security, prominent political scientists such as Thomas Homer-Dixon narrowed the emphasis to study the relationship between environmental change and violent conflict, justifying the move as a way to define a more tractable research question. Compared to security, he wrote, "Violence is easier to define, identify, and measure; this focus helps bound our research effort."[14]

While Homer-Dixon restricted the focus to violent conflict, others sought to broaden the agenda under the umbrella concept of "human security."[15] As I have noted, efforts to inject a human security frame into the climate discussions culminated in a chapter on human security in the 2014 IPCC report. There, human security was defined "as a condition that exists when the vital core of human lives is protected."[16] The "vital core" of human security extends beyond material well-being to include "culturally specific" nonmaterial factors that people require to fulfill their interests. This broad definition of security has its detractors. As Roland Paris argued, "human security seems to encompass everything from substance abuse to genocide."[17] Moreover, the definition makes causal analysis challenging since factors that could cause human security are part of the definition.[18] While I largely agree that "human security" may

---

[11] Art 2003, 3.   [12] Ullman 1983, 129.   [13] Ibid., 133.
[14] Homer-Dixon and Levy 1995b, 189.
[15] United Nations Development Programme 1994; Dalby 2009. Since 1996, the Global Environmental Change and Human Security (GECHS) program has been one of the leading voices on human security and the environment; Barnett, Matthew, and O'Brien 2010, 18.
[16] Adger et al. 2014, 759.   [17] Paris 2004, 371.   [18] Ibid.

conceptually stretch the concept of security too far, the attention to individual well-being has some salutary properties, emphasizing the safety and well-being of individuals and not just the territorial integrity of states.

Despite these efforts, the narrower research agenda on the environment and violence has dominated and been the primary focus of criticism. Dan Deudney's critique has continued resonance. He saw efforts to securitize the environment – that is, to label the environment as a security issue – as a strategic ploy by advocates to generate more attention. While national security issues typically command higher priority and resources, securitizing the issue has risks, including the tendency for countries to interpret responses to security problems in terms of national self-interest rather than the collective good.[19]

These concerns notwithstanding, there now exists a well-developed literature focused on the links between climate change and violent conflict. While I agree that the tight emphasis on conflict makes research easier, it narrows research possibilities in ways that the literature has already found confining. For example, as I discuss in the section The Second Debate, the links between climate-related processes and civil wars have been somewhat inconclusive and contested. As a consequence, scholars have expanded their research agenda to focus on lower-level communal conflicts as well as other forms of social conflict including riots and strikes. It is arguable whether these constitute threats to national security or are merely security-related because they involve violence.[20]

As I argued in Chapter 1, security threats can be defined more expansively without broadening the scope to encompass all harms to human welfare. The test is whether the damage potential from climate effects are commensurate to the threat posed by a traditional armed external attack. Maximally, security threats include risks to the survival of the country but also more limited dangers such as the vulnerability of the seat of government, the survival of the regime, threats to critical infrastructure, and large-scale loss of life.[21] What these have in common is the risk of damage on a scale that would create unacceptable losses. While both large-scale damage and unacceptable losses are both necessarily ambiguous in their meaning, they help assess the stakes in whether or not a specific episode rises to the level of a security threat, what elevates a fire, flood, drought, or

---

[19] Deudney 1990, 467. For more recent applications to the securitization of climate change, see Rothe 2017; Diez, Lucke, and Wellmann 2016; Hardt 2017.
[20] This section builds on Busby 2008.
[21] My approach has some similarity to the work of Price-Smith 2001.

hurricane from a disaster emergency risk to something more significant, given the seriousness of the threat.

For some low-lying island nations, climate change constitutes an existential threat to the country's survival. An invasion by a foreign government might mean your state ceases to exist as an independent entity. By the same token, if sea-level rise and storm surge require Vanuatu or Kiribati to be abandoned, that would be as consequential as an external invasion. Even if an entire country was not threatened, its seat of government might be highly exposed to climate hazards. A government might believe this to be a security threat, since a devastating weather emergency could pose the kind of dangers posed by a decapitation strike on the country's capital from an external aggressor. Even large-scale losses of life and large-scale damage to critical infrastructure could rise to the level of security challenges. Once again, we can imagine the reaction if a neighboring country bombed one's refineries or depopulated a city. A country would surely consider these national security threats. By the same logic, a major storm, like Hurricane Katrina, that killed and made homeless thousands and damaged and destroyed critical infrastructure would raise a similar level of concern.

While we might assume that what makes armed attacks security challenges is the agency of the external party, there are several reasons to believe that the scale of damage to life and property is more important to the definition. First, there are direct implications of climate effects for the military. Military bases themselves are often in harm's way from climate-related events,[22] and bases depend on wider civilian infrastructure for power, transport interconnections, and their workforce.[23] Humanitarian emergencies often require the mobilization of the military for response and rescue. In the context of the United States, we can think of the mobilization of 50,000 members of the US National Guard after Hurricane Katrina in 2005 as an illustration.[24] Thus, security-oriented agencies in a state have to have contingency plans for these actions at home (and, in the case of powerful countries like the United States, for overseas deployment as well). Even if one rejects the view that the deployment of security forces for domestic humanitarian operations constitutes a security problem, the opportunity costs of diverting security

---

[22] Department of Defense 2019; US Department of Defense 2018; Banerjee 2018; Department of Defense 2014.
[23] Busby 2019a. [24] Orrell 2010.

professionals from defense of the nation from external threats to internal disaster relief has to be considered.

Second, humanitarian emergencies often compromise a state's monopoly on the use of force for a limited period of time over a specific geographic area. Local police stations may be damaged, and in the face of downed power lines, interrupted water and waste services, and closed shops and banks, desperate people and/or unscrupulous individuals may take advantage of post-emergency chaos to loot property and/or engage in violence. Climate-related emergencies, particularly for swift-onset hazards like cyclones, may thus require a pulse of external security to restore order. Here, we can think of the portraits of localized looting after Typhoon Yolanda struck the Philippines in 2013.[25]

In rare circumstances, these threats could constitute threats to the regime in power, not from the direct physical effects of the weather emergency itself, but as a result of calls for resignations and regime change for failed government responses to emergencies. Governments are frequently blamed for weak responses to climate-related emergencies, such as US presidents George W. Bush after Hurricane Katrina and Donald Trump after Hurricane Maria, as well as Pakistan's president Asif Ali Zardari after the 2010 floods. There is an emergent literature on whether democracies or autocracies face differential punishment pressures from publics for perceived failures of preparation and response.[26]

Arguably, here the threats to regimes are largely internal. While in many functioning democracies, peaceful protest will be considered merely an expression of democratic liberties, collective expressions of discontent in other countries may be considered threats to state security, even if mobilization is peaceful. In some country contexts, already riven by conflict and with active militant groups, these events can serve as a source of grievance, recruitment, and mobilization or can facilitate the emergence of such groups. This is, at least, the stylized narrative for protest activity that emerged in the wake of the Syrian droughts of the 2000s. Here too, there is a literature on whether natural disasters make conflict more likely. Like much of the climate–conflict literature, the research on disasters and conflict has produced mixed findings as discussed in the section The Second Debate. For present purposes, it is enough to lay out the logic that might elevate these threats to security challenges.

---

[25] Yap 2013.   [26] Smith and Flores 2010; Quiroz Flores 2015.

Another pathway I discuss in more detail is that climate hazards may undermine the long-term economic growth of countries and thus deprive the state of resources required to exercise a monopoly on the use of force over its territory. Indeed, given the importance of low economic growth in many empirical studies of conflict onset, this rationale of state capacity undermined by climate hazards is an important one that I will return to later in the chapter, also in the section The Second Debate.

What this discussion means is that the study of climate and security rightfully ought to be extended beyond an exclusive focus on climate and conflict. One can both broaden and still reasonably bound one's climate security aperture to include conflict as well as large-scale climate-related disasters. What qualifies as a large-scale disaster?[27] What matters less is the specific threshold but whether local officials consider these disasters to have created negative consequences for security. This section has provided a preliminary defense for why climate hazards may constitute security challenges beyond their potential contribution to conflicts. In the next section, I review the arguments put forward by scholars on the explicit links to conflict in the two generations of scholarship on environmental security.

THE FIRST GENERATION OF ENVIRONMENTAL SECURITY

As the Cold War wound down in the late 1980s and early 1990s, the demand to broaden the definition of security gained more traction. There was considerable optimism the environment could finally get the attention it deserved. Jessica Mathews captured this perspective in 1989, writing: "Man is still utterly dependent on the natural world but now has for the first time the ability to alter it, rapidly and on a global scale."[28] The dystopian underpinning of environmental threats loomed large in this assessment. The journalist Robert Kaplan captured the zeitgeist in his 1994 essay "The Coming Anarchy" in which he suggested the

---

[27] What distinguishes a large-scale climate-related disaster from smaller events may be somewhat context specific, given that states face differential exposure to different hazards and vary in size and by population. Existing natural disaster databases like the EM-DAT International Disaster Database have thresholds for damage and other criteria by which exposure to physical hazards merit inclusion in the dataset in the first place. Datasets like these include a number of minor events. To be more precise, we can establish some arbitrary thresholds for large-scale events such as those that cause loss of life, affect populations, or cause monetary damages in the top 90th or 95th percentile of the distribution for that country.

[28] Mathews 1989, 177; for a similar view, see Myers 1989.

environment would be the defining national security issue of the early twenty-first century.[29]

That essay made Thomas Homer-Dixon more well known. Homer-Dixon is a Canadian scholar, at the time based at the University of Toronto. He and his collaborators in the Environmental Change and Acute Conflicts Project (ECACP) delivered an ambitious and complex portrait of the links between the environment and conflict, drawing on the case studies of Rwanda, South Africa, and other places.[30] Homer-Dixon's scholarship and Kaplan's cruder version helped catapult environmental security onto the agenda of the Bill Clinton administration.[31]

Homer-Dixon foresaw a future of environmentally driven scarcity potentially leading to violence, particularly within developing countries.[32] While inspired by the eighteenth century cleric Thomas Malthus, Homer-Dixon sought to avoid criticism of being seen as an "environmental determinist."[33] He wrote that environmental factors were neither necessary nor sufficient for conflict.[34] Moreover, understanding the environmental contribution to conflict was complicated given a tangled chain of causation, interactions between environmental and social causes, effects that only occur above certain thresholds, and feedback loops.[35] While he despaired of assessing the relative causal importance of environmental factors, Homer-Dixon argued that some conflicts cannot be understood without including environmental scarcity.[36]

He distinguished three different kinds of environmental scarcity that could, when coupled with social and political factors, lead to conflict. The first was *supply-induced scarcity* due to environmental degradation, the second *demand-induced scarcity* due to population growth, and the third *unequal resource-based distribution*.[37] Whether situations lead to violence depends on the capacity for societies to innovate and overcome scarcity.[38] Here, it is important to note that Homer-Dixon focused on renewable resources, such as fisheries and timber or processes like the

---

[29] Kaplan 1994.
[30] Homer-Dixon 1991, 1994, 1999; Homer-Dixon and Blitt 1998; Percival and Homer-Dixon 1998.
[31] Peluso and Watts 2001, 4.   [32] Homer-Dixon 1991, 78.
[33] Malthus thought the rate of population growth would inexorably exceed the capacity of food production to expand, leading to boom–bust cycles of population growth and famine; Malthus 1798.
[34] Homer-Dixon 1999, 7.   [35] Homer-Dixon 1991, 86, 107; 1999, 105–106, 174.
[36] Homer-Dixon 1999, 7–9.   [37] Ibid., 15; Homer-Dixon 1994.
[38] Homer-Dixon 1999, 1994.

hydrological cycle and the climate. Nonrenewable resources like oil and minerals, which scholars of the resource curse think of as important drivers of conflict, are not part of his framework.[39]

Homer-Dixon generated three hypotheses of conflict types: (1) simple scarcity between states; (2) group identity-based conflicts within states affected by internal migration; and (3) relative deprivation conflicts where economic decline disrupts social institutions and leads to domestic strife. He found little support for the first hypothesis but stronger support for the other two.[40]

By virtue of its visibility and far-reaching claims, the work of Homer-Dixon and collaborators was always a likely target for vigorous criticism.[41] As I noted in Chapter 1, one of the most potent lines of critique was the absence of paired cases in the research. Without side-by-side cases of conflict and nonconflict, it was not possible to tease out the scope conditions for when environmental scarcity would lead to violence. For his part, Homer-Dixon's rejoinder was that early in a research program it is appropriate to select on the dependent variable (that is, choose only cases of conflict) and most likely cases (those where there was environmental degradation) to trace the specific role environmental factors played: "The aim is to determine if the independent and dependent variables are actually causally linked and, if they are, to derive inductively from a close study of many such cases the common patterns of causality and the key intermediate and interacting variables that characterize these links."[42] While that might have been defensible two decades ago in the emergence of the environmental security literature, we have yet to see that critique manifest in case study work on climate and security.

The climate and security literature, for its part, has largely been dominated by quantitative work. The case studies that have been written,

---

[39] Ross 2015.
[40] Homer-Dixon 1994, 18–25. Along with Homer-Dixon's research, the Swiss scholar Günther Baechler's initiative, the Environmental Conflicts Project (ENCOP), is also recognized. Baechler and his collaborators produced a multivolume set, initially in German, of forty qualitative case studies of "environmental conflict." They identified different pathways, types, and syndromes where environmental stress leads to conflict (and the threat of violence). They hypothesized that environmental conflicts manifest if and when actors "instrumentalize" cleavages such as ethnic differences; Baechler 1999a, 108, 1998.
[41] For a trenchant, quasi-Marxist critique, see Peluso and Watts 2001.
[42] Homer-Dixon and Levy 1995a, 194. Homer-Dixon and collaborators defended the project elsewhere with a call for methodological pluralism and the virtues of single case studies in exploring causal mechanisms; Schwartz, Deligiannis, and Homer-Dixon 2001.

many of them in policy publications, typically suffer from the same methodological critique offered by Levy in the 1990s of single cases that trace the role played by climate factors in various conflicts in South Sudan, Syria, and North Africa.[43] Other case study work by regional studies experts is often more speculative, seeking to assess what climate change might mean for security in particular countries in the future.[44]

As Nils Petter Gleditsch argued in 1998 in his critique of Homer-Dixon, quantitative studies have the virtues of generalizability, of being able to capture correlations over a wide range of cases.[45] However, until the last decade, time-series, spatially disaggregated environmental data was limited in availability. This meant that early studies that sought to leverage quantitative evidence to assess the claims of environmental security with crude measures produced mixed findings.[46]

THE SECOND DEBATE: CLIMATE CHANGE AND SECURITY

A new literature on climate and security emerged in the mid-2000s and leveraged high-resolution, time-series data on environmental indicators made possible by improved satellite and geo-referenced coverage.[47] This revolution facilitated statistical tests of connections between proxies for climate change related processes (i.e. droughts, temperature change, and rainfall volatility) and security outcomes, namely, the onset and incidence of violent conflict within states.

However, after more than a decade, we are left with a large body of research that has produced mixed findings. As the 2014 IPCC chapter on human security concluded: "The evidence on the effect of climate change and variability on violence is contested. Although there is little agreement about direct causality, low per capita incomes, economic contraction, and inconsistent state institutions are associated with the incidence of violence."[48] Where there are reasonably robust correlations between climate hazards and conflict, such as for temperature, there has been insufficient exploration of causal mechanisms to understand when climate

---

[43] Faris 2007; Ki-moon 2007; Werrell and Femia 2013; Gleick 2014; Kelley et al. 2015.
[44] Moran 2011.   [45] Gleditsch 1998.
[46] Hauge and Ellingsen 2001; de Soysa 2000; Esty et al. 1999.
[47] There is a different literature on water and conflict that finds inter-state water wars have almost never occurred; Wolf 1998. See also my discussion on security and water; Busby 2017a.
[48] Adger et al. 2014, 758. Several authors have carried out periodic stock-taking exercises on this literature, including Nordås and Gleditsch 2007; Gleditsch 2012; Salehyan 2014.

factors lead to violence and other security-related outcomes. To understand this assessment, it helps to walk through a number of studies.

The connections between climate and security emerged in the mid-2000s in the policy community.[49] Debates accelerated after the release of several US think tank reports around 2007, including one by the CNA Corporation, a joint Center for a New American Security and Center for Strategic & International Studies (CNAS–CSIS) effort, and a paper I wrote for the Council on Foreign Relations.[50] These emphasized the potential role of climate change as a threat multiplier in the exacerbation of security problems, with a particular focus on US national security.[51] These discussions culminated in high-level attention to climate and security by the US government and the United Nations Security Council.[52] As I discussed in Chapter 1, some policy-oriented research in this space sought to identify the potential connections between climate factors and specific conflicts such as in Darfur, Sudan, and Syria.

A parallel academic discussion emerged contemporaneously with the policy debates.[53] This research focused largely on whether proxies for climate change were correlated with conflict, with rainfall and Africa receiving particular attention. In much of this work, it is not absolute scarcity of rainfall per se that is thought to cause conflict, but the extent to which rains deviate from expected levels, with the emphasis on lower than normal rainfall.

Early studies looking at rainfall found promising results. In 2004, Edward Miguel and collaborators found, using rainfall variation as a proxy for economic growth, that negative growth shocks of 5 percent increased the likelihood of civil conflict in Africa by more than 12 percent in the following year. They argued that lower economic growth would both increase individual incentives to engage in conflict and undermine state capacity to repress violence.[54] A second study by Cullen Hendrix and

---

[49] Nigel Purvis and I wrote a study for the United Nations in 2004, in which we emphasized climate-driven humanitarian emergencies as the most proximate concern; see Purvis and Busby 2004.

[50] CNA Corporation 2007; Campbell et al. 2007; Busby 2007. See also my 2008 paper in *Security Studies* for a more theoretical account of the ways climate change could pose a threat to US national security; Busby 2008.

[51] Other countries like Germany and the UK also carried out similar efforts; WBGU 2007; Mabey 2008.

[52] Busby 2016a.   [53] Barnett 2003; Barnett and Adger 2007.

[54] Miguel, Satyanath, and Sergenti 2004. Ciccone provided a critique of this methodology and suggested these findings disappear if one uses rainfall levels rather than growth rates in rainfall; Ciccone 2011.

Sarah Glaser also focused on civil conflict in Africa. They examined the contribution of long-term trends (including a location's climate suitability for agriculture and freshwater availability) to conflict onset. They also assessed the contribution of interannual deviations from normal rainfall to the triggering of conflicts. They found that higher than normal rains and land suitable for agriculture were negatively correlated with conflict, but only when controlling for other social, political, and economic factors. Hendrix and Glaser argue that good rains in a single year reduce the incentives for engaging in conflict because farming is more attractive. At the same time, areas that are amenable to agriculture over the long term have higher economic returns, also diminishing the likelihood of conflict.[55] In a third study, Marc Levy and collaborators assessed the connections between rainfall anomalies and conflict outbreaks worldwide using spatially disaggregated data. They found rainfall anomalies were correlated with high-intensity civil conflicts but not low-intensity ones. They argue that rainfall variability affects the economy and state capacity to manage conflicts.[56]

However, other studies produced discrepant results. Researchers associated with the Peace Research Institute of Oslo (PRIO) found no association between drought and civil wars in Africa.[57] In Theisen's study of Kenya, lower than normal rains were actually correlated with reduced conflict.[58] In other articles, it appeared that abundance might be a more potent mechanism that triggers conflict as groups have more reason to clash in time of plenty. Better rains might give raiding parties engaged in communal conflict more cover to conceal attacks.[59] Clionadh Raleigh and Dominic Kniveton found this pattern of rainfall abundance accentuated communal conflict (such as between herders and farmers) while anomalously dry conditions enhanced civil conflict (between rebel movements and governments).[60]

Still other studies have emphasized political variables over environmental ones. In their assessment of range wars between pastoralists and farmers in East Africa, Christopher Butler and Scott Gates argued that asymmetric property rights rather than resources per se fuel banditry by poorer parties.[61] Similarly, in their examination of similar conflicts in the

---

[55] Hendrix and Glaser 2007.   [56] Levy et al. 2005.
[57] Theisen, Holtermann, and Buhaug 2012.   [58] Theisen 2012.
[59] Meier, Bond, and Bond 2007; see also Hendrix and Salehyan 2012; Salehyan and Hendrix 2014.
[60] Raleigh and Kniveton 2012. Fjelde and von Uexkull 2012 found the opposite – that large negative deviations in rainfall in Africa were associated with more conflict.
[61] Butler and Gates 2012.

The Second Debate: Climate Change and Security

Sahel, Tor Benjaminsen and collaborators attributed violence to agricultural encroachment that impeded mobility by pastoralists, opportunism in rural areas with the decline of the state, and rent-seeking behavior by elites.[62]

The climate–conflict literature has generated a sharp dispute between quantitative scholars aligned with PRIO[63] and California-based scholars Edward Miguel, Marshall Burke, and Solomon Hsiang.[64] PRIO scholars, for the most part, have not found strong correlations between climate-related variables and conflict, while Miguel and coauthors, by contrast, have. At the risk of oversimplification, their disputes have largely been based on model specification and differences over methodology.[65]

A 2009 Burke et al. paper found that for every 1 degree increase in Celsius, there was a 4.5 percent increase in the incidence of violent conflict.[66] Halvard Buhaug found the results did not hold up when one included additional data, used alternative model specifications, or included other variables such as political exclusion.[67] A 2013 meta-analysis by Solomon Hsiang and co-authors fueled the debate further. They estimated the average effects of a variety of climate indicators (temperature increases, positive deviations in rainfall, negative deviations in rainfall) on violence across sixty different studies, examining both "personal violence" (which included studies of baseball pitchers beaning more batters on hot days) as well as "inter-group" violence (which included studies of state collapse, civil wars, and other measures). Their provocative claim was that for every 1 standard deviation of climate

---

[62] Benjaminsen et al. 2012.
[63] Buhaug 2010; Theisen, Holtermann, and Buhaug 2012; Buhaug et al. 2014; Buhaug 2014; Nordås and Gleditsch 2007; Buhaug, Gleditsch, and Theisen 2008; Gleditsch and Nordås 2014; Gleditsch 2012; Gleditsch, Nordås, and Salehyan 2007; Theisen 2008.
[64] Miguel, Satyanath, and Sergenti 2004; Burke et al. 2009; Hsiang, Meng, and Cane 2011; Hsiang, Burke, and Miguel 2013; Hsiang and Meng 2014.
[65] Other prominent scholars include Marc Levy who, after earlier contretemps with Thomas Homer-Dixon, became more convinced of the links between climate change and conflict/national security. Another prominent scholar is John O'Loughlin whose publications are in line with the PRIO school. Idean Salehyan, Cullen Hendrix, and Clionadh Raleigh are important, more idiosyncratic scholars. Other eclectic researchers include Jürgen Scheffran at the University of Hamburg and collaborators in the Research Group Climate Change and Security (CLISEC). This group, along with Hans Günter Brauch, has contributed multivolume books on climate and security through the Hexagon Series on Human and Environmental Security and Peace. Three special issues – a 2007 issue of *Political Geography*, a 2012 issue of the *Journal of Peace Research*, and a 2014 issue of *Political Geography* – included many leading figures.
[66] Burke et al. 2009.   [67] Buhaug 2010.

indicators the frequency of interpersonal violence increased by 4 percent and intergroup conflict by 14 percent.[68] Buhaug and coauthors raised various objections – about model specification, choice of control variables, and other arcana – that resulted in a back and forth with Hsiang and his collaborators.[69]

Leaving aside which side is right in these disputes, the California research on temperature as well as the wider meta-analysis is largely silent on the question of causal mechanisms. While they identify some plausible mechanisms and pathways, they acknowledge more work needs to be done: "To date, no study has been able to conclusively pin down the full set of causal mechanisms, although some studies find suggestive evidence that a particular pathway contributes to the observed association in a particular context."[70] While their research focused on finding correlations between climate phenomena and conflict, the field has moved toward identifying discrete causal pathways between specific climate phenomena (such as too much rain) and particular kinds of conflict (such as communal violence).

Heretofore, most studies of climate and conflict tested direct relationships between physical hazards and conflict rather than indirect pathways through food prices, effects on agriculture, migration, disasters, and economic growth. Research has started to address these lacunae. Scholars have begun to focus on the causal pathways to negative security outcomes through their effects on agriculture, economic growth, disasters, and migration as well as the mediating role of institutions. I summarize some of the findings here.[71]

In terms of agriculture, there are two dimensions, one focusing on food production and another on food prices. Emily Meierding urged scholars to study the indirect pathways, focusing on the agricultural sector and food prices because those parts of the economy are the most tightly coupled to climate processes.[72] Depressed agricultural production (and lower income) makes joining a rebellion more attractive, and higher food prices might serve as a source of grievance for consumers.

One study by Nina von Uexkull and colleagues focused on growing season droughts. They found conflict incidence in Africa and Asia to be more likely when droughts occurred in agriculturally dependent areas

---

[68] Hsiang, Burke, and Miguel 2013.
[69] Buhaug et al. 2014; Hsiang and Meng 2014; Buhaug 2014.
[70] Hsiang, Burke, and Miguel 2013, 7.   [71] Busby 2018b; Busby 2019d.
[72] Meierding 2013.

with high levels of political exclusion. This work focused on the more contextual and contingent factors that led to conflict and examined climate data from periods most consequential for farming.[73] That study informed subsequent research to identify countries at risk in the wake of severe water deficits; this examined countries which had high agricultural dependence, a history of conflict, and discriminatory institutions.[74] Another paper in this vein by Maystadt and Ecker connected drought to civil conflict in Somalia through the effects on depressed livestock prices that, in turn, make recruitment by rebel groups more attractive and conflict more likely.[75]

Research by Ore Koren underscores the complex role food production plays in sustaining armies. One study, using crop yield data on wheat and corn in Africa, concluded that food abundance, rather than scarcity, was correlated with political violence. Food-rich regions may draw in a variety of actors seeking control of harvests for their own gain.[76] Koren argued that food-rich regions become sites of contention as a means of denying opponents sufficient food to field forces.[77] In previous work, he and his coauthor found that food insecurity was also highly correlated with conflict.[78] We might reconcile these discrepant findings by noting that while some may enjoy abundance, others may not. In areas experiencing declining resources, areas with local abundance might become sites of contestation.[79]

Other research has examined the effects of food price shocks on social conflict. The role of food price shocks in the Arab Spring looms large, the argument being that the increase in global food prices – emanating from weather-related harvest reductions in Russia, Argentina, and other grain producers – spurred protest activity.[80] Lagi claimed that global food price shocks in the lead up to the Arab Spring were highly correlated with "food riots" in the Middle East and North Africa.[81] Smith, however, noted that many countries insulate their publics from global food price shocks through domestic subsidies. Using rainfall as a driver of domestic food price increases, he found that protests and riots become more likely if domestic food prices increase.[82] Hendrix and Haggard further showed that regime type mattered in whether food price hikes led to riots or protests. They found that democracies were more likely to experience

---

[73] von Uexkull et al. 2016. [74] Busby and von Uexkull 2018.
[75] Maystadt and Ecker 2014. [76] Koren 2018. [77] Koren 2019b.
[78] Koren and Bagozzi 2017. [79] Kahl 2006. [80] Werrell and Femia 2013.
[81] Lagi, Bertrand, and Bar-Yam 2011. [82] Smith 2014.

protests than authoritarian regimes, in part because authoritarian regimes tend to subsidize food to insulate urban consumers. In authoritarian regimes, food price shocks may serve as important drivers of protests which can lead to coups and regime turnover.[83]

A second related and understudied pathway is through the effects of climate and conflict on economic growth. Here, climate changes and variability could depress economic growth (perhaps through the effects on agriculture or as a result of disasters), either making it more attractive for people to rebel and/or by undermining state capacity to suppress violence and to provide services. Early work in this space has been inconclusive.[84] There is also a vigorous empirical debate in economics on the effects of natural disasters on long-term economic growth.[85] This research has not been connected to that on conflict, but if it can be established that disasters have a negative impact on economic growth, then the well-established link between economic growth and conflict would likely be operative, with disasters having an impact on conflict through economic growth.[86]

A third underexplored pathway to conflict is through migration. Reuveny argued that climate-related migration could lead to interethnic conflict over resources, distrust, and rivalry between socioeconomic groups.[87] Research by Idean Salehyan and Kristian Gleditsch suggests that refugees can bring newcomers into conflict with longtime residents over limited resources and government programs, with conflicts spilling over to neighboring polities.[88] Clionadh Raleigh and coauthors suggested climate migrants, to the extent that this is an identifiable category, might be different from refugees. They argued that many environmental migrants' movements are likely to be temporary; their departures might be seen as forced by acts of nature, making them more sympathetic to receiving locations. Moreover, environmental migrants might be so vulnerable that they are less likely to engage in violence.[89] With some locations, low-lying island nations in particular, becoming inhospitable to human settlement, it is little less clear if many population movements *will* be temporary. Early empirical work on migration and conflict by Koubi et al. was inconclusive.[90] In specific cases, migration has been

---

[83] Hendrix and Haggard 2015.   [84] Koubi et al. 2012; Koubi 2017.
[85] Shabnam 2014; Cavallo et al. 2013.   [86] Collier 2007.
[87] Reuveny 2007; see also Reuveny and Moore 2009.
[88] Salehyan and Gleditsch 2006; Reuveny 2007; Reuveny and Moore 2009.
[89] Raleigh, Jordan, and Salehyan 2008.   [90] Koubi et al. 2016a; Freeman 2017.

The Second Debate: Climate Change and Security

identified as a driver of conflict – for example, in the Syrian civil war.[91] As I explain further below, several scholars have, however, contested the links in this case.[92] This is an area that is difficult to study. Using data on asylum applications, a recent study creatively sought to assess the effects of climate change on conflict, and the effect of conflict on migration, in Western Asia.[93] Whether migration leads to conflict, climate migration itself may be a security concern in its own right, given the sensitivity of the topic. Teasing out whether people moved because of climate change or due to other factors is a challenge.

A fourth channel is the effects of climate disasters on security. Disasters may lead to conflict through the effects on economic growth or, potentially, when failed disaster response leads to grievances among affected populations. Nel and Righarts showed the effects of disasters on conflict to be the most severe in low- and medium-income countries with high inequality, low economic growth, and mixed political regimes (either partially democratic or partially authoritarian). While the effects were stronger for earthquakes and volcanoes, the results held up for climate-related disasters.[94] However, Slettebak found climate-related disasters actually made conflict less likely.[95] Other studies have examined connections between disasters and regime survival.[96] A separate small literature outside of security studies has examined the correlates of disaster mortality; poverty, population exposure, and government effectiveness loom large across different accounts and multiple hazards.[97]

The findings here are ambiguous, partially a function of whether we distinguish between hazards (as physical phenomena) and disasters (as social outcomes that represent failures of preparation and response). We may also need to distinguish between swift-onset hazards such as cyclones and storms and slow-onset ones such as drought. Some work suggests that disasters may precipitate peace rather than conflict, when groups rally around the common challenge of survival, rebel movements have been too weakened by the disaster to continue the fight, or where the disaster makes a conflict ripe for resolution with targeted and well-distributed aid

---

[91] Gleick 2014; Werrell and Femia 2013; Kelley et al. 2015.
[92] Fröhlich 2016; de Châtel 2014; Selby et al. 2017a; Hendrix 2017b; Gleick 2017; Kelley et al. 2017; Selby et al. 2017b.
[93] Abel et al. 2019. For a critique, see Koren 2019a.   [94] Nel and Righarts 2008.
[95] Slettebak 2012; for similar results, see Bergholt and Lujala 2012.
[96] Quiroz Flores 2015; Flores and Smith 2010.
[97] Peduzzi et al. 2009; Peduzzi et al. 2010; Yonson, Gaillard, and Noy 2016.

flows.[98] There is also good related work by Alejandro Quiroz Flores and Alastair Smith on disasters and leader survival, that is, whether failed responses to disasters lead to leadership challenges in certain regimes.[99]

The role of institutions is another mediating factor between climate phenomena and security outcomes. Institutions, both local and international, may diminish or exacerbate the likelihood of conflict, depending on how they are designed and implemented. Institutions affect the distribution of services, the capacity of response, and whether disputes escalate. As noted above, von Uexkull et al. focus on how political institutions that exclude certain groups exacerbate the risk of conflict in agricultural societies experiencing growing season droughts.[100] Linke et al. draw attention to both official government and customary domestic institutions and how rules over natural resource management potentially amplify or moderate conflict.[101] For transboundary river basins, well-designed institutions diminish the risks of conflict by allocating water, planning for shocks, and facilitating dispute resolution.[102]

Scholars have exploited better geo-referenced datasets to examine subnational conflict patterns and a variety of kinds of conflict. We are also seeing scholarship on regions other than Africa, including Asia and the Middle East. The best work seeks to specify the conditions under which climate-related hazards lead to particular kinds of conflict, distinguishing between kinds of states (e.g. between exclusive and inclusive institutions, states with stark group cleavages), kinds of contexts (e.g. between urban and rural areas), and kinds of hazards (e.g. swift onset versus slow onset).[103] While this disaggregated analysis is a productive step forward, there still may be room for a more unifying theoretical framework on the causal pathways between climate change and security outcomes that includes but is not limited to conflict.[104] What follows in Chapter 3 is an effort to do just that.

---

[98] Kelman 2006; Egorova and Hendrix 2014.
[99] Quiroz Flores 2015; Flores and Smith 2010.   [100] von Uexkull et al. 2016.
[101] Linke et al. 2018b.   [102] De Stefano et al. 2012; Tir and Stinnett 2012.
[103] For a similar take, see Hendrix, Gates, and Buhaug 2016.
[104] Several papers have tried to synthesize what we know about the links between climate and conflict, including Mach et al. 2019; Koubi 2019; Theisen 2017.

3

The Argument, Method, and Mechanisms

State Capacity, Institutional Inclusion, and International Assistance

As discussed in Chapter 2, the climate security literature has paid insufficient attention to causal mechanisms, but it has moved in a positive direction to unpack various pathways to security outcomes – through agricultural output, food prices, migration, disasters, and economic growth. Scholarship has also differentiated explanations for different kinds of security outcomes; that is, the climate-related processes that contribute to civil conflict might be different from those that contribute to smaller-scale communal conflict.[1]

Some scholars have made a call to move beyond conflict onset as the primary concern of research to the dynamics of conflict, whether and how climate processes might affect the timing and duration of conflicts.[2] While these appeals to narrower, more context-specific arguments offer important nuance, there is the risk of a proliferation of arguments and the loss of some generalized understanding about climate and security. Although it is important to specify the conditions that lead from climate phenomena to security outcomes, can we still sketch a general pathway that can help us understand a range of phenomena, including violent conflict but also humanitarian emergencies?

The premise of this chapter and, indeed, the wider book, is that we can identify general scope conditions for when climate changes are likely to lead to negative security outcomes. The argument in its reduced form is as follows: In the wake of exposure to climate hazards, negative security outcomes, which include conflict but also humanitarian emergencies, are most likely to occur when states have (1) limited capacity; (2) exclusive

[1] Raleigh and Kniveton 2012. [2] Hendrix, Gates, and Buhaug 2016.

political institutions; and (3) where international assistance is absent or distributed in a one-sided manner that reinforces the resources available to some groups rather than others. Where state capacity reflects a government's ability to help groups in times of need, inclusive political institutions capture their willingness to do so. International assistance can partially compensate for weak state capacity. In this chapter, I unpack the importance of each concept individually and consider what metrics we might use to evaluate them before synthesizing their collective contribution.

Governance looms large in many accounts of why climate effects lead to negative security outcomes in certain contexts. This informed the choice of specific factors in this book. A 2005 study identified a number of variables that correlated with climate-related disaster mortality. Among the factors correlated with disaster mortality, a subsequent expert elicitation identified government effectiveness and voice and accountability as the two most important.[3] In recent years, a number of articles have found that political exclusion is a key factor contributing to conflict in a number of empirical studies of climate and conflict.[4] A 2019 expert elicitation on climate and conflict risk factors identified low socioeconomic development, low state capability, and intergroup inequality as the most important factors in driving conflict.[5] These findings helped inform my choice of state capacity and political exclusion as part of a synthetic, relatively spare argument, to which I added a dimension as yet underexplored in the climate security literature – international assistance.

State Capacity

States that lack effective governments, that lack a monopoly of force over their territory, will be less able mobilize resources and deploy them for public purposes, including service delivery and suppression of violence. Andrew Price-Smith defines state capacity as follows: "[T]he capability of government, and its level determines the state's ability to satisfy its most important needs: survival, protection of its citizens from physical harm as

---

[3] Brooks, Adger, and Kelly 2005; Adger et al. 2004.
[4] Theisen, Holtermann, and Buhaug 2012; Raleigh 2010; von Uexkull et al. 2016; Ide et al. 2020. Wider conflict studies also focus on political exclusion and intergroup inequality, including Cederman, Gleditsch, and Buhaug 2013.
[5] Mach et al. 2019.

a result of internal and external predation, economic prosperity and stability, effective governance, territorial integrity, power projection, and ideological projection."[6] To allow for multiple goals that states may have, Hanson provides a pithy definition of state capacity: "[I]t is the ability of state institutions to effectively implement official goals."[7]

My expectation is that countries with lower state capacity, all else being equal, will do a worse job in preparing for and responding to climate hazards. For paired cases in subsequent chapters, I should observe that countries that do better relative to others vis-à-vis climate hazards should have more state capacity than those that experience worse outcomes (so, Ethiopia compared to Somalia; Lebanon compared to Syria; Bangladesh and India compared to Myanmar). My expectation is that states with very bad security outcomes like Somalia, where large numbers of people died in the wake of the 2011 drought, ought to be among the worst performers globally in terms of state capacity. States that improve their performance over time, such as Ethiopia and Bangladesh, ought to show improved capacity.

Scholars such as Homer-Dixon and Price-Smith note that capacity itself can be affected by processes of environmental change and public health outbreaks. In his pathbreaking work *The Health of Nations*, Price-Smith used state capacity as the outcome of interest to be explained.[8] While I acknowledge climate hazards can affect state capacity, I am more interested in how the two intersect at moments in time in the wake of episodic exposure to extreme weather events that are of relatively short duration, though clearly climate change will have more long-lived effects that may transform state capacity over time. To be sure, slow-onset hazards like droughts unfold over weeks and months, so their effects do not manifest as quickly. Still, I am using climate hazards, both swift-onset and slow-onset ones, as causal triggers for security outcomes rather than long-term trends.[9]

This may seem curious for those interested in the long-term effects of climate change on societies. Others, notably scholars such as Jared

---

[6] Price-Smith 2001, 27; see also, Price-Smith 2009.   [7] Hanson 2018, 19.
[8] Drawing on Homer-Dixon, Price-Smith provides another definition of state capacity as "one country's ability to maximize its prosperity and stability, to exert *de facto* and *de jure* control over its territory, to protect its population from predation, and to adapt to diverse crises"; Price-Smith 2001, 25.
[9] Hendrix and Glaser 2007. In the health space, Price-Smith distinguishes between short-term outbreak events and long-term attrition processes. My focus here is more on the short-term events; Price-Smith 2001, 15.

Diamond, have explored how environmental change has led to societal collapse in the historical record.[10] However, it is unclear what inferences from ancient civilizations are applicable to states in the modern era. While it is possible to study the effects of long-lived processes on social orders, the causal chains are likely to be longer, making it more difficult to establish that environmental change rather than social or political processes have some responsibility for the observed outcomes. As a consequence, I am interested in the political effects and security consequences of relatively short-term manifestations of climate change, namely, extreme weather events that create emergency conditions that states have to respond to (or not as the case may be). I come back to the challenges of studying the causal role of climate trends in Chapter 8.

While community level responses are the first line of defense to such extreme weather events, I agree with Price-Smith that states remain the most important actors in the international system, and that the ability of the territorially bounded state to exercise control over its entire territory, both for the provision of security and other public services, is of enormous consequence.[11] Nonstate actors are important service providers, and they can partially fill gaps and create local coping capacity where a state presence is weak or absent, particularly when paired with international assistance. Nonetheless, climate hazards may pose such extreme risks to publics that civil society actors cannot effectively respond without state assistance – such as lift capability to deliver emergency supplies and carry out rescue operations. For this reason, the focus in this book is still state-centric.

States vary in their capacities to address climate change, and this requires us to define more precisely what we mean by it and how to measure it. Homer-Dixon and collaborators identify nine dimensions of interest including four intrinsic elements of the state – its *human capital* (the competence of the bureaucracy); what they call *instrumental rationality* (its ability to make use of information); its *coherence* (the degree to which actors share the same ideological bases); and *resilience* (the ability of the state to respond to shocks). They identify five dimensions that reflect state–societal relations including *state autonomy* (whether the state is able to act independently of external forces); *legitimacy* (the state's moral authority); *reach* (the state's ability to get things done over its entire territory); *responsiveness* (whether the state meets, or tries to meet, the

---

[10] Diamond 2004; see also, Hsiang, Burke, and Miguel 2013.
[11] Price-Smith 2001, 27, 2009.

needs and grievances of its people); and *fiscal strength* (whether the state can finance its programs).[12] Price-Smith operationalizes five proxies in his work and generates a composite index to measure state capacity. Some of his proxies – for example, gross national product (GDP) – better reflect other underlying constructs, and this gets to the difficulty of measuring state capacity independently of other properties.[13]

Hendrix identifies three broad understandings of state capacity, one focusing on the ability to project military power over territory, another based on bureaucratic and administrative capacity, and a third focusing on the quality and coherence of institutions.[14] Of these various dimensions, the second is most germane. Here, I am influenced by Francis Fukuyama on governance as policy execution or the ability to "make and enforce rules." He sees capacity as a function of professionalization and available resources.[15] Drawing on Michael Mann, Fukuyama writes of the importance of "infrastructural" power, which he distinguishes from "despotic" power. For Mann, despotic power is the scope of actions a state is empowered to take without "routine, institutionalized negotiation with civil society."[16] While despotic power is seen as power over civil society, infrastructural power is the amassed power of the state to enforce rules in coordination with civil society. It is this rule-making and enforcement capacity over territory that is most relevant to this book.

With respect to climate hazards, state capacity requires the ability to mobilize and deploy resources in planning, preparation, and response. Administrative capacity to protect the citizenry from harm requires the ability to make and enforce rules that would discourage human settlement in highly vulnerable areas, establish rules over construction and resource use to diminish vulnerability in the event of hazard exposure, develop and implement plans for how to respond to climate shocks, and mobilize resources to aid recovery in the wake of exposure. At the broadest level, state capacity thus requires the ability to project the state's power over an entire territory, but it also involves the capacity to mobilize resources to pay for plans, preparations, and responses.

The cases of interest in this book date back at their earliest to the 1970s and extend forward to the present. While some of the cases compare states

---

[12] Barber 1997.   [13] Price-Smith 2001, 28.
[14] Hendrix 2010. Drawing on Skocpol, Hanson similarly distinguishes between three dimensions: coercive capacity, administrative capacity, and extractive capacity (the latter reflecting the state's ability to raise revenues to support its operations); Hanson 2018, 19.
[15] Fukuyama 2013, 349.   [16] Mann 1984, 188.

at the same or similar moments in time, we are also interested in longitudinal data to compare Bangladesh, India, and Ethiopia to themselves over time. Thus, even if there is no single authoritative source of state capacity measures, there may be measures that capture much if not all of this time span.

Operationalizing state capacity is difficult.[17] Perhaps one of the best-known sets of governance measures comes from the World Bank. The Worldwide Governance Indicators capture six dimensions of governance: Voice and Accountability; Political Stability and Lack of Violence; Government Effectiveness; Regulatory Quality; Rule of Law; and Control of Corruption. They are based on surveys and expert appraisals from companies, public officials, and NGO assessments. Diverse data sources comprise each of these individual measures. Of these, Government Effectiveness, Regulatory Quality, and Control of Corruption are perhaps the most relevant to the concept of state capacity to prepare for and respond to climate hazards.[18]

One limitation of the Worldwide Governance Indicators is that they are only available dating back to 1996, which precludes comparison with earlier periods of interest here, from the 1970s and 1980s. However, one measure of bureaucratic quality from the PRS Group, a political risk firm, dates back to 1984 as part of its International Country Risk Guide. This measure is incorporated into the World Bank's metric of government effectiveness. Like the Worldwide Governance Indicators, the PRS Group's measure of bureaucratic quality is based on expert assessment; its work is geared toward informing private sector actors about expropriation risk and policy uncertainty. As a consequence, bureaucratic quality focuses more on policy instability, making it an imperfect proxy for a state's ability to implement policy, but useful nonetheless: "The institutional strength and quality of the bureaucracy is another shock absorber that tends to minimize revisions of policy when governments change.

---

[17] Fukuyama 2013, 350.
[18] Kaufmann, Kraay, and Mastruzzi 2010; World Bank 2019g. The Bank defines Government Effectiveness as "the perceptions of the quality of public services, the quality of the civil service and the degree of its independence from political pressures, the quality of policy formulation and implementation, and the credibility of the government's commitment to such policies." Regulatory Quality captures "perceptions of the ability of the government to formulate and implement sound policies and regulations that permit and promote private sector development." Control of Corruption reflects "perceptions of the extent to which public power is exercised for private gain, including both petty and grand forms of corruption, as well as 'capture' of the state by elites and private interests"; World Bank 2019g.

3 The Argument, Method, and Mechanisms

Therefore, high points are given to countries where the bureaucracy has the strength and expertise to govern without drastic changes in policy or interruptions in government services."[19]

V-Dem also has a narrower measure of fiscal capacity based on expert review that also has longitudinal coverage back to the early twentieth century. However, this measure mostly captures shifts in the source of taxation ranging from states unable to raise any tax revenue to those that can derive revenue from a variety of economic transactions, including sales, income, corporate profits, and capital. This metric may inadequately reflect the state's capacity to translate tax revenue into service delivery. Moreover, for some of the countries of interest such as Somalia and Myanmar, the number of experts consulted for some years is three or fewer, making valuations subject to swings based on reviewer availability. V-Dem suggests dropping point estimates where there are three or fewer experts.[20]

Fukuyama raises a related issue. He worries that aggregate survey-based measures like these are problematic since experts might evaluate countries differently, making cross-country and temporal comparisons suspect.[21] What alternatives are there to perception-based indicators? As Fukuyama notes, measures such as tax collection potentially tell us something, but they are inadequate as taxes may be collected but not transformed efficiently into services. Some states may not be reliant on tax revenues since they capture natural resource rents from oil, diamonds or other sources. Another potential indicator of state capacity is the degree of professionalization and education of the bureaucracy. Because there could be considerable variation in governance capacity across departments and geographically over a country's territory, in an ideal world we would disaggregate our measures of state capacity to assess what functions the government performs well and where governance within a state is better or worse.[22] Comparative tax collection data is often hard to come by, let alone more fine-grained, function-specific, or subnational assessments of state capacity.

Recent studies have tried to use lights at night, that is, satellite images of electricity coverage in countries, as a proxy for state capacity, the advantage being that such a measure can show geographic variation in the reach of the state over a country's territory.[23] This is potentially attractive as it can allow for some closer matching between the geography of climate

---

[19] PRS Group 2012.   [20] V-Dem 2019.   [21] Fukuyama 2013, 353–354.   [22] Ibid.
[23] Koren and Sarbahi 2018.

hazards and subnational state capacity, with the expectation that a state with lower subnational capacity that is affected by climate hazards in that same region ought to experience worse security outcomes in that area. Given that other scholars are using the same indicator as a proxy for economic development,[24] I am reluctant to rely on this particular measure, though I acknowledge the value of disaggregated subnational metrics of state capacity.

Moran et al. developed a composite index of government effectiveness, drawing on many of the World Bank's indicators of government effectiveness but including some others. Like the accompanying legitimacy index, the effectiveness score included political, security, economic, and social components, each of which was comprised of three indicators.[25] These scores are available for 2000–14, providing another robustness check as a point of comparison.

Another dataset developed by Hanson and Sigman seeks to measure three dimensions of state capacity – extractive capacity, coercive capacity, and administrative capacity – with coverage for the period 1960–2009. They, too, draw on Mann's theories of infrastructural and despotic power and derive twenty-four indicators of state capacity from different sources, some of which are surveys. One of their capacity measures is highly correlated with the World Bank's measure of government effectiveness. This approach has the advantage of including measures beyond taxation capacity to analyze other aspects, namely, the ability to exercise a monopoly of force over territory and the ability to implement policy, not merely leverage revenue from the citizenry.[26]

There are also efforts to evaluate state capacity with respect to disaster risk reduction efforts, an indicator directly relevant to the subject of this book. INFORM, a collaboration of various international organizations, has evaluated this capacity of states since 2012. Its index now includes data on 191 countries and it has generated an overall risk measure based on exposure, socioeconomic vulnerability, and lack of coping capacity.[27]

---

[24] Noor et al. 2008; Xie et al. 2016; Keola, Andersson, and Hall 2015.
[25] The *political* indicators included quality of public service, the number of successful coups d'état in the last five years, and government tax revenue as percentage of GDP. The *security* indicators included the intensity of ongoing armed conflict, the size of displaced populations, and the proportion of the country affected by conflict. The *economic* indicators included GDP per capita, the poverty head-count ratio, and primary commodity exports as percentage of the total. The *social* indicators included infant mortality rate, child immunization rates, and the percentage of the population with access to improved water sources. Moran et al. 2018.
[26] Hanson and Sigman 2013.   [27] Marin-Ferrer, Vernaccini, and Poljansek 2017.

3 The Argument, Method, and Mechanisms 45

The lack of coping capacity measure is the closest proxy of interest and it has an institutional and an infrastructural component.[28] The institutional component has three parts: a measure of government effectiveness drawn from the Worldwide Governance Indicators; a measure of corruption perception from Transparency International; and a measure of disaster risk reduction drawn from country self-assessments of progress toward the United Nations Hyogo Framework for Action, an action plan to address risk reduction that came out of the 2005 World Conference on Disaster Reduction.

The Hyogo self-assessment indicator is the most distinctive contribution relevant to disasters. Countries were asked to evaluate their progress on disaster risk reduction encompassing: (1) the degree of institutional commitment to disaster risk reduction; (2) their capacity for early warning; (3) their ability to use knowledge to enhance resilience; (4) the strength of efforts to reduce underlying risks; and (5) the extent to which disaster preparedness was being strengthened at all levels. Some forty countries including Somalia did not report on this measure, which could itself reveal lack of state capacity, though richer countries such as Russia also did not report. Countries conducted these self-assessments at different times, and there is limited longitudinal coverage.[29] As the report notes, countries might also inflate their own grades; however, the creators regarded this as useful information in the absence of anything better. They are experimenting with new approaches, including subnational analysis. Despite its limitations, this index provides another data point on state capacity. The purpose of these various indicators is to triangulate different measures of state capacity. In Chapters 4, 5, and 6, I reference several if not all of these measures to compare countries to each other, and to themselves over time. That said, they represent proxy indicators for more complex and specific processes relative to protecting a populace from climate hazards and responding to extreme events.

Here, there are more specific functions that states have to carry out, to plan, prepare, and respond to discrete kinds of climate hazards. For countries that face perennial problems of drought or cyclones, state capacity in this arena includes detection and early warning systems,

---

[28] The infrastructural component includes a variety of indicators on access to communication, physical infrastructure, and access to health systems. These are less properties of the state, so the institutional metric is of greatest relevance here.

[29] The INFORM report includes the most recent data. In the 2018 report, data for the Hyogo Framework for Action (HFA) indicator is from 2015 at the latest and 2009 at the earliest. Thow et al. 2018.

communication systems to share information with the public, policies to protect the populace from damage (such as food and water reserves and emergency shelters), and response capacities such as lift and transport to take people out of harm's way and other measures to restore services. Separate from the outcome (large numbers of people dying or suffering in the wake of a climate hazard), we can ask:

(1) Does the government have natural hazard monitoring and early warning capability?
(2) Is there a plan in place to prepare for and respond to such hazards?
(3) What is the level of staffing of that effort?
(4) How much funding is allocated?
(5) For particular hazards, what stockpiles/levels of food, water, transport, and emergency shelter existed prior to the hazard events?

When scaled to account for population differences, these indicators can tell us something more specific about the adequacy of preparation efforts between countries at similar moments in time, as well as within countries over time.

Political Inclusion

While state capacity determines whether states *can* respond to climate hazards, other factors shape whether they *will*. Following work by Colin Kahl, Daron Acemoglu and James Robinson, and Douglass North among others, the second dimension of relevance is the *degree of political inclusion*.[30] According to Acemoglu and Robinson, inclusive institutions are characterized by "power broadly distributed in society and [institutions that] constrain its arbitrary exercise."[31] If we think of state stability as underpinned by elite pacts between groups to share power and resolve differences through law and politics rather than by force, then political inclusion implies incorporation of all politically and militarily relevant subgroups in government decision-making and "fair" apportionment of resources and programs.[32]

---

[30] Kahl 2006; Acemoglu and Robinson 2013; North, Wallis, and Weingast 2009. This approach also has some affinities with selectorate theory where exclusive regimes have small selectorates that provide few public goods; Bueno de Mesquita et al. 2003.
[31] Acemoglu and Robinson 2013, 82.
[32] For similar thinking, see North, Wallis, and Weingast 2009.

Polities with exclusive political institutions are likely to respond to climate-related hazards with measures largely limited to the political base of the regime, such as coethnics or the leader's home region.[33] This leaves less favored regions with meager to no access to resources that would protect them from harm or provide them with the ability to respond to climate shocks, such as emergency provisions, food aid, water, shelter, medical attention, transport, and cash. All else being equal, my expectation is that states with more inclusive governments will be more willing to come to their citizens' aid in the wake of exposure to climate hazards. I come back to ways to operationalize this concept below. Here again, I should observe that countries that experience very bad outcomes in the wake of hazard exposure (such as Myanmar in 2008 and Syria in 2011) should have worse indicators of political inclusion than states that experience better outcomes (such as Bangladesh, India, and Lebanon).

In 2006, Colin Kahl made an important argument about when demographic and environmental stress (DES) would lead to conflict. While much of Thomas Homer-Dixon's work presumes that the origins of conflict emerge from below, Kahl focuses on *state exploitation*, where elites "capitalize on scarcities of natural resources and related social grievances to advance their parochial interests."[34] Kahl argues that states with exclusive institutions and stark cleavages (what he calls "groupness") are more likely to be vulnerable to environmental scarcity related conflicts.

Inclusive political institutions, Kahl argues, insulate states from these kinds of pressures since a variety of groups have some opportunity to have their voices incorporated into government and they offer some protection for weaker groups from exploitation by others. Exclusive political institutions offer few such protections. In societies where people have overlapping identities and ties or in homogenous societies, the division into socially separate groups is limited. By contrast, in societies where group identities are divided starkly along racial, religious, or ethnic lines, the potential for group mobilization and conflict increase.[35]

In his call for a renewed research agenda on the relationship between environmental change and mass atrocities, Cullen Hendrix writes,

---

[33] Other climate security scholars have started to assess the role played by political exclusion. These include: von Uexkull et al. 2016; Busby and von Uexkull 2018; Theisen, Holtermann, and Buhaug 2012; Seter, Theisen, and Schilling 2018.
[34] Kahl 1998, 82.   [35] Kahl 2006; Kahl 2002; Kahl 1998.

"Institutional inclusiveness matters for whether groupness translates into the exclusionary, winner-take-all politics that legitimates this kind of othering and implies unsparing competition for resources, with a winner-take-all logic that legitimates the use of violence."[36] In practice, institutional inclusiveness thus includes federalism, efforts to devolve power regionally, an independent judiciary, and checks on executive power such as votes that require supermajorities and policies that give minorities voice opportunities. These constraints on leaders and institutional practices to resolve conflicts are two reasons why inclusive institutions are less likely to suffer from top-down violent oppression or the emergence of conflicts that bubble up from dissatisfied groups.[37]

Related ideas about institutions animate Acemoglu and Robinson's work. In their book, *Why Nations Fail*, they distinguish between inclusive and extractive political institutions. Inclusive political institutions, in their view, have two characteristics – some degree of central authority and pluralism: "We will refer to political institutions that are sufficiently centralized and pluralistic as inclusive political institutions. When either of these conditions fails, we will refer to the institutions as extractive political institutions."[38]

Acemoglu and Robinson try to explain the different economic trajectories of countries rather than the emergence of violence. They saw an intimate connection between politics and economics. Extractive political institutions are inconsistent with inclusive economic institutions, and, by the same token, combinations of extractive economic institutions and inclusive political institutions are unstable. Institutions that concentrate power for the benefit of the few are seen as inimical to inclusivity: "If the distribution of power is narrow and unconstrained, then the political institutions are absolutist."[39] In absolutist systems, they argue, "those who can wield this power will be able to set up economic institutions to enrich themselves and augment their power at the expense of society."[40]

Inclusive economic institutions permit majority participation in the economic life of the country through measures that secure private property rights, provide for the unbiased rule of law, where public services are delivered without favoritism, and markets are open to new entrants.[41] In the long run, inclusive institutions are more likely to generate sustained economic growth than extractive ones, because measures to protect the economic interests of a narrow elite will ultimately lead to calcified economic interests that are resistant to new ideas and practices.[42]

---

[36] Hendrix 2016, 8.   [37] Ibid., 3.   [38] Acemoglu and Robinson 2013, 82.   [39] Ibid., 80.
[40] Ibid.   [41] Ibid., 74–75.   [42] Ibid., 86.

Acemoglu and Robinson also argue that insufficient central state authority is also a danger. Writing about Somalia, they elaborated: "[P]olitical power in Somalia has long been widely distributed – almost pluralistic. Indeed there is no real authority that can control or sanction what anyone does." They continued, "at the root of it is the Somali state's lack of any kind of political centralization, or state centralization, and its inability to enforce even the minimal amount of law and order to support economic activity, trade, or even the basic security of its citizens."[43]

Here, I think Acemoglu and Robinson conflate inclusivity with state capacity. I prefer to keep those properties separate, as a state may have capacity, that is, central authority to enforce rules, but not be inclusive. It is arguable whether a state lacking capacity to enforce the rule of law over its territory can be inclusive.

This quality of inclusiveness is similar to the approach in *selectorate theory*, which argues that autocratic regimes have small winning coalitions and can stay in power only by generating private goods for the few rather than broadly based public goods for the mass of the population.[44] There are also some parallels with North et al.'s discussion of *limited or natural access orders* and the ways these societies control violence. Limited access orders are essentially elite bargains where violence is limited through "the manipulation of economic interests by the political system in order to create rents so that powerful groups and individuals find it in their interest to refrain from using violence."[45]

Both of these approaches are premised on societies being organized for a winning or dominant coalition, with implications for how out-groups are treated, how they respond, or what happens if there are breakdowns in coalition dynamics, particularly in the wake of shocks. As North et al. note, climate shocks, among other jolts to the political and economic order, can require renegotiation of the basic bargain between elites:

Externally, unpredictable changes in relative prices, climate disasters, bumper crops, technological change, and newly hostile neighbors are part of the world. All societies are subject to random and unexpected shocks. In natural states, the changes may affect the distribution of violence potential and require a renegotiation of the distribution of privileges and rents within the dominant coalition as well as changes in the membership of the coalition as new powerful interests arise and old interests weaken.[46]

---

[43] Ibid., 80. [44] Bueno de Mesquita et al. 2003. [45] North et al. 2012, 3.
[46] North, Wallis, and Weingast 2009, 21.

Open access orders differ from limited access orders by restructuring relations between elites to become more impersonal, rather than rely on patronage ties, short-lived organizations associated with their founders, and private security networks.[47]

In limited access orders, recurrent violence in the wake of coalition adjustment failures is a strong possibility, particularly in what North et al. call "fragile natural states" like Somalia and Afghanistan that can barely function as they confront both internal and external threats.[48] This concept of fragility is a reflection of state capacity.

With an emphasis on inclusive institutions, we can also see the intersection with the role played by political exclusion as a source of grievance and conflict accelerant in divided societies. For a number of years, one of the enduring puzzles in civil war studies has been the absence of findings in the quantitative literature on the association between ethnic fragmentation and civil war.[49] Indeed, these results seemed to strengthen the claims by Collier and others that greed and opportunities for self-enrichment explain the emergence of conflict rather than grievances.[50] While the world witnessed high profile episodes of seeming ethnic conflict, such as between the Tutsi and Hutu in Rwanda and Burundi, the absence of statistical correlation left analysts puzzled. More recent studies by Lars-Erik Cederman and his collaborators have explained this nonfinding by identifying and developing new metrics of ethnic political exclusion that better track the circumstances in which ethnic-based grievances might emerge. This approach suggests groups marginalized from political power might resent their political and economic exclusion and therefore organize and ultimately resort to violence if their demands for inclusion are not reciprocated. The argument finds that ethnic political exclusion as well as horizontal economic inequality between groups can explain conflict emergence and is relevant to how long conflicts last.[51] This work has pioneered new datasets on Ethnic Power Relations (EPR) to identify politically relevant groups that are discriminated against by the ruling coalition in different country contexts or are powerless.[52]

A number of studies in the climate security space have found that political exclusion in the wake of climate hazard exposure can contribute to conflict. For example, in von Uexkull et al.'s study of conflict incidence

---

[47] North et al. 2012, 17.   [48] North, Wallis, and Weingast 2009, 42.
[49] Fearon and Laitin 2003.   [50] Collier and Hoeffler 2004.
[51] Buhaug, Cederman, and Gleditsch 2014; Cederman, Gleditsch, and Buhaug 2013.
[52] Cederman, Min, and Wimmer 2009.

in Africa, they found conflict to be more likely when growing season drought affected excluded groups in agriculturally dependent countries.[53]

Operationalizing political exclusion can be challenging. For example, Bretthauer uses the Ethnic Power Relations (EPR) dataset and identifies as a threshold for high political exclusion countries that have 20 percent of their population excluded from power-sharing in government.[54] EPR rankings are available from 1946 to 2017 and change over time if ethnic groups gain or lose political power as part of regime transitions. The data have also been georeferenced to the range of where ethnic groups are located within countries; this is useful since climate hazards affect specific places within countries.

While the EPR approach has its virtues, some countries, such as Ethiopia and Cameroon, which have struggled with problems of political exclusion in recent years, do not meet the threshold because groups like the Oromo in Ethiopia and the Anglophone groups in Cameroon were technically "junior partners" in the government, although they were dissatisfied.[55] In other countries, like Somalia, for instance, all the populace is technically Somali, but there are finer forms of clan-based identity that are not captured by EPR. In other countries, such as Haiti,[56] class may be as or more of a potent source of division between groups, suggesting that EPR should not be the sole arbiter of whether a country has exclusive political institutions.

A metric which captures other forms of political exclusion has been developed by the V-Dem project, which charts political inclusion by gender and socioeconomic grouping dating back to 1789. These data are based on expert evaluations so the usual caveats about the potential subjectivity of the experts applies, though the project has a number of robustness checks and the methodological sophistication to generate some confidence in the approach. In particular, its measure of social group access to public services is perhaps the best indicator to gauge the inclusivity of service provision as it relates to the prevention of disasters and the response to climate hazards. While its indicators of political inclusion focus on parties, its social groups measure focuses on other sources of distinction such as caste, language, race, religion, and ethnicity.[57]

---

[53] von Uexkull et al. 2016. Other studies have also found support for political exclusion, but less or mixed support for climate/environmental drivers; see Theisen, Holtermann, and Buhaug 2012; Seter, Theisen, and Schilling 2018.
[54] Bretthauer 2015.   [55] Busby 2018b.   [56] ETH Zurich 2015.
[57] Lührmann et al. 2018.

3 *The Argument, Method, and Mechanisms*

The importance of inclusive political institutions is not a new observation. Amartya Sen's landmark book *Development as Freedom* argued persuasively that "there has never been a famine in a functioning multiparty democracy."[58] Leaders in democracies have to be attentive to their citizens' needs in order to stay in office. Moreover, democracies tend to have vibrant press freedom which means that food shortages and climate-related crises are publicized and can become politically salient. In authoritarian regimes, the absence of electoral incentives insulates leaders from responding to the needs of their citizenry. Furthermore, the lack of a free press means that leaders may not hear about humanitarian emergencies, particularly if their retinue of advisers are sycophants and fearful of the consequences of reporting bad news. This portrait of authoritarian regimes is compelling, though it may lack some nuance. For example, because protest in urban centers may pose a challenge to regime stability, authoritarian leaders may respond selectively with subsidies for cities, at the expense of rural populations.[59]

Some democracies may also have groups that are marginalized if not fully excluded from political power. In 2005, Hurricane Katrina's impact on New Orleans led to claims that the Bush Administration's desultory response was because of racial bias, epitomized by hip-hop artist Kanye West's complaint, "George Bush doesn't care about black people." In 2017, the Trump Administration's lackadaisical response to Hurricane Maria raised concern that Puerto Rico's status as a territory of the United States (and the racial makeup of the population) marginalized the island from the mainland.[60] Despite these caveats, Sen's core observation about democracies compared to authoritarian regimes is, in general, valid for democracies. To triangulate the degree of political inclusion, we can also look to their degree of democratization as a proxy, using Polity IV scores. These rank the continuum from full democracies (+10) to full autocracies (−10), a composite index of three parameters − the degree of political participation, the openness and competitiveness of the chief executive, and constraints on the chief executive. Data are available from 1800–2017, which allows us to observe variation between countries at the same moment in time but also within-country variation over time. In Chapter 4, where I explore Ethiopia's approach to drought over time, the latter is important.[61] In countries like Ethiopia that experienced improved

---

[58] Sen 2000, 178.
[59] Hendrix 2013; Hendrix and Haggard 2015; Smith and Flores 2010.
[60] Nixon and Stevens 2017. [61] Marshall and Gurr 2016.

3 The Argument, Method, and Mechanisms

climate-security outcomes over time, one of the drivers could be more inclusive governance systems.

As mentioned above, my argument has some affinity with selectorate theory. Countries with large winning coalitions ought to be more responsive to their citizens in the provision of public goods, and those with smaller winning coalitions should provide fewer public goods.[62] The original data from the *Logic of Political Survival* tracked winning coalition size through 1999 and was extended to 2005 by Cao and Ward.[63] As an alternative indicator of exclusion, countries in the bottom 25 percent of the distribution, relative to the size of the winning coalition, can be coded as having high political exclusion. However, scholars of autocracies have questioned the usefulness of these measures, noting that the size of the winning coalition is often not known with precision, given the opacity of decision-making processes.[64] Calculations of winning coalition size also rely on some of the same indicators that comprise Polity IV data on regime type. Despite these concerns, winning coalition sizes could provide some additional indicators from which we can judge political exclusion.

An alternative approach, building on a United States Agency for International Development (USAID) methodology, tracks state legitimacy using an index approach. While an imperfect proxy for political exclusion, the two are related. It is the lack of political inclusion that creates a legitimation problem for countries and also creates disparities in service provision for groups. In a report for USAID, Moran et al. tracked state legitimacy between 2000 and 2014, which permits some triangulation for cases in the contemporary period as well as changes over time.[65]

The index of legitimacy includes four dimensions: political, security, economic, and social legitimacy. Each dimension is comprised of three additional indicators:

- *Political* – competitiveness of political participation, citizen participation in selecting government, and asylum requests as percentage of population;
- *Security* – state use of political terror, presence of militant groups against the state, and the number of rival military organizations;
- *Economic* – control of corruption, rule of law and property rights protection, and number of days to start a business;

---

[62] The argument is actually a bit more complicated in that they evaluate the size of the selectorate, the group that chooses the leader, and compare the ratio of the winning coalition to the selectorate, to identify countries.
[63] Cao and Ward 2015.   [64] Gallagher and Hanson 2015.   [65] Moran et al. 2018.

- *Social* – military expenditures as percentage of GDP, percentage of parliamentary seats held by women, and life expectancy at birth.

Where these diverse indicators are available but produce discrepant findings, it is also useful to interrogate a case and ask whether there are important groups in society that lack representation in government or perceive themselves to be underrepresented. This would allow us to identify country cases like Ethiopia where groups such as the Oromo who are numerically large – around 30 percent of the population – have felt marginalized from power from the early 1990s through the 2010s, while the minority Tigrayans held key positions of authority.

Of these first two dimensions, my expectation is that state capacity may matter more than political exclusion in shaping whether climate changes produce humanitarian emergencies or social conflict. States with limited state capacity will likely produce widespread suffering with or without inclusive governance systems. However, states with some state capacity but exclusive political institutions will likely generate biased policies and outcomes, with in-groups the beneficiaries of government policies that insulate them from the damaging effects of climate hazards and out-groups disproportionately harmed.

International Assistance

Responses to climate change may be as, or more, important than the hazards themselves in determining whether a drought, flood, or other climate hazard becomes a disaster or generates other negative security outcomes. This is the main lesson of vulnerability studies. While state responses are hugely important, international assistance can help compensate for capacity constraints by building capacity over time and by responding in emergency settings with food aid, humanitarian response, conflict mediation, peacekeepers, or other measures to address human suffering, to extend state authority to restore order, or to quell conflicts.[66] In studies of disaster risk reduction, there are fears that international aid might encourage moral hazard, where states rely on external assistance rather using their own resources to prepare for climate hazards. In work on sub-Saharan Africa, however, Bussell and her collaborators found these fears were overblown.[67]

---

[66] On the role of aid in conflict mitigation, see de Ree and Nillesen 2009; Findley 2018.
[67] Bussell 2014.

3 The Argument, Method, and Mechanisms

However, in the study of aid and civil wars, moral hazard has been found to be more of a concern: Aid may extend civil wars as warring parties fear the end of conflict will see those resources dry up. Moreover, in that literature, aid flows can extend civil wars by providing lootable assets to one side of a conflict. Aid flows may also pose a threat to rebel groups by providing the state with assets to strengthen its power. As a consequence, aid projects themselves may become targets if perceived as a threat to rebel authority.[68]

For the purposes of the case studies in this book, my main theoretical expectation is that international assistance can compensate for state weakness in the lead up to, and in the wake of, hazard exposure. By international assistance, I include traditional overseas development assistance (ODA) that precedes a crisis, but this measure also includes emergency humanitarian assistance in the wake of hazard exposure, which is often counted by donor governments separately from ODA, though there is some double-counting that occurs. Hereafter, when I refer to assistance or aid I mean both investments in disaster risk reduction and emergency international response assistance. Moreover, diaspora groups, foundations, and non-governmental organizations may also provide significant quantities of assistance, beyond that which is offered by intergovernmental organizations or bilateral donors. Furthermore, militaries often provide emergency humanitarian assistance which may not come from traditional foreign aid budgets or be counted as part of emergency response flows. I am mostly interested in the period just after a climate hazard has been recognized to be of grave concern. I am somewhat interested in whether anticipatory foreign aid investments in disaster risk reduction have taken place, though this is likely already captured in metrics of state capacity.

I temporally focus on the measures taken when a hazard is imminent or has just happened, in the days leading up to and after a swift-onset hazard like a cyclone or in the weeks and months after a slow-onset hazard like a drought unfolds. However, in some cases, emergency foreign assistance may not be allowed in, which would mean local populations have to rely on the state or community resources. In other cases, the state or local actors will seek to channel or capture aid for their favored groups. Diaspora groups may also channel their support to friends, relatives, and coethnics. To the extent that assistance patterns in the wake of a hazard are one-sided to favor some groups at the expense of others,

[68] Findley 2018.

we should observe differential suffering based on groups that get aid and those that do not.

In states where virtually no assistance is allowed in at all, the effects of a climate hazard will depend on a state's capacity and, secondarily, on its degree of political inclusion. In weak capacity states, the effects will likely be broadly negative for most groups, with the effects depending on the severity of the hazard. In higher capacity states with inclusive institutions, states that do not allow assistance in will experience some negative outcomes, geographically concentrated in high hazard exposure areas but without discriminatory impacts based on ethnicity, class, or other sources of social division. In higher capacity states with exclusive institutions, states that do not allow aid in should experience biased outcomes like those that channel aid in a one-sided manner. Regime supporters should also suffer but not nearly as much as discriminated ethnic or social groups.

While I recognize that the quality of international assistance matters, here my rough measures (i.e. whether it is provided and, if provided, whether aid is distributed in a one-sided manner) provide some traction on the role of aid in reducing negative security outcomes. That said, tracking aid flows and their distribution in the wake of hazard exposure is a challenge.

ODA measures are available from the Organisation for Economic Co-operation and Development (OECD) Development Assistance Committee, with additional donors captured by the AidData project. The Financial Tracking Service (FTS) of the United Nations Office for the Coordination of Humanitarian Affairs (OCHA) provides one measure of emergency assistance.[69] As mentioned, funds for disaster assistance are often but not always different from flows of finance for ODA. The reporting of both disaster and aid flows is often opaque, subject to some double-counting and lag effects, so the timeliness and overall levels of funding may be difficult to observe, and the distribution more difficult still. In the wake of a hazard event, the line between emergency response and ODA will likely be blurry, as some relatively short-term sources of funds may also come from ODA budgets. Still, as a first cut, we can review ODA levels in the lead up to a hazard event, and OCHA FTS flows in the wake of the event. If the United Nations has issued an emergency appeal, we can assess both the volume of funds and how quickly the appeal was met, as some appeals do not receive an adequate or timely response.

---

[69] OCHA Financial Tracking Service. n.d.a.

3 The Argument, Method, and Mechanisms

Some forms of assistance, such as military emergency aid, may not be picked up by OCHA FTS, and it is not clear that governments adequately report all forms of emergency assistance to OCHA FTS. The US military, for example, has shared a catalogue of its emergency disaster aid for the period 2005–10, and USAID's Office of Foreign Disaster Assistance has shared its data for the period 2000–9. For events that occurred outside of this time period, news reports and press releases provide some metrics of assistance levels.

Perhaps the easiest indicator to observe is whether aid is allowed in by domestic actors. Two cases in this book, namely, Somalia in the wake of the 2011 drought and Myanmar in the wake of Cyclone Nargis in 2008, provide evidence of local actors denying external actors entry for the delivery of international assistance, though some aid was ultimately permitted. This makes observing the effects of aid denial difficult to track. We have to ask the counterfactual: What levels of international aid would have been delivered by when, were it not for governmental interference? The challenge is that if aid is provided, how can we assess whether it is delivered in a biased manner? This is a difficult area of study for foreign assistance. Here again, we may have to rely on news reports and scholarly after-action reviews of aid capture by the regime.

How then do climate hazards, state capacity, political inclusion, and foreign assistance relate to each other in leading to negative security outcomes? In the next section, I briefly recapitulate what I mean by security outcomes. I then sketch out several causal pathways that build upon the basic conjunction of hazards, capacity, inclusion, and aid.

Climate Hazards, Causal Chains, and Security Outcomes

As I argued in the Chapter 2, I conceive of negative security outcomes as including but not limited to conflict. Humanitarian emergencies constitute security threats, not least because of human security outcomes with the risk of large-scale loss of life. They also pose national security risks in so far as they often require the mobilization of military assets for humanitarian rescue, which, if nothing else, creates opportunity costs for militaries diverted from their primary purposes of national defense.

Humanitarian emergencies wrought by climate hazards, particularly swift-onset hazards, may also lead to temporary loss of law and order in the wake of a flood, fire, or storm, creating local security challenges – potentially made worse if there is already weak governance in a region facing irredentist movements or lawlessness.

Conflict, both rebel movements and communal conflicts between groups, is another potential risk associated with climate hazards, though the causal chain is longer than for humanitarian emergencies, particularly with respect to slow-onset hazards such as droughts. When comparing cases, I wanted to find two cases that were similarly affected by a climate hazard but produced different results.

For a climate hazard, I look to extreme weather events, such as swift-onset hazards like floods, cyclones, and wildfires. Existing data platforms – for example, the United Nations Environment Programme's (UNEP) Global Risk Data Platform – can serve to identify hazard events.[70] Drought is a more difficult concept to operationalize as there is no agreed upon definition of the concept and many different ways to operationalize it. Do we mean changes in rainfall patterns (meteorological drought) or are we talking about impacts specifically for crops (agricultural drought)?[71]

The climate and security field, with its emphasis on Africa where rainfed agriculture is dominant, has often looked at deviations in rainfall from normal levels. However, there are different ways to operationalize what constitute normal baseline levels, depending on whether one is looking at aggregate national yearly rainfall data, disaggregated rainfall data specific to particular geographies within a state, monthly rainfall compared to previous monthly means, or seasonal rainfall for growing season months. There is a proliferation of metrics like the Standardized Precipitation Index, the Palmer Drought Severity Index, and other alternatives that have tried to operationalize the concept of drought in different ways. The challenge is made more difficult where rainfed agriculture is supplemented by irrigation and/or groundwater withdrawals, potentially dampening the effects of negative rainfall shocks on agriculture. This difficulty is made more severe still when we recognize that droughts may be a product of a confluence of factors, including temperature, evapotranspiration, groundwater withdrawals, and topography. Recognition of these complications has led researchers to develop more complex water balance models, such as one pioneered by the consultancy ISciences.[72]

In Chapters 4, 5, and 6, I identify countries that faced similar exposure to climate hazards during the same time period, ideally in the same year, but sometimes within a few years of each other. For drought-related

[70] Global Risk Data Platform 2013.   [71] Lyon 2011.   [72] ISciences n.d.

3 *The Argument, Method, and Mechanisms* 59

measures, I try to triangulate with different sources to get at similar phenomena. I seek to compare hazards of relatively equal magnitude such as Category 5 hurricanes that affect two different countries but have differential effects.[73] Where possible, I seek geographically disaggregated data on hazard events to locate both when and where they occur in a country.

To identify paired cases, I seek cases where one country experienced a negative security outcome and the other did not. For negative security outcomes, I rely on the EM-DAT International Disaster Database for humanitarian emergencies, which gives a rough account of the severity of the impacts of natural hazards with estimates of the number of fatalities and the number of people affected. EM-DAT includes some geographic markers, so one can compare the physical hazards to the geography of disaster impacts.[74]

For conflict measures, I rely on indicators of civil war (i.e. conflicts between a rebel group and the government) or communal conflict (i.e. conflict between groups in society). I draw from the Uppsala Conflict Data Program which includes measures of armed conflict, nonstate conflict, and battle deaths.[75]

My approach to case selection has been to choose regional neighbors. Regional neighbors may face similar climate phenomena but have domestic political differences and differing orientations to external assistance that materially shape why some places experience negative security outcomes and others do not. The approach of case selection is roughly Mill's method of difference.[76] I try to identify countries that face very similar circumstances but produce different outcomes because of key differences on one (or a few) parameters. The challenge is rough matching since countries may differ on characteristics beyond those I have identified as theoretically important. Countries in the same region may have some cultural, climatic, of economic similarities that make them apt comparisons, though neighbors potentially run the risk of contagion between countries from migration or other externalities.

While paired cases have been selected based on the premise that climate hazards contributed to negative security outcomes in one of the cases, that causal chain has to be demonstrated not assumed. In some of the cases, notably the Syrian civil war, the causal role of climate hazards has been

---

[73] Walker et al. 2018. I recognize that there are efforts to revisit the cyclone Saffir–Simpson scale to incorporate other dimensions beyond wind speed.
[74] CRED 2019.   [75] UCDP 2019.   [76] Mill 2002.

disputed by scholars.[77] In cases where there is a lengthier causal chain, like the Syria conflict case, I therefore also pursue process tracing as a complement to comparative case methods. Process tracing unpacks the temporal causal sequence between cause and effect, including intervening variables.

The literature on process tracing has identified different tests of stringency that evidence can provide. Information that is consistent with an explanation but is neither necessary nor sufficient to explain an outcome is regarded as a *straw-in-the-wind*, a relatively weak test. For example, did a drought precede the crisis in Syria? Harder tests are so-called *hoop tests*: evidence that would be necessary to confirm a hypothesis but would not on its own confirm it. Did the drought in Syria lead to population movements and protests in cities that received large numbers of migrants? Stronger still is a *smoking gun* test where evidence would be sufficient to confirm an argument. Did the protesters identify the drought as one of their reasons for protesting? Finally, *doubly decisive* tests would provide both necessary and sufficient evidence to support a case and rule out competing explanations.[78] Were the protests across Syria almost exclusively about drought?

Below, I connect my argument and sketch the putative causal pathways to different security outcomes, starting with the simplest (humanitarian emergencies), extending to more complex outcomes (conflict), through lengthier causal chains and the indirect pathways discussed in Chapter 2.

While I recognize that climate hazards and conflict can have complex feedback loops over longer time horizons, this project focuses on relatively short-term responses to climate hazard events – the days, months, and first years following a hazard event. The diagrammed pathways represent streamlined and relatively simple causal diagrams like those in Colin Kahl's research rather than more complex pathways developed by Thomas Homer-Dixon in the first generation of environmental security scholarship.[79]

THE CAUSAL PATH TO LOSS OF LIFE

Perhaps the most direct path to negative security outcomes is if a climate hazard puts large numbers of people in harm's way. For swift-onset hazards in particular, where people do not have much warning, this can

---

[77] Selby et al. 2017a. [78] Punton and Welle 2015; Collier 2011.
[79] Kahl 2006; Homer-Dixon 1991.

lead to large-scale death from direct exposure. In time, many more may die from thirst, hunger, and disease. The death toll can be especially high when governments are either unwilling or unable to help their citizens (see Figure 3.1). The Myanmar cyclone of 2008, discussed in Chapter 6, is illustrative of this causal pathway.

The causal path starts with a hazard event and ends in potential loss of life. Along the way, governments can respond to the hazard event, which is, in turn, shaped by their capacity and degree of political inclusion. Weak capacity states and regimes with high political exclusion would have central tendencies to be vulnerable to adverse outcomes in the event of shocks.[80] While those structural parameters shape the likely response, there is agency by governments. Similarly, international actors can respond and, here again, the level and nature of their response provides scope for agency. While both are represented as following hazard events in Figure 3.1, some measure of advanced preparation, early warning systems, planning, building codes, and other policies will have been enacted prior to the hazard (possibly supported by international assistance) and will shape how serious the outcomes are.

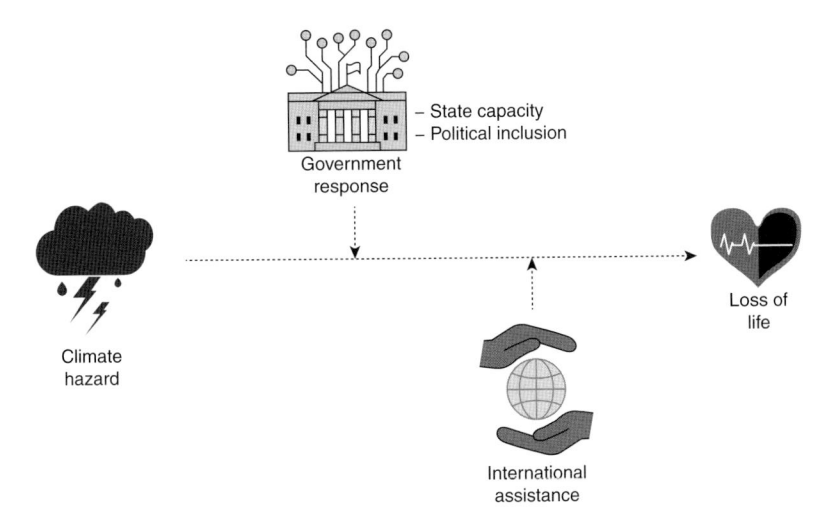

FIGURE 3.1 Simple pathway to loss of life

---

[80] On vulnerability, see Busby et al. 2018; Busby, Smith, and Krishnan 2014; Busby et al. 2013.

The potential for large-scale loss of life is not merely a function of responsiveness and capacity of the government in the aftermath of exposure, but it also includes preparations (or the lack thereof) in the lead up to exposure. If governments have not prepared for such contingencies through early warning measures or investment in risk reduction, then outcomes are likely to be worse, even if they display a willingness to respond after a crisis occurs. As in response, both capacity and institutional inclusion matter for whether a state is able and willing to prepare for a crisis.

Another piece of this causal chain is the role played by international assistance. The death toll in a humanitarian emergency can be dampened if broad-based international assistance is permitted to respond to the crisis. By international assistance, I mean both short-term humanitarian relief and more long-term ODA. When I use the terms assistance and aid hereon, it is shorthand for both. However, in some instances, governments (think of Myanmar after cyclone Nargis) or other political actors (Al-Shabaab after the 2011–12 drought) might be unwilling to allow international assistance in or they may impose a variety of impediments that block the flow of assistance to a trickle. International aid donors might also refuse to provide aid to a country and/or discourage others from assisting (as is observed in both the Somalia and Syria cases). In other circumstances, local actors may be able to divert aid so that it is one-sided and groups marginal or hostile to the regime cannot access resources while coethnics or those closer to the ruling elite are supplied. While the denial of aid outright might lead to broad suffering in the population, one-sided aid delivery ought to show discrete outcomes between populations that receive aid and those that do not.

As discussed in the section International Assistance, the line between short-term emergency relief and longer-run development aid has blurred in practice. Not only do such emergencies potentially shift development priorities, but efforts intended to get people back in their homes and up on their feet economically in the immediate wake of an emergency have lasting legacies for development. Beyond the immediate effects on human well-being, these responses to climate-related emergencies can prove contentious if certain groups get special treatment while others feel left out or underserved. This observation is in keeping with what scholars call the "backdraft" potential of responses to climate change – that the responses may prove to be as, if not more, contentious than the direct physical exposure itself.[81]

---

[81] Dabelko et al. 2013.

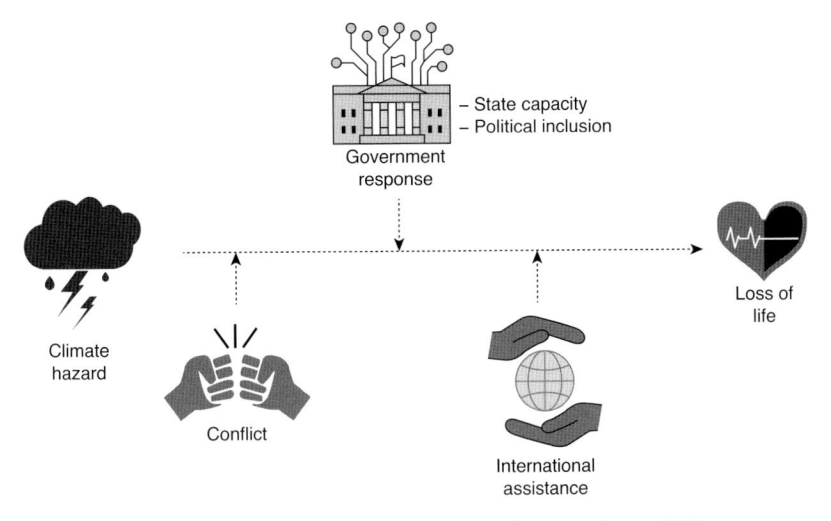

FIGURE 3.2 A complex emergency pathway to loss of life

A variation on this pathway is when a climate hazard affects a country already experiencing conflict – what has sometimes been referred to as "complex emergencies."[82] Here, the likelihood of famine or drought is made more likely because the presence of active conflicts interrupts the flow of humanitarian assistance. Over the longer run, the presence of armed conflict can also undermine institutional capacity and accentuate exclusive institutions. Those processes, in turn, can make conflict itself more likely (see Figure 3.2). The 2011–12 famine in Somalia discussed in Chapter 4 is an exemplar of this pathway.

THE CAUSAL PATH TO CIVIL/COMMUNAL CONFLICT

A secondary effect of exclusive institutions and one-sided aid delivery can be emergence or activation of grievances among marginalized groups that can translate into demands for redress through public protests and formal requests for assistance. Protests can escalate into riots and violence if groups have limited discipline, are spontaneous, and where participants are overwhelmingly young men. These episodes, in turn, can lead to violence if the state reacts to those demands with violent repression. In situations where there is already an active insurgency or exclusive

---

[82] The concept does not have a uniform definition but often reflects the concatenation of climate hazards and political instability, including armed conflict; see Macias 2013.

institutions, groups may not wait for repression to engage in violence (see Figure 3.3). This is a longer causal chain. The causal path begins with the hazard and passes through societal impacts which, in turn, lead to demands for redress or local contention. While government response and international assistance are represented at the tail-end of the path, they can enter the picture in the immediate aftermath of exposure. Indeed, preparations and assistance can also predate the hazard event itself.

Here, one example of societal effects of climate hazards is decreased agricultural production and lower economic output. These impacts can affect the motives for people to engage in conflict since lower returns from agriculture or legal economic activity can encourage people to join rebellions or engage in illegal raids on livestock, grazing lands, or watering holes of rival groups.[83] Over the long run, a decline in agricultural production (through the effects on tax revenue from commodity exports) could deprive the state of resources to support and contain internal security and exercise a monopoly of force over its terrain.

Another societal impact of hazard exposure could come through impacts on food prices rather than economic growth. Here, the contention associated with a climate hazard is driven by feelings of dissatisfaction among citizens who face higher food prices in the wake of deteriorating

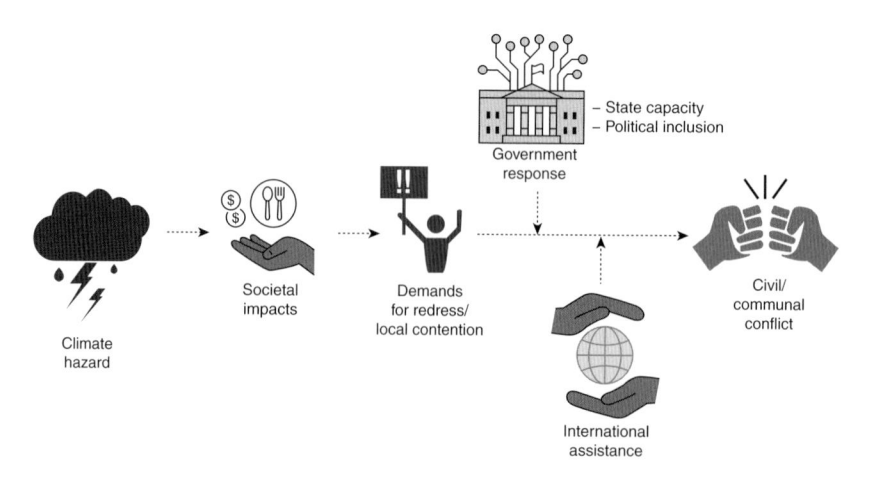

FIGURE 3.3 Pathway to civil/communal conflict

---

[83] Koubi 2017; Koubi et al. 2012.

local or global agricultural conditions that depress the production of basic foodstuffs such as grains. While social protest on its own is not a negative security threat – and indeed may be quite salutary in advancing human progress – protests can turn violent spontaneously or be repressed violently. These kinds of dynamics are more likely in countries with significant urbanization such as South Africa and Egypt, where a large proportion of the population no longer grows its own food and thus is more sensitive to price shocks to foodstuffs.[84] Here, climate shocks would intersect with weak state capacity, exclusive political institutions, and absent or one-sided aid in a different way. Weak states might not be able to buffer populations from food price shocks, leading to humanitarian emergencies and protest activity unless international assistance can help compensate with food aid. States with exclusive political institutions but some capacity might deliver services and international aid to politically connected groups, leading to suffering by underserved groups and potentially creating a source of grievance.

State capacity, exclusive institutions, and one-sided international aid may be relevant even where the target of violence is not the state. In response to deteriorating agricultural conditions, groups may engage in local contention over resources, as has occurred between farmers and herders over grazing lands and watering holes. In weak capacity states, communal conflict between societal groups may erupt because the state lacks both repressive capacity to clamp down on violence as well as adequate conflict resolution mechanisms, such as functioning court systems, that can facilitate nonviolent mediation of disputes. The state may even side with one societal group at the expense of other groups, perhaps escalating grievances between the groups or in ways that violently repress one group in a conflict.

Another societal impact of exposure to a climate hazard is human migration. Here, a climate shock may directly force large numbers of people to migrate internally or across borders if their homes have become uninhabitable due to storm damage or flooding. A slower process may unfold during an extended drought, which may ultimately lead farmers to migrate to towns or cities, as food reserves are depleted, water sources dry up, and livestock are sold and/or die off. Gradual salinization of agricultural lands or sea-level rise (punctuated by periodic storm surge events) may also induce similar calculations to migrate. While swift-onset disasters may trigger large flows of people all at once from affected areas, these

[84] Smith 2014.

movements may be temporary; once the hazard has passed there may be opportunities and support for rebuilding and recovery.[85] Slow-onset hazards like droughts may be associated with more long-term if not permanent movements, as it may take some time for the drought process itself to pass, and the loss of assets may make it challenging for migrants to return home and pick up their livelihoods again.

What then is the mechanism by which migration leads to security consequences? We can think about the immediate humanitarian and security challenge of protecting and providing for displaced persons in emergency centers such as stadiums, religious houses of worship, warehouses, or improvised or existing camps. In such situations, establishing rule of law and the security of residents and their property can be challenging in a group setting, particularly for women. That is all of a piece with the immediate proximate challenge outlined above with respect to a humanitarian emergency, which is both temporally and often geographically close to the site of occurrence of the original hazard.

The longer route to security consequences from migration is temporally and geographically more distant, when those who have left because of a climate hazard are not integrated into a wider population with whom they interact and potentially compete for services and resources. This pathway may be fairly direct in the wake of a swift-onset hazard to force migration, or it may contain more steps if the process goes through declines in agricultural production. Interaction with existing populations can be a source of friction. Here, there is some work to suggest that the migrants themselves are less likely to initiate conflict, in part because they are vulnerable newcomers subject to reprisals from locals in terms of lost employment, services, expulsion, or worse if they are politically active or engage in law-breaking.[86]

What may be as, if not more, important is how local groups receive the migrants and how government actors serve them and the wider community. Here, the local community may be more willing to absorb migrants into the population if they are seen to be displaced by visible acts of god like swift-onset cyclones or floods. It may be difficult, on the other hand, to disentangle the motivations for migrants displaced by slow-onset droughts. While in some circumstances, such as in situations where famine risk has been declared, there may be recognized suffering, in other contexts prolonged drought might be among the mixed motives people have for leaving an area.

[85] Raleigh, Jordan, and Salehyan 2008. [86] Ibid.; Koubi et al. 2016b.

The Causal Path to Civil/Communal Conflict

Although governments may provide migrants with some resources to settle in new locations, those benefits can become a source of jealousy if local actors perceive them to be one-sided, with longtime residents not eligible for such generosity or having to foot the bill for such measures. To the extent that the newcomers represent different groups based on ethnicity, religion, or livelihood, that can serve as a source of friction too.

Where people migrate to a region populated by those from the same social group, their migration may swell the ranks of people potentially disaffected with the government in a particular locale. Here, the conflict risk may be less of a function of friction between locals and new migrants, but from the combined population of locals and migrants who together are alienated from the government. Figure 3.4 shows the pathway to conflict from climate hazard through migration. The Syria case discussed in Chapter 5 is an exemplar of this pathway.

For each of these plausible causal pathways between climate shocks and security outcomes, we can see how state capacity, inclusive governance, and broadly distributed international aid together diminish the risks of humanitarian suffering as well as social conflict. States with weak capacity, exclusive political institutions, and absent or one-sided international aid are the most likely cases for humanitarian emergencies and social conflict. In between, there are a range of other cases where we can make some predictions about likely outcomes, since cases may not neatly fit into these ideal types. We can simplify each of these three dimensions as dichotomies, though in practice they are more likely to be continuums. Together, this gives us eight possible configurations. As I suggested earlier,

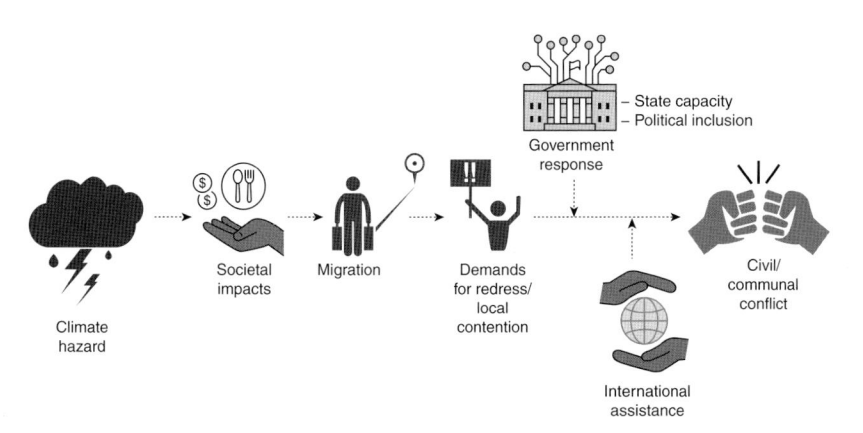

FIGURE 3.4 Extended pathway to conflict through migration

it may not matter as much whether states with weak capacity have inclusive or exclusive institutions since the state is likely incapable of service delivery, even to preferred groups. The first two configurations of weak capacity states (Types 1 and 2) thus produce humanitarian emergencies and possibly social conflict where aid is absent or delivered in a one-sided fashion. The next two configurations (Types 3 and 4) produce limited humanitarian emergencies and reduced risks of conflict because international aid is broadly distributed (see Table 3.1).

The next four configurations of strong capacity states produce different outcomes depending on both the degree of political inclusion and how foreign aid is delivered (see Table 3.2).

Type 6 is the least likely case for negative security outcomes because the state has strong capacity, is inclusive, and aid is delivered in a broad-based manner. Type 5 produces limited negative outcomes only if the severity of the hazard exceeds government capacity. However, a capable, inclusive state distributes aid broadly to all affected social groups. In these cases, a state might not receive international aid, but if it does, it would unlikely allow aid to be delivered in a one-sided fashion.

Type 7 are cases where there should be disparities of suffering because a capable, exclusive state distributes aid to some social groups at the expense of others and either forbids the delivery of international assistance or channels it to some groups at the expense of others.

Type 8 are cases where international actors distribute aid broadly to compensate for a government's favoritism. These may be cases where international actors come into conflict with a capable state that seeks to

TABLE 3.1 *Weak capacity states – configurations 1–4*

	Capacity	Institutions	International assistance	Predicted outcomes
Types 1 and 2 Weak, no/one-sided aid	Weak	Exclusive/inclusive	Absent or one-sided	Large-scale suffering or conflict, perhaps one-sided if aid supports some groups at the expense of others
Types 3 and 4 Weak, broad-based aid	Weak	Exclusive/inclusive	Broad-based	Reduced humanitarian emergency, no or minimal conflict

TABLE 3.2 *Strong capacity states – configurations 5–8*

Type	Capacity	Institutions	International Assistance	Predicted Outcomes
Type 5 Strong, inclusive, absent or one-sided	Strong	Inclusive	Absent or one-sided	Limited negative outcomes because the state has the capacity and will to distribute aid broadly.
Type 6 Strong, inclusive, broad-based	Strong	Inclusive	Broad-based	Reduced humanitarian emergency, no or minimal conflict.
Type 7 Strong, exclusive, no/one-sided aid	Strong	Exclusive	Absent or one-sided	Disparities of suffering and conflict, perhaps one-sided if aid supports some groups at the expense of others
Type 8 Strong, exclusive, broad-based	Strong	Exclusive	Broad-based	International aid is delivered to compensate for exclusionary state policies. Friction between international actors and the government.

divert aid for preferred groups. It is unclear whether there will be cases where a strong, exclusive state will permit broad-based forms of aid delivery, as this might pose a threat to the regime's legitimacy.

In each of the paired cases in Chapters 4, 5, and 6, I code the countries at different moments in time in terms of each of the eight types and evaluate whether the outcomes are consistent with my prediction for those configurations. My aim is two-fold: (1) to explain why some countries at similar moments in time experience different outcomes; and (2) to explain why the same country over time experiences negative security outcomes at some moments but not others. States that experience improved capacity and inclusion ought to show the most improved results. States that have weak capacity ought to show the worst results, particularly if no foreign aid is forthcoming to compensate for weak state capacity. States that have improved capacity but not inclusion ought to have biased results with some groups insulated from climate shocks while others suffer, even if foreign aid is allowed in. Foreign aid, in such circumstances, may be delivered broadly, though states with strong capacity and exclusive regimes may jealously protect their sovereignty and prefer to keep out aid rather than allow it to be distributed to all needy groups.

The next chapter on Somalia and Ethiopia contrasts their experiences with drought and famine in the 2010s, but also shows variation within Ethiopia over time, as a severe drought in the mid-1980s led to famine but did not in the 2010s.

4

Droughts and Famine in Somalia and Ethiopia

> Famine stops at the Somali border. I assure you this is not a political manipulation of the data – it is the data we have. Basically, the people without a functional state and collapsing markets are being hit much harder than their counterparts in Ethiopia and Kenya, even though everyone is affected by the same bad rains, and the livelihoods of those in Somalia are not all that different than those across the borders in Ethiopia and Kenya.
>
> — *Edward Carr*[1]

In 2011, the Horn of Africa experienced a devastating drought, described as "the worst in 60 years."[2] In July 2011, the United Nations declared a famine in two regions of southern Somalia. When it faced a severe drought in 2015, Ethiopia experienced food insecurity but no famine.[3] A report for the United Nations Food and Agricultural Organization (FAO) estimated that between 2010 and 2012 nearly 260,000 Somalis died above and beyond expected levels of mortality.[4] Ethiopia's 2015 drought affected nearly ten million people, but Ethiopia escaped famine.[5] Alex de Waal went so far as to claim that the death count was "near zero."[6]

What accounts for the differential impacts of drought on mortality between Somalia and Ethiopia? That is the central question animating this chapter. Amartya Sen seemingly has an answer to the question "Why famine?" in his magisterial work, *Development as Freedom* – democracy. The virtues of democracy are two-fold: politicians who have to run for reelection are responsive to the concerns of their citizens, and societies

---

[1] Carr 2011. [2] BBC 2011. [3] United Nations 2011. [4] Checchi and Robinson 2013.
[5] Climate and Development Knowledge Network 2017. [6] de Waal 2018.

with free presses bring to light problems that otherwise might remain hidden from authoritarian leaders.[7] However, while Ethiopia has had elections, it is not a democracy. According to Polity V data, Ethiopia ranked a −3 between 2005 and 2017, which is technically a closed anocracy, neither fully democratic (+6 and higher) nor authoritarian (−6 to −10).[8] Thus, we need to have a fuller, or different, explanation to understand the outcomes in Somalia and Ethiopia.

The first section of this chapter provides an overview of the criteria for case selection and reviews the cases covered and how these relate to my theoretical expectations. The second section contrasts the Somalia and Ethiopia experiences with each other and uses within-case variation in Ethiopia to show how the country has reduced its vulnerability to famine. This section reviews hazard exposure and presents data on capacity, inclusion, and assistance. The third section puts the pieces together in a synthetic narrative of why Somalia and Ethiopia experienced different outcomes after drought in the 2010s, and how Ethiopia was able to escape its past history of famine. The fourth section explores alternative explanations.

EXPECTATIONS AND CASE SELECTION

In each of these case study chapters (Chapters 5 and 6, as well as this chapter), I pair country cases that have faced similar climate exposure but experienced different security outcomes. I also take advantage of variation over time within countries to explain why some hazards generated negative security events in some periods and not others. Each country case begins with some documentation on climate hazard exposure. It may not be possible with current scientific attribution techniques to isolate whether anthropogenic climate change caused the particular hazard event, but it should be a plausible proxy, that is, the kind of hazard that scientists expect to see accompanying climate change. This will usually be reflected in some deviation from normal rainfall or temperature patterns, or an extreme weather event such as a cyclone or drought. Here, I select extreme weather events that were plausibly caused by climate processes,

---

[7] Sen 2000.
[8] Center for Systemic Peace 2020. In 2018, Ethiopia transitioned to an open democracy, coded as 1. From 2002 to 2011 Somalia was coded as −77, which reflects cases of "interregnum" or anarchy. In 2012, Somalia was coded as −66 which is a foreign interruption, after which Somalia transitioned to 5 or open anocracy.

though these are not in and of themselves events that can be said to be caused by, or made more likely by, anthropogenic climate change. Some are associated with natural processes like El Niño or La Niña events, though those processes too may be accentuated by climate change.[9]

Neighbors that are similarly physically exposed provide excellent paired cases for exploration as the agroecological zones and livelihoods of the populations may be similar, though there are risks of political effects across borders. Ideally, both countries face severe climate exposure at the same moment in time, though even neighbors facing a drought might not be equally affected. In the case of Ethiopia and Somalia, I found comparing the 2011 drought in Somalia to the 2015 drought in Ethiopia to be a better case comparison than using the effects of the 2011 drought on Ethiopia, which had less consequential effects on the country's agriculture.

We also need to identify proxy measures to reflect state capacity, political inclusion, and foreign assistance. Since no single source necessarily is authoritative on these dimensions, it may help to triangulate using multiple sources. For foreign assistance, the challenge is two-fold. First, whether aid is allowed in is a first test, as some governments or subnational actors often refuse help in the wake of hazard exposure (or are denied help by external actors), as we shall see with respect to Somalia (in the section Putting the Pieces Together). Second, we need to have some information about the distribution of aid if it is allowed in, to ascertain whether it is delivered in a one-sided manner to regime supporters and denied to opponents and disfavored groups. That task is harder. We also need to assemble a narrative for how these pieces fit together, to lead to negative security outcomes in one instance but not the other. Ideally, research should also be open to alternative explanations for these outcomes.

While paired country cases are one way to approach the question, we can also exploit within-case variation to observe how changes within a country may reduce or exacerbate vulnerability to security outcomes over time. For example, as I argue in the section Evidence of State Capacity, improvements in state capacity in Ethiopia alongside more (but incomplete) political inclusion diminished the risk of famine over time.

---

[9] Cho 2016. While the science of attributing specific climate events to anthropogenic climate change is improving, few of the cases explored in this book have been subject to that nascent form of inquiry.

Between countries, I should observe differences in state capacity, inclusiveness, and/or aid provision. The better performing country should have stronger state capacity, more inclusive institutions, and possibly more international assistance. The worse performing state should have some combination of weak state capacity, exclusive political institutions, and no or one-sided aid delivery. A state that improves its performance over time may develop higher state capacity, more inclusive institutions, and/or more broad-based aid delivery. In other cases, a state that faced deteriorating or inconsistent performance may face fluctuations in capacity, inclusiveness, or aid. The outcomes in these cases ought to match up with changes in some of the proxy indicators for these phenomena.

More importantly, I ought to find support for the causal narratives connecting climate hazards and outcomes in news reports and scholarly accounts of the cases, both the sequence and the combination of factors. This chapter and Chapters 5 and 6 are meant to build on the excellent scholarly work of regional studies experts. Given the geographic diversity of the cases covered and the physical danger associated with some of these country contexts, I elected to draw from deep dives that country experts have already carried out.

In this chapter, Somalia and Ethiopia are compared to each other and, in Ethiopia's case, to itself over time. While Somalia has had numerous drought episodes in the latter years of the twentieth century and in the early twenty-first century, I focus on the 2011 drought that led to a famine and such severe outcomes. In Ethiopia, I use the 2015 drought as a point of contrast since the country was able to mobilize to avoid famine. I also compare Ethiopia to itself and consider the 1984 drought that led to famine, which claimed perhaps as many as 600,000 lives. The 2015 drought severely tested the Ethiopian government but loss of life was limited. While the central question animating this chapter is to explore why, in the 2010s, Ethiopia and Somalia experienced different outcomes in the face of severe droughts, I also seek to exploit within-case variation to understand how Ethiopia has reduced mortality in the face of droughts.

Here, I am setting aside the question of the relationship between climate hazards and conflict. The famine contrast between Somalia and Ethiopia provides for a somewhat cleaner comparison. However, the causal relationship between drought and conflict is relevant. In Somalia's case, Maystadt and Ecker have examined the connections between drought and civil war, finding that drought diminishes the returns from pastoralism by lowering the prices of livestock, thus making engaging in conflict activities more attractive. Even though Somalia's

Expectations and Case Selection

violence predated drought episodes, using conflict event data from the Armed Conflict Location and Event Dataset (ACLED), Maystadt and Ecker were able to show a spike in conflict events that followed drought periods.[10] As I discuss in the section Alternative Explanations, one could argue that the differences in famine mortality between the countries may be driven by differential levels of violence rather than capacity and inclusion. I suggest that comparatively higher violence in Somalia is a reflection of weaker capacity and less inclusive governance, but also over time exacerbates both of these parameters. Indeed, one could make the case that Ethiopia's experience in the 1980s and in 2015 is similar: Weak administrative capacity and high political exclusion made violence more likely while a more capable and inclusive state diminished the risks of violence decades later.

Somalia could also be subject to within-case analysis – though I am constrained by space and unable to explore this here. Somalia also experienced multiple years of severe drought, but there were only two famines. The country has faced weak state capacity throughout the entire period, ever since a coup toppled Siad Barre in 1991. That means that variation in performance is largely a function of the other two parameters. Here, the provision of aid at some junctures but not others would have to explain the differential outcomes, since even a weak state with inclusive institutions is unlikely to be able to respond effectively to a severe climate hazard. Thus, if Somalia experienced a famine in 1992 and 2011, my argument anticipates that delayed, absent, or blocked aid flows explain the outcomes.[11] The implication is that permissive flows of aid in other drought periods prevented famine events. This chapter focuses on the failed response to the 2011 Somalia drought, leaving other episodes in Somalia's history to be explored in subsequent research and by other scholars.

Ethiopia, for its part, experienced famine in 1984 but none since. In 1984, Ethiopia was a very poor country with a communist military dictatorship that withheld aid from suffering populations and did little to alert the world to its suffering. While a massive aid response was ultimately triggered by that drought, it came too little too late. Thus, Ethiopia was a most likely case for famine in 1984: a weak state, with politically exclusive institutions, and unable to elicit international help.

---

[10] Maystadt and Ecker 2014.
[11] EM-DAT International Disaster Database lists a number of drought disasters in the 2000s in Somalia including 2000, 2004, 2005, 2008, and 2010. Curiously, the 1992 famine is not listed, and the death toll for the 2011–12 famine is only 20,000; CRED 2019.

The severe drought episode in 2015 that avoided famine was a function of improved state capacity, more inclusiveness, and steady support from international donors.

These case dynamics are summarized in Table 4.1 (the droughts discussed in this chapter are in bold). Building on the discussion on causal pathways in Chapter 3, I also specify which predicted pathway the outcome corresponds with.

SOMALIA AND ETHIOPIA

I examine the paired cases of Somalia and Ethiopia, comparing them to each other and, in Ethiopia's case, to itself over time. Before providing evidence in support of my claims, I lay out my general argument. In 2011, Somalia suffered a devastating famine in which nearly 260,000 people were estimated to have died above and beyond normal rates of expected mortality. By contrast, Ethiopia, which faced similar exposure, did not experience a comparable famine in 2015 when the country was buffeted by an extreme drought. This was quite a different outcome from the mid-1980s when Ethiopia faced a drought that took the lives of hundreds of thousands, which, in turn, became an international cause célèbre with the 1985 Live Aid concert.

What set these two countries apart from each other in the 2010s and how was Ethiopia able to avoid its past fate in 2015? Between 1992 and 2016, Somalia languished largely without a functioning central government. Meanwhile, Ethiopia had developed considerable government capacity since the 1980s, in the decades after the Ethiopian People's Revolutionary Democratic Front (EPRDF) seized power from the socialist Derg government in 1991. Despite arguably having an authoritarian government, Ethiopia has increased state capacity and been amenable to foreign aid and humanitarian assistance, which has helped to alleviate famine risk. In 2015, Ethiopia faced an extreme drought, and the government was up to the challenge, avoiding large-scale loss of life.[12]

Substantiating the argument requires several pieces of confirmatory evidence. First, we need to identify periods of extreme drought in Somalia and Ethiopia. Second, we need to show differences in state capacity between Somalia and Ethiopia in the 2010s as well as differences in Ethiopia itself in the 1980s and the 2010s. Third, we need to be able to

---

[12] Since the drought, some of the gains have been undone by resurgent ethnic conflict, discussed later in this chapter in the section Concluding Thoughts.

TABLE 4.1 *Comparing Somalia and Ethiopia*

Country	Hazard events	Capacity	Institutions	International assistance	Outcomes
Somalia	Droughts 1992 2000 2004 2005 2008 2011–12	Weak	Exclusive	Limited access 1992, 2011 Broad-based aid delivery	Two famines 1992, 2011–12 (Type 1) No famine (Type 3) 2000, 2004, 2005, 2008
Ethiopia	Droughts 1984 2000 2002 2010 2015	Weak 1984 → Improved state capacity 2000, 2002, 2010, 2015	Exclusive → Increasing inclusivity but still somewhat exclusive	Limited access 1983–5 Broad-based aid delivery 2015	Famine 1984 (Type 1) No famine 2000, 2002, 2010, 2015 (Type 6 or Type 8)

show that Ethiopia had exclusive political institutions in the 1980s with improvement by the 2010s, though there were still some questions about political inclusion, compared to other countries. Somalia should largely have exclusive political institutions throughout this period, though arguably weak state capacity may be more important than the degree of political inclusion. Fourth, we should be able to show that Somalia experienced an aid blockage in 2011.[13] In the case of Ethiopia, we should be able to show limited or one-sided aid delivery in the 1980s to favor preferred groups but more broad-based aid delivery in the 2010s. Finally, while it is important to be able to show that the indicators are consistent with my expectation, I also draw on area studies experts of both countries to substantiate that my interpretation is consistent with scholars and practitioners with more ground-level expertise.

Evidence of Hazard Exposure

Somalia's main rainy season, *gu*, is April to June. The second rainy season, *dayr* or *deyr*, lasts from October to December.[14] Ethiopia's weather patterns are somewhat different. The main crop seasons are the *belg*, a short rainfall season from February to May, and the *kiremt or meher*, with rains from June to October. The *kiremt/meher* season produces more than 90 percent of the country's cereals output[15] and is an important season in most of the country except for the south and southeast.[16]

How can we assess that these countries experienced comparable levels of drought? There is no universal standard for measuring drought. Scholars distinguish between meteorological drought (in terms of changes in precipitation or temperature from "normal" levels), agricultural drought (focusing on changes in soil moisture in areas and periods important for agriculture), and hydrologic drought (which emphasizes low water levels in streams, lakes, reservoirs, or subsurface sources). Drought may be more severe when lower than normal rainfall or low groundwater levels combine with high temperatures to affect evapotranspiration of plants.[17]

There are different ways to operationalize drought including the Palmer Crop Moisture Index, the Palmer Drought Severity Index, and

---

[13] An extension of the argument would be an expectation to see delayed aid delivery to Somalia in 1992 and more permissive aid delivery in other drought episodes in the country.
[14] Encyclopedia Britannica n.d.b.   [15] USDA 2008b; Encyclopedia Britannica n.d.a.
[16] Ethiopia National Meteorological Agency n.d.   [17] Lyon 2011.

the Standardized Precipitation Index, among others. Some are more data intensive than others. Some rely on weather station data, which became increasingly challenging to collect in the postcolonial era and amid conflict in a number of countries.[18] Newer data sources available since the late 1990s rely on satellite data to observe patterns from space. Some datasets combine station and satellite data.[19] In areas where countries have access to irrigation, groundwater withdrawals may compensate for lack of rainfall. Measuring drought under such circumstances may demand more complex water balance models.[20]

Ethiopia and Somalia are both overwhelmingly reliant on rainfed agriculture. Irrigation has not featured prominently in either country. In 2011, the World Bank estimated that 0.5 percent of agricultural land in Ethiopia was irrigated.[21] There is some irrigation in southern Somalia in the Shabelle and Juba riverine areas; the FAO estimated that in 2003 some 5 percent of cultivated land in Somalia was equipped for irrigation.[22] I am primarily interested in the effects on agriculture; smallholder farmers tend to be those most vulnerable to famine in Ethiopia and Somalia. Therefore, it makes sense to focus on drought during their growing seasons.[23]

National-level data may provide a first indication of trends in countries, though it may mask considerable subnational variation. Some such as Maystadt argue that temperature anomalies, particularly in Africa, are a better measure of drought, particularly given the relative scarcity of rainfall weather station data and that rainfall data may have more errors.[24] Whether we use rainfall, temperature or some combination, we are interested in deviations from some "normal" reference period. Scholars operationalize this in different ways. Some compare rainfall in the present period to rainfall levels across an entire period such as 1980–2009. Thus, rainfall in 2005 would be compared to the average rainfall in that particular year or month across the entire period 1980–2009.[25]

This seems a curious way to think about drought since it includes future periods in the calculation of what constitutes normal rainfall. But, agriculturalists are not passive and will move herds, change planting cycles or crops, sell livestock, consume seed, or migrate to other areas in the face of droughts.

---

[18] Ibid. This may also affect weather station data on temperature; Schultz and Mankin 2019.
[19] Sun et al. 2018.   [20] ISciences n.d.   [21] World Bank 2019a.
[22] Food and Agriculture Organization 2005.
[23] This is in keeping with state-of-the-art climate security scholarship and recommendations from other scholars; von Uexkull et al. 2016; Koubi 2019; Meierding 2013; Busby 2018b.
[24] Maystadt and Ecker 2014, 1167.   [25] Maystadt and Ecker 2014.

To make judgments about what to do under such circumstances, they will reflect on their own personal experience of normal rains or temperature. Thus, another way to get at deviations from normal rainfall is to compare rainfall or temperature in the present period to the average in some number of previous periods based on the experience of farmers. What is a reasonable period of memory to imagine that farmers will have? Many of them likely start helping their parents in the fields at a young age, though their memory of planting cycles may develop later in life. In previous collaborations (and similar to work by other scholars), I used a twenty-year rolling average on a monthly basis and looked at deviations from normal rainfall (and temperature) during those months.[26]

Here, I operationalize rainfall a little differently by focusing on aggregate deviations during growing season months, specific to Ethiopia and Somalia. Thus, I compare March rainfall to previous average March rainfall for the previous twenty years (excluding the current March). I do the same for each month. I then calculate the size of the deviation of the current March from the average. Then I look at the aggregate rainfall deviations across each growing season. So, for Somalia, that means the net rainfall difference from normal for the *gu* rains (May, June, July) and the net rainfall difference from normal for *dayr* rains (October to December). For Ethiopia, that means the size of cumulative rainfall deviations for *belg* rains (February to May), as well as the size of total rainfall anomalies for the *kiremt* rains (June to October).

I examine rainfall seasons separately to see if the sequence of rainfall shows consecutive low rainfall seasons and/or multiple years of low rainfall. To address Maystadt's argument that temperature is a more reliable indicator of drought, I also do the same for temperature to see if there are large heat wave events during the rainy season – which presumably would be another contributor to drought. Data on rainfall and temperature come from the Climate Research Unit (CRU) at the University of East Anglia by way of the World Bank Climate Knowledge Portal.[27] CRU data are derived from station data dating back to 1901. These are available in gridded form at 0.5 degree spatial resolution but have been aggregated here up to the national level.

The figures below show the cumulative rainfall deviations from the previous twenty years for each agricultural season in Somalia and

---

[26] Busby et al. 2018; Busby, Smith, and Krishnan 2014; Smith 2014. See also Koubi et al. 2012. I thank Todd Smith for STATA code to generate the twenty-year rolling average.
[27] World Bank n.d.

Ethiopia. The Somalia data show that the largest drop in *gu* rains accompany the 1992 famine. There were a number of other years of low rainfall in the mid-1990s and early 2000s. In the lead up to the 2011 famine, Somalia faced lower than normal rains for three consecutive planting seasons. *Gu* rains were much lower than normal in 2009 and were slightly below normal in 2010. *Dayr* rains were also below normal in 2010 (see Figures 4.1 and 4.2).

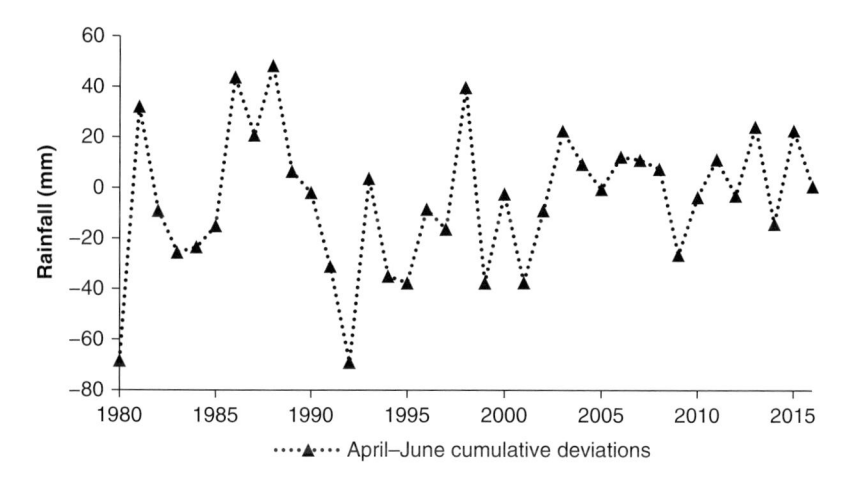

FIGURE 4.1 Somalia *gu* rainfall deviations, 1980–2016
Source: Author's calculations using CRU data

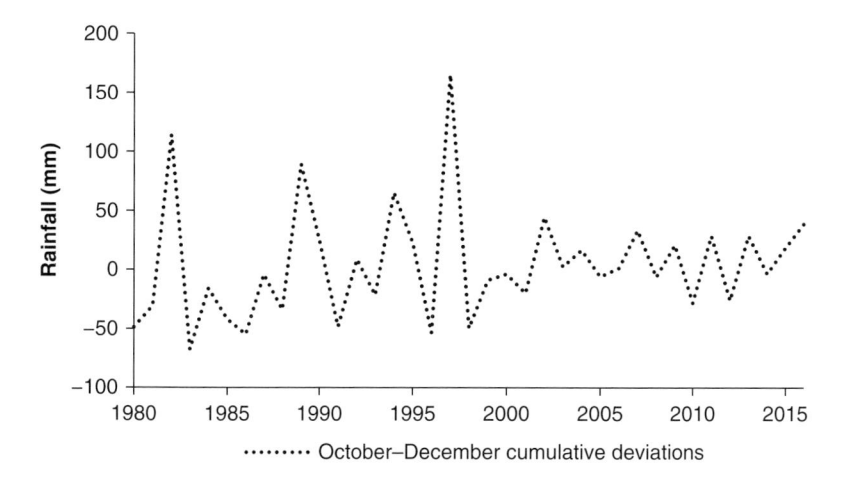

FIGURE 4.2 Somalia *dayr* rainfall deviations, 1980–2016
Source: Author's calculations using CRU data

In the case of Ethiopia, the data show a precipitous decline in *belg* rainfall leading up to the 1984 drought. There were other notable nationwide *belg* below normal rainfall periods in 1988, 1992, 1994, 1999, 2001, and 2009. While the 2015 rainfall deviation was not especially low by historic standards, it followed several years of good rains (and does not reflect subnational variation). The *kiremt* rains showed an especially large drop in 1987, as well as below normal rainfall years in 2002, 2004, and 2013. These national level observations roughly correspond to de Waal's record of major food crisis years, 1984–5, 1999–2000, 2002–3, and 2015–16, though the rainfall declines leading up to 2015 were not especially severe by historic standards (see Figures 4.3 and 4.4).[28]

As a robustness check, I also calculated temperature deviations for the agricultural months for both countries. For rainfall, cumulative deviations are relevant as multiple months of below average rainfall may negatively affect agricultural production. For temperature, the average across months rather than the cumulative difference is more germane.

In Somalia, 1998, 2003, and 2009 were the most notable years when *gu* season temperatures were above normal – more than 1.6°F in 2009 in the lead up to the famine. *Dayr* temperatures were also above normal in 2009 and 2010, 1.2°F and 1.4°F, respectively. Like the *gu* years, other hot *dayr*

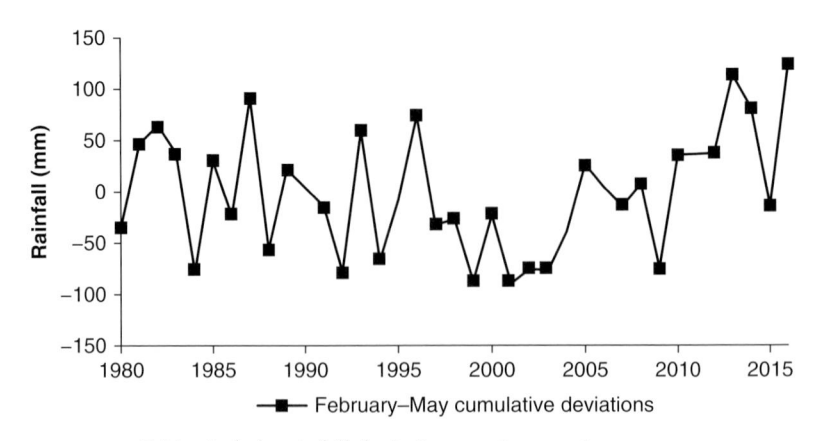

FIGURE 4.3 Ethiopia *belg* rainfall deviations, 1980–2016
Source: Author's calculations using CRU data

[28] de Waal 2018.

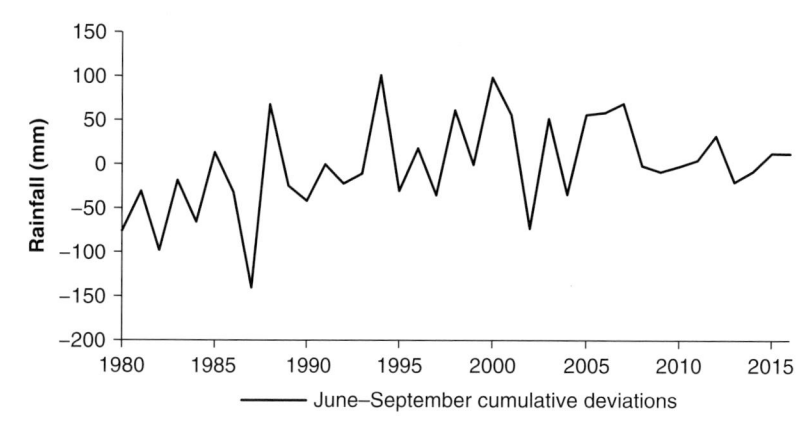

FIGURE 4.4 Ethiopia *kiremt* rainfall deviations, 1980–2016
Source: Author's calculations using CRU data

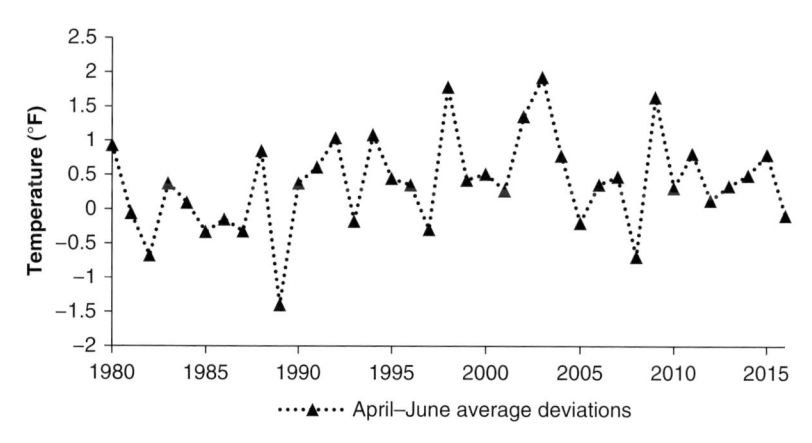

FIGURE 4.5 Somalia *gu* average temperature change, 1980–2016
Source: Author's calculations using CRU data

years were 1998 and 2003. These temperature findings confirm the rainfall data, suggesting that the period leading up to the 2011 famine was both drier and hotter than normal (see Figures 4.5 and 4.6).

In Ethiopia, *belg* high temperature years included 1980, 1998, 2003–4, 2009–10, and 2016. *Kiremt* high temperature years included 1983, 1987, 2002, 2009, and 2015. A number of these years correspond to de Waal's periods of food crisis in Ethiopia, including 1984–5, 1999–2000, 2002–3, and 2015–16. Ethiopia is perennially subject to drought so other years such as 2009 were also dry years, though not flagged by de Waal as a food

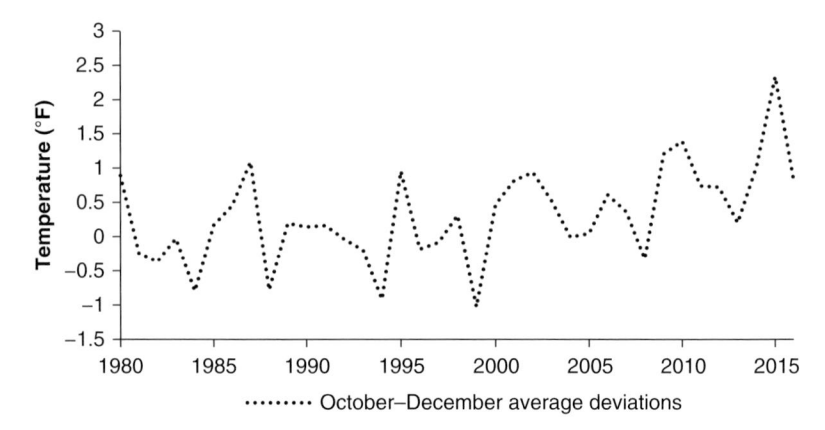

FIGURE 4.6 Somalia *dayr* average temperature change, 1980–2016
Source: Author's calculations using CRU data

crisis year. Thus, 2015 rainfall was below normal, but not exceptionally so. However, *kiremt* temperatures were more than 1.7°F above the norm for that season compared to the previous twenty years (see Figures 4.7 and 4.8).

Because these are national data, they may obscure subnational drought patterns. Other data sources provide leverage on subnational rainfall. Map 4.1 shows rainfall in East Africa for late 2010 from the CHIRPS project, a weather monitoring project supported by the United States Agency for International Development (USAID).[29] We observe large weather deviations from normal rainfall in both southern Somalia and throughout much of Ethiopia for the period October to December 2010. Throughout Somalia and much of eastern Ethiopia, rains were more than 55 percent below normal compared to the average for that time of year for the period 1981 to 2010.[30]

In Map 4.2, we can see that the rainfall anomaly for the *dayr* rains from October to December 2010, with rains more than 100 mm below normal in much of southern Somalia.[31]

---

[29] The data available here is from USGS and USAID 2019. According to the website, "the Climate Hazards Group InfraRed Precipitation with Station data (CHIRPS) is a 30+ year quasi-global rainfall dataset. Spanning 50°S–50°N (and all longitudes), starting in 1981 to near-present"; See Climate Hazards Group n.d.
[30] USGS and USAID 2018.
[31] Here, the data tool calculates anomalies based on data for the entire period from 1981 through early 2019 when the figure was generated; USGS and USAID 2019.

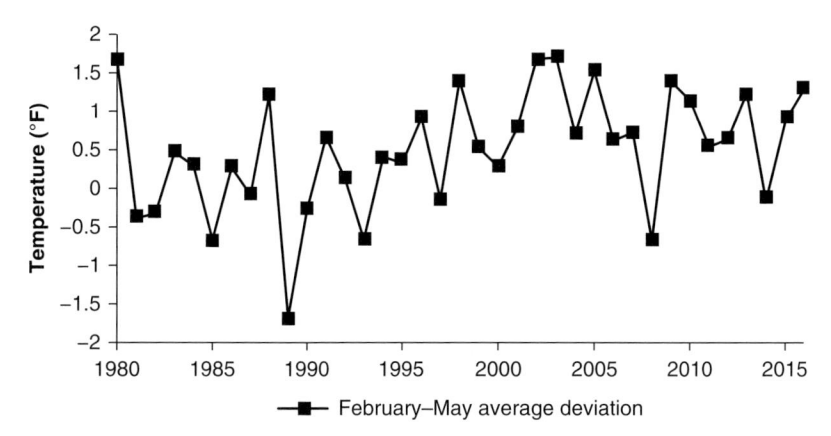

FIGURE 4.7 Ethiopia *belg* average temperature change, 1980–2016
Source: Author's calculations using CRU data

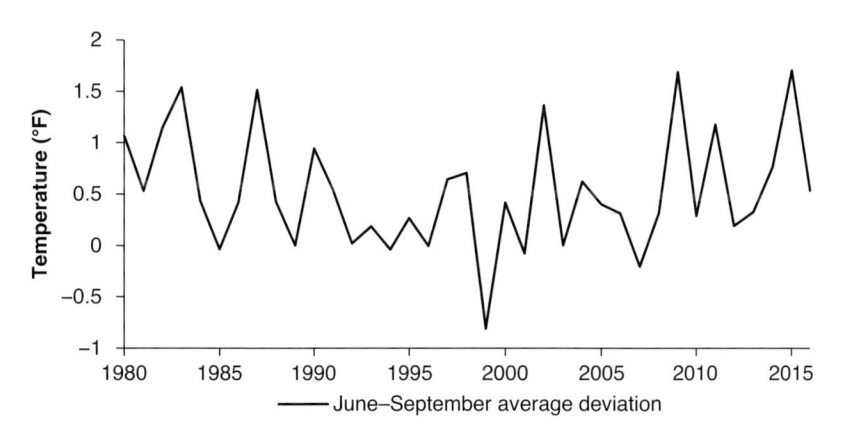

FIGURE 4.8 Ethiopia *kiremt* average temperature change, 1980–2016
Source: Author's calculations using CRU data

Map 4.3 shows that that the rains were not all that much better in the 2011 *gu* growing season, April to June 2011.

In Figure 4.9, we observe the interannual deviation in rainfall for what would become one of the most affected regions in Somalia, Lower Shabelle, for both of those rainy periods. These show lower rainfall for the 2010 *dayr* and 2011 *gu* compared to the previous year.[32]

---

[32] The Lower Shabelle region is located on the coast in southern Somalia, an L-shaped region that kinks in to the west, the third region in the maps above from bottom (as shown in Map 4.2).

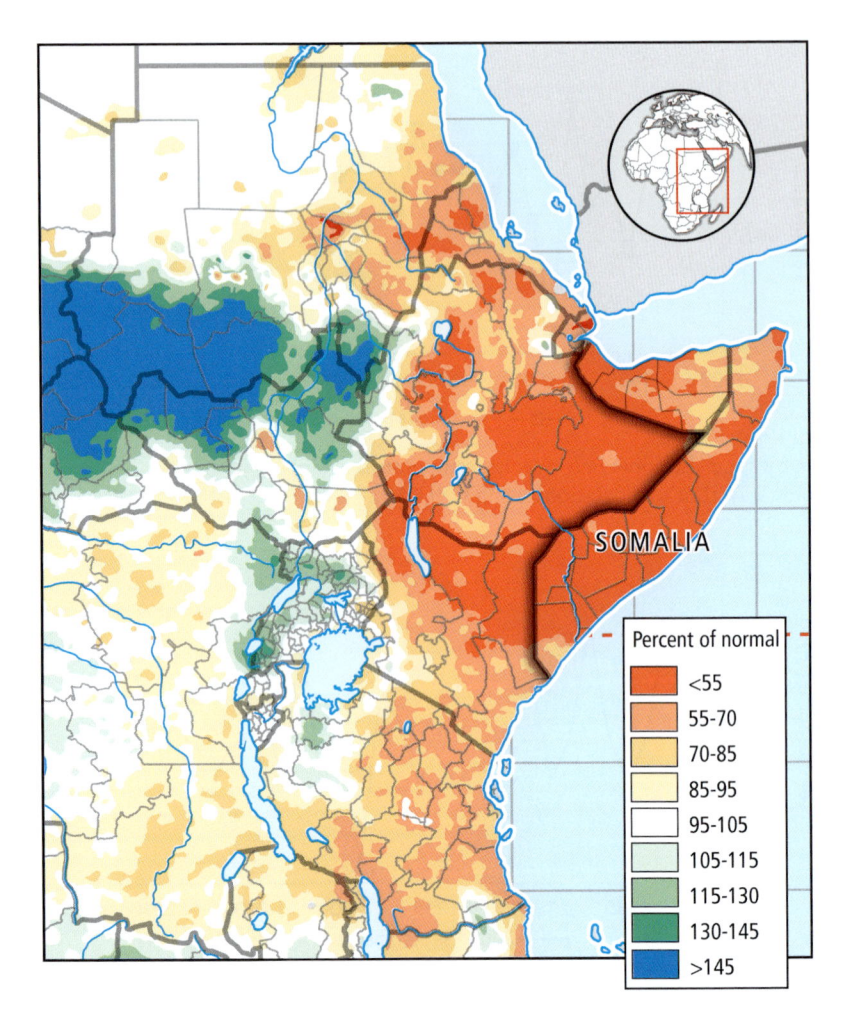

MAP 4.1 Rainfall deviations in East Africa, October–December 2010
Source: United States Geological Survey (USGS) and USAID, https://earlywarning.usgs.gov/fews/product/599

What were the impacts of those failed rains on harvests in Somalia? The early 2011 *dayr* harvests were less than 20 percent of the average, and the July 2011 *gu* harvests were expected to be no better than 50 percent of the 1995–2010 average.[33] These harvest failures put some 3.7 million in

[33] FEWSNET 2011; Salama et al. 2012.

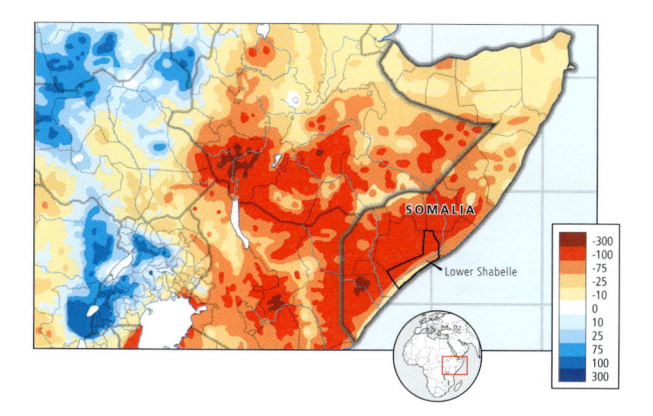

MAP 4.2 *Dayr* three-month anomaly (October, November, December 2010) (mm)
Source: CHIRPS, http://chc-ewx2.chc.ucsb.edu/

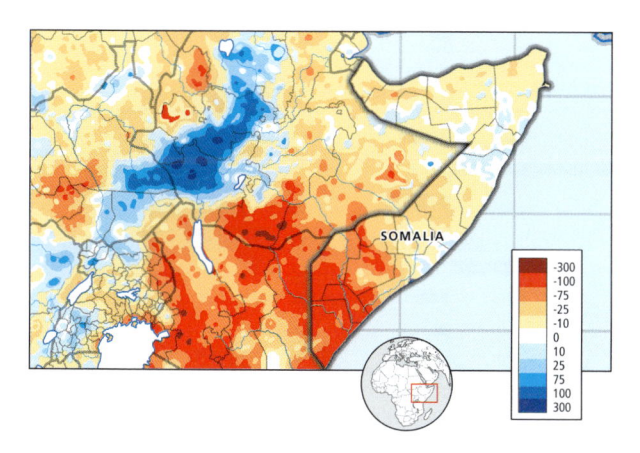

MAP 4.3 *Gu* three-month anomaly (April, May, June 2011) (mm)
Source: CHIRPS, http://chc-ewx2.chc.ucsb.edu/

crisis, most of them requiring "immediate, lifesaving assistance."[34] Food prices rose, the value of livestock declined, labor prices declined, and vast numbers of people experienced food deficits which led to malnutrition on a massive scale. In desperation, people sought food by migrating to major

[34] FEWSNET 2011.

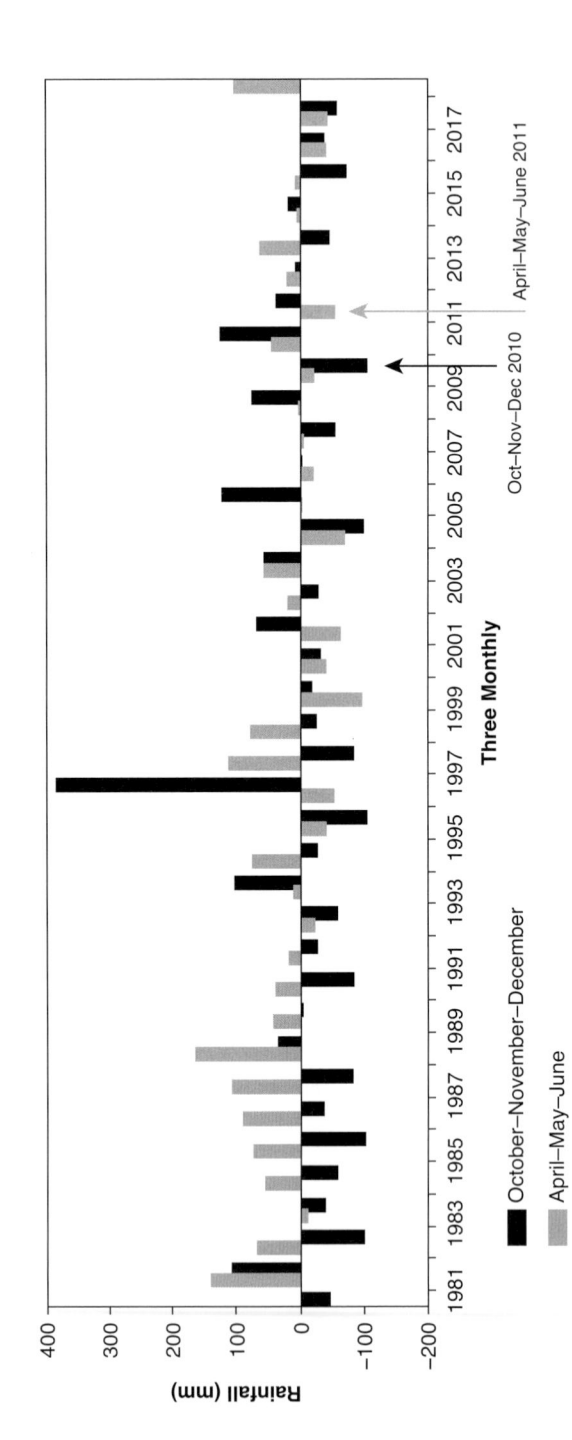

FIGURE 4.9 Interannual variability in rainfall, Lower Shabelle, Somalia
The vertical arrows indicate the rainy season periods of interest.
Source: CHIRPS, http://chc-ewx2.chc.ucsb.edu/

cities and neighboring countries. Nearly 1.5 million were internally displaced by the end of 2010;[35] perhaps as many as 150,000 emigrated to Kenya and another 120,000 to Ethiopia.[36] It was not inevitable that the failed harvests would result in famine since effective governance and food aid can typically ensure that people do not face starvation and disease in the wake of harvest disruptions. What then led to such extreme loss of life?

Before turning to this question, let's look at Ethiopia, an important contrast case. While Map 4.1 provides some suggestive evidence that the countries could be compared at the same moment in time, differences in rainfall and harvest schedules between the two countries and analysis from regional experts suggests that 2015 was a more significant drought episode in Ethiopia, affecting the north and central highlands of the country. This El Niño-influenced drought started to emerge in early 2015 when the *belg* rains began about a month late.[37]

As Map 4.4 shows, from February to September 2015, rainfall totals were estimated to be about half to three-quarters as much as normal. It is important to note that scientists have not been able to say whether climate change made this particular drought more likely. This was an extremely rare drought, between a 1 in 60 or a 1 in 260-year drought. An assessment of different climate models produced inconclusive results, with some models suggesting climate change had a role while other models showed the drought was consistent with natural variability.[38] Whether this particular event can clearly be linked to climate change is not all that consequential for the argument of this book: The science of attribution studies is a young one, so we may have better methods in time to ascribe a role for climate change in having made the drought in Ethiopia more likely.[39] Moreover, in much of the climate security literature, scholars are using climate hazards such as droughts as proxies for the kinds of outcomes that climate change is expected to influence, in terms of incidence and severity.

The problems begin to show up in rainfall anomalies for *belg* season rains in the Oromia and Somali regions of northeastern Ethiopia in February to March 2015 (see Map 4.5).

By the long *kiremt* rains from June to October, the negative rainfall deviation would become even more pronounced. Map 4.6 shows the rainfall anomaly for the beginning of *kiremt* from June to August.

---

[35] Maxwell and Fitzpatrick 2012, 8.   [36] Maxwell and Majid 2016, 70.
[37] Climate and Development Knowledge Network 2017, 2.   [38] Ibid., 4.
[39] A study of the 2011 drought for the UK Met Office did find that climate change made that drought more likely. Climate Home News 2013.

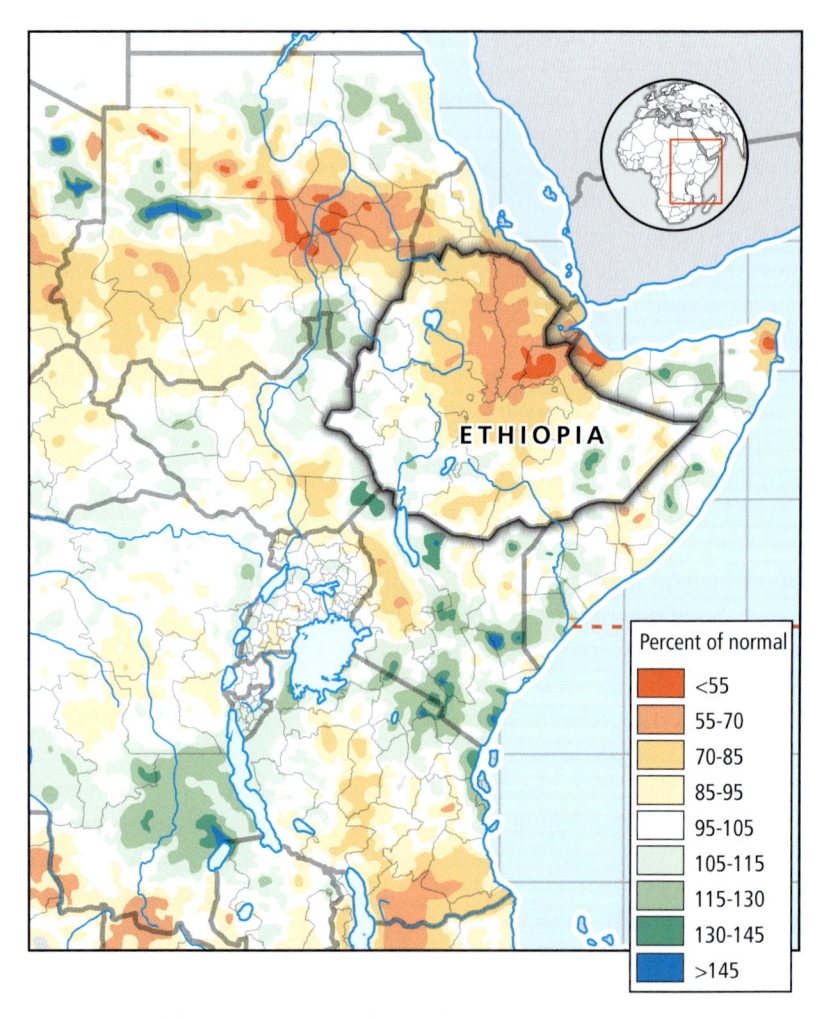

MAP 4.4 Rainfall deviations, East Africa, February–September 2015
Source: USGS and USAID, https://earlywarning.usgs.gov/fews/product/596

We can see the deviations in normal rainfall for the Oromia region for 2015, dating back to *kiremt* rains which were already low compared to the previous year (see Figure 4.10).

In a December 2015 assessment, USAID described this drought as Ethiopia's worst in fifty years, captured by the assessment in Figure 4.11 of March to September rainfall totals dating back to 1960 for central/eastern Ethiopia. This figure also captures the drought in the

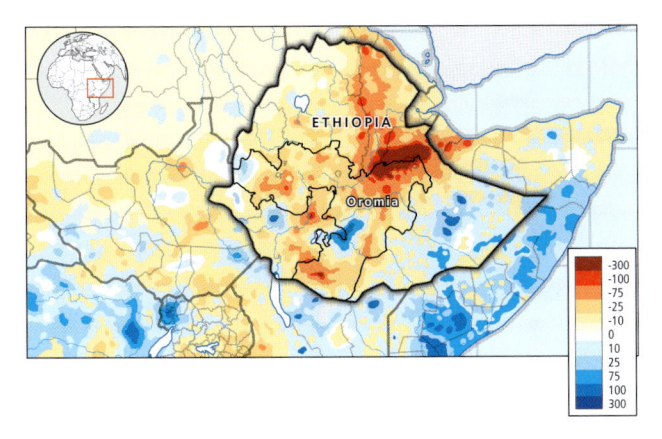

MAP 4.5 *Belg* three-month anomaly (February, March, April 2015) (mm)
Source: CHIRPS, http://chc-ewx2.chc.ucsb.edu/

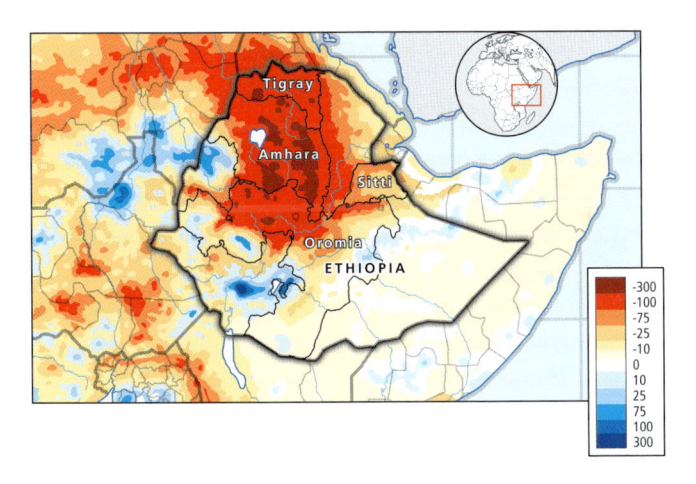

MAP 4.6 *Kiremt* three-month anomaly (June, July, August 2015) (mm)
Source: CHIRPS, http://chc-ewx2.chc.ucsb.edu/

mid-1980s that became a global cause through the 1985 Live Aid concerts.

What effects did these rainfall deficits have on livestock and harvests? USAID suggested that the effects on northern pastoral areas in Afar and the Sitti zone of the Somali region were especially harsh. Some 200,000 animals died with some 13,000 families in the Sitti zone alone losing their livestock and forcing them to migrate to camps for the internally displaced.

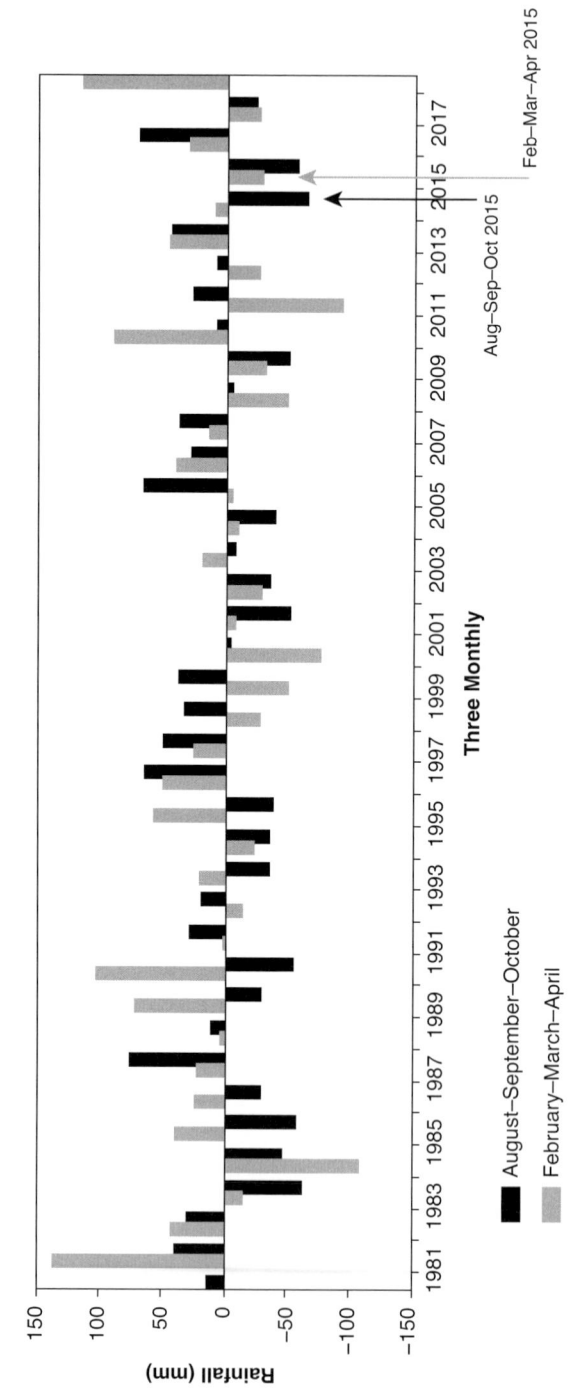

FIGURE 4.10 Interannual variability in rainfall, Oromia region, Ethiopia
a The vertical arrows indicate the rainy season periods of interest.
Source: CHIRPS, http://chc-ewx2.chc.ucsb.edu/

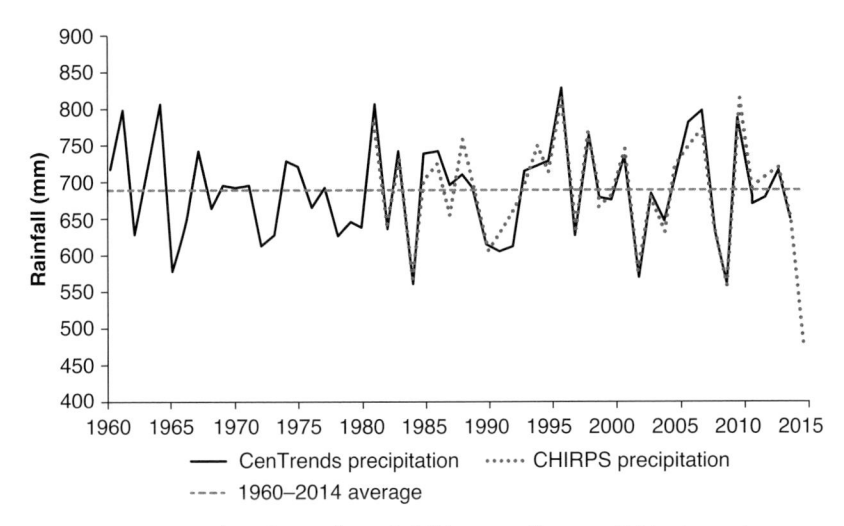

FIGURE 4.11 March to September rainfall in central/eastern Ethiopia, 1960–2015 CenTrends is the Centennial Trends Greater Horn of Africa precipitation dataset. See Funk et al. 2015.
Source: FEWSNET/USGS, Florida State University, https://bit.ly/3x65td4

The *belg* harvests failed in highland regions of northern Amhara, central Tigray, and eastern and central Oromia. The average harvest decline was estimated to be 25 percent in assessed zones, with as much as one-third of *woredas* (districts) experiencing harvest declines of more than 70 percent. As a result, some fifteen million Ethiopians were estimated to require food assistance.[40] As noted in the opening paragraph of this chapter, while some 260,000 additional Somalis died beyond normal rates of mortality in 2011, the mortality associated with the 2015 drought in Ethiopia was near zero. What accounts for the difference between the two countries?

Background on Governance in Somalia and Ethiopia

Having demonstrated that there were serious rainfall anomalies in the lead up to the famine in Somalia in 2011 and Ethiopia's food crisis in 2015, how different was their state capacity? It is helpful to put both countries' governance in context. Somalia has been unstable since Siad Barre was deposed from power in 1991, having subsequently experienced almost

---

[40] FEWSNET 2015, 2.

continuous civil war.[41] After the Union of Islamic Courts was driven from its hold on southern Somalia by Ethiopian troops in 2006, it was succeeded by a more violent Islamist group, Al-Shabaab, which formed in 2007. Al-Shabaab has intermittently controlled large swathes of territory in southern and central Somalia, including parts of Mogadishu, with peak political control between 2009 and 2011.[42] While the government showed some signs of more consolidation in the 2010s, particularly after the 2011 famine, it has variously been described as a "fragile" state or lacking a "functioning" government for much of the period since 1992.[43]

Its neighbor, Ethiopia, has had its own share of political volatility. Its authoritarian feudal leader Haile Selassie had little concern for the welfare of Ethiopia's citizens. He was driven from power in 1974 after he failed to address an unfolding famine that killed some 200,000 people. The Derg, the communists who overturned Selassie, proved equally murderous. In the face of a drought and rebellions from the Eritreans and Tigrayans in 1984–5, the Derg responded with forced displacement, crop destruction, and livestock seizures in rebel held territories. All told, as many as 600,000 people died. The failure delegitimized the Derg who continued to rule until 1991, when they were militarily defeated by the coalition of parties that formed the EPRDF, which has ruled the country thereafter. Intermittently, Ethiopia has continued to face armed conflict, including a war from 1998 to 2000 with breakaway Eritrea, which seceded in 1993. Ethiopia has also continued to face armed opposition from a number of internal groups. Before synthesizing the role played by the intersection of state capacity, inclusion, and foreign assistance, I summarize these indicators for both countries.

Evidence of State Capacity

As discussed in Chapter 3, scholars have developed a variety of measures of state capacity, none of them a perfect proxy. For the purposes of this chapter, I begin with the World Bank's Worldwide Governance Indicators and their indicators of Government Effectiveness, Regulatory Quality, and Control of Corruption which get at different dimensions of

---

[41] Somalia has had an armed conflict in which at least one party is the state for every year between 1991 and 2017, save 1997–9 and 2003–5; UCDP/PRIO 2018.
[42] International Crisis Group 2018.   [43] UCDP/PRIO 2017b, 2017a.

infrastructural power – the administrative capacity to enact and implement policy.[44]

Table 4.2 shows that compared to the beginning of the dataset – in 1996 – Ethiopia experienced considerable relative improvement in Government Effectiveness and Control of Corruption and, to a lesser extent, Regulatory Quality. Meanwhile, Somalia lurked at the bottom of the global rankings for all three dimensions. Of these, Government Effectiveness is closest to a proxy for administrative capacity, "capturing perceptions of the quality of public services, the quality of the civil service and the degree of its independence from political pressures, the quality of policy formulation and implementation, and the credibility of the government's commitment to such policies."[45]

As Figure 4.12 shows, Ethiopia's measure of Government Effectiveness, in particular, shows some signs of having stalled and even declined subsequent to 2015, perhaps a reflection of the political turmoil that emerged in the wake of the 2015 drought. Nonetheless, the general picture in Ethiopia is of improvement relative to 1996. And, if the data were to go back further, it would likely show an improvement relative to the Derg regime of the 1980s that presided over the famine that killed hundreds of thousands.

To get a longer longitudinal record, I drew on data from the PRS Group on bureaucratic quality that goes back to the mid-1980s, with scores ranging from zero (no quality) to 4 (maximum bureaucratic quality).[46] Records start for both countries in 1985, with Somalia actually having higher bureaucratic capacity than Ethiopia in that year (the latter had none). With a civil war in Somalia erupting in the late 1980s, its bureaucratic quality would evaporate and never recover. In 1995, Ethiopia's bureaucratic quality would tick up for the first time in the record and then hover at 1 from 1996 to 2006 before increasing again to 1.5 in 2008, where it has remained ever since. These

---

[44] Kaufmann, Kraay, and Mastruzzi 2010.
[45] World Bank 2019g. Regulatory Quality captures "perceptions of the ability of the government to formulate and implement sound policies and regulations that permit and promote private sector development." Control of Corruption reflects "perceptions of the extent to which public power is exercised for private gain, including both petty and grand forms of corruption, as well as 'capture' of the state by elites and private interests."
[46] Bureaucratic quality "measures institutional strength and quality of the civil service, assess[es] how much strength and expertise bureaucrats have and how able they are to manage political alternations without drastic interruptions in government services, or policy changes. Good performers have somewhat autonomous bureaucracies, free from political pressures, and an established mechanism for recruitment and training." Kaufmann, Kraay, and Mastruzzi 2010.

TABLE 4.2 *Governance indicators for Ethiopia and Somalia*

Indicator	Country	Year	Percent ranking
Government Effectiveness	Ethiopia	1996	7.1
		2001	39.34
		2015	29.33
	Somalia	1996	1.09
		2001	0
		2015	0
Regulatory Quality	Ethiopia	1996	9.78
		2001	18.01
		2015	12.98
	Somalia	1996	0
		2001	0.47
		2015	0.96
Control of Corruption	Ethiopia	1996	18.82
		2001	28.44
		2015	40.87
	Somalia	1996	5.38
		2001	0
		2015	1.44

Source: World Bank 2019g

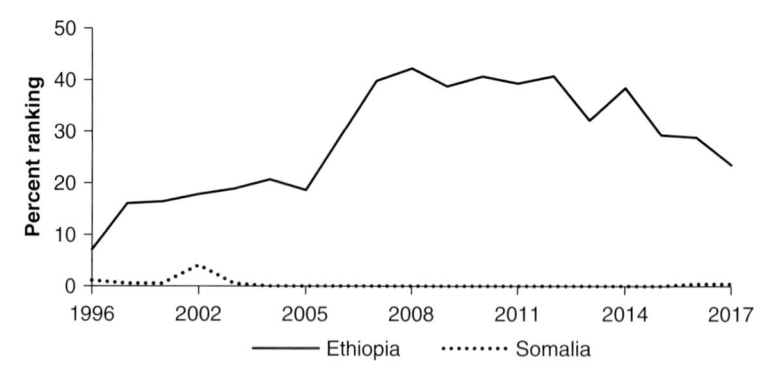

FIGURE 4.12 Government effectiveness in Ethiopia and Somalia, 1996–2017
Source: World Bank 2019g

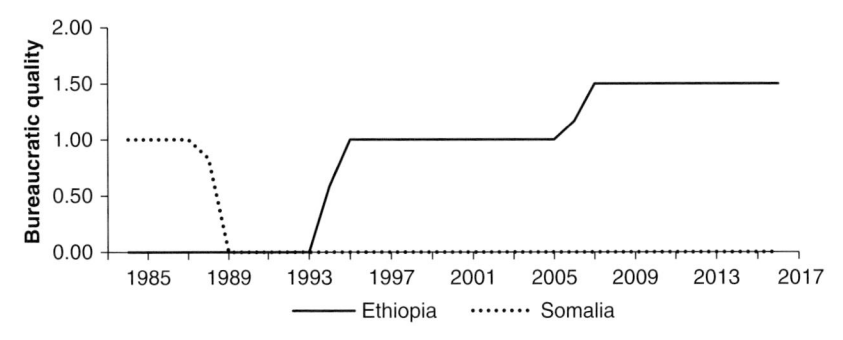

FIGURE 4.13 Bureaucratic quality in Ethiopia and Somalia, 1985–2017
Source: PRS Group 2012

observations are consistent with the World Bank data and my theoretical expectations (see Figure 4.13).

We can also look for indicators of state capacity specific to drought preparedness, including the scientific capacity to predict major drought events and carry out early warning, and the capacities to prepare for such events through anticipatory measures that would diminish their impact such as grain storage, irrigation, insurance schemes, income, and food support. We can also evaluate whether a government has response measures in place to ensure adequate nutrition such as food-for-work schemes and food donations as well as credit and inputs to help farmers replant and/or restock livestock.

Whether the government has an agency tasked to handle drought and wider natural disasters is the first point of departure, followed by a review of staffing, budgets, and the plans of those agencies. We can then evaluate some metrics of their preparations such as the size of food stocks and the number of people covered by various entitlements programs. Some of these government programs are financed by foreign assistance so it can be difficult to disentangle where capacity actually lies. For example, both Somalia and Ethiopia likely rely on the USAID Famine Early Warning Systems Network (FEWSNET), an early warning and famine system that was founded in 1985, partially in response to the Ethiopian drought. Nevertheless, we can review some of these measures to buttress wider government capacity indicators.

As discussed in Chapter 3, the INFORM index evaluates the risk reduction capacity of 191 countries. Its lack of coping capacity measure includes both an institutional and an infrastructural component. The institutional dimension is of most relevance to this project; it gets at state capacities to reduce risks before and after hazard exposure. In INFORM's 2018 index, Somalia was the worst performing country on the institutional measure, which captures perceptions of government effectiveness and corruption, with 9.2 (out of 10). Somalia did not submit a self-assessment of its risk reduction efforts as per the 2005 Hyogo Framework (discussed in Chapter 3 and further below). Ethiopia, for its, part, was ranked 113 out of 191 in the world for the overall institutional measure. In its 2015 self-assessment under the Hyogo Framework, Ethiopia rated 2.9 on a 1 to 5 point scale (with lower numbers being better), which put it alongside France and the United States in terms of its self-perception of preparedness efforts. The other two indicators average 6.5 (on a 10-point scale with 10 being worse), putting the country alongside countries like Kenya and Gabon. The average of the self-assessment and the other two governance metrics for Ethiopia is 4.7, which places the country alongside Ecuador and Botswana. Whatever the precise estimate, it is clear that Ethiopia and Somalia are far apart in terms of risk reduction capacity.

A 2014 comparative evaluation by Bussell and her collaborators assessed ten African countries for their disaster risk reduction preparations, also using the Hyogo Framework but in a more expansive way.[47] Each of the five dimensions of the Hyogo Framework was evaluated on a 5-point scale, with five being the highest and best score. Dimensions included: (1) prioritization of risk reduction with strong institutional support; (2) capacity for early warning; (3) efforts to build a culture of resilience; (4) programs to reduce underlying risk factors; and (5) strengthening preparation and response at all levels. On three dimensions, Ethiopia received the highest marks: prioritization (4.3), early warning (4.3), and reducing underlying risk factors (3.7). It was below average for the culture of resilience (2.7) and near the top scores for strengthening efforts at all levels (3.3).

Bussell and her colleagues noted the authoritarian tendencies of the Ethiopian government to exercise control over international aid agencies and NGOs, but that these mostly served to channel aid effectively rather than thwart implementation: "Ethiopian bureaucracy was found to be reasonably well-developed and able to implement disaster preparedness

---

[47] In addition to Ethiopia, other countries included Kenya, the Gambia, Ghana, Malawi, Mozambique, Senegal, Togo, Zambia, and Zimbabwe. The studies were based on fieldwork and desk review.

activities well and without substantial interference from politicians."[48] This evaluation is something of a static snapshot, though the authors did note that capacity had improved over time, as the government moved from a heavy emphasis on emergency response to focus on long-term recovery and resilience.[49]

Ethiopia's initial disaster management institutions date back to 1973 when the Relief and Rehabilitation Commission was established in the wake of a famine. In 1993, not long after the EPRDF came to power, that institution was renamed and restructured as the Disaster Preparedness and Prevention Commission. In 2004, it was split into two – the Disaster Preparedness and Prevention Agency and the Food Security Coordination Bureau (for longer-term development needs). In 2009, the Disaster Risk Management and Food Security Sector (DRMFSS) replaced these two agencies.[50] Further reforms were approved in 2013 as part of a comprehensive overhaul and updating of procedures, which fed into a new strategic framework with donors the following year.[51]

Bussell and her coauthors credit the Ethiopian government with having a strong early warning system, though with room for improvement. They note that the Ethiopian government, the NGO Save the Children, the World Food Programme, and other entities carry out risk assessments twice a year, with staff deployed to collect field level data for two months at a time. Early warning and monitoring are performed both by FEWSNET and the National Meteorological Agency among other partners. Bussell and colleagues suggest further refinements beyond biannual assessments were in process to provide more ground level detail and to be more proactive before crises emerge.[52] Among the initiatives was an effort to improve *woreda* (district level) risk profiles.[53]

In terms of prevention and response capacity, perhaps the best-known program that these agencies administer is the Productive Safety Net Programme, an entitlement program coordinated by the government that seeks to improve the food security of the population. It was initially implemented as a temporary measure in 2005 to serve five million people[54] but has endured and expanded to serve between eight and eleven million per year.[55] This program features prominently in the later discussion of the 2015 drought in the section Putting the Pieces Together.

---

[48] Bussell 2014, 18.   [49] Ibid., 106.   [50] Ibid., 101.
[51] Global Facility for Disaster Risk Reduction n.d.; Government of Ethiopia 2013, 2015.
[52] Bussell 2014, 104–105.   [53] Africa Climate Change Resilience Alliance 2015.
[54] de Waal 2018.   [55] Bussell 2014, 107.

Bussell and her colleagues concluded that Ethiopia had high state capacity in the disaster risk reduction space, though they conflate capacity and inclusion:

> The combination of strong government leadership to shift DRM [disaster risk management] strategy from reactive to a more sustainable risk reduction, combined with an effective mechanism for inclusive, multi-stakeholder engagement and action, as well as a strong level of commitment and resources from the donor and NGO community, suggests a high level of capacity in Ethiopia to deal with natural hazards.[56]

However, this has not always been the country's history. The Derg regime's response to the drought of the 1980s is a challenging one to assess in terms of state capacity. de Waal writes that there were ample warnings of the drought in Ethiopia – some twenty-one warnings between March 1981 and October 1984 – and that the government was aware of them.[57] The government's response to the crisis was a program of forced resettlement from the highlands to the lowlands, which was imposed for ideological reasons and subjected populations to government control of food aid that could be withheld and used as a weapon against them for disloyalty. While the government was capable of forcibly relocating more than 500,000 in the first fifteen months of the program, it proved incapable of providing them with adequate inputs for agriculture in their new locations.[58] Lemma and Cochrane blame the failures of this program on "poor planning."[59] As discussed in Chapter 3, a state can possess capacity in some respects but not others. I am more interested in infrastructural power of service delivery than despotic power to coerce the populace.

de Waal sees the program as more of a deliberate outcome rather than a negative by-product. He blames the program of villagization and political control of aid as a "second degree famine-crime" intended to deliver a military and political victory over opponents of the regime as part of a counterinsurgency strategy. That makes rendering a judgment of capacity harder since if the intent was to deny people ready access to land to grow food, or food itself, the program was successful in the short run, though it triggered wider political opposition that would lead to the government's ultimate demise.[60] Still, if the intent of the regime was to use its forced resettlement scheme to reorient agriculture around a new model that would actually produce food, the effort was a spectacular failure since it undermined rather than enhanced agricultural production.[61]

---

[56] Ibid., 118.   [57] de Waal 1991, 360.   [58] Gill 2012, 46–47.
[59] Lemma and Cochrane 2019.   [60] See de Waal 2018, chapter 8.   [61] Pankhurst 1992.

Somalia, for its part, has largely faced a prolonged period of state incapacity without central institutions that extend over the country's entire territory.[62] Not only has the government faced almost continuous challenges to its monopoly of force from militia like Al-Shabaab, but two northern regions – Somaliland and Puntland – have pursued autonomy (and, in Somaliland's case, secession). As the Global Facility for Disaster Risk Reduction concluded: "The legacy of conflict in Somalia has weakened the ability of government institutions to address natural hazard and climate-related challenges."[63]

Menkhaus has written extensively about the failures of external support for state-building in Somalia since the early 1990s. This initially focused on supporting the capacity of civil society institutions until the early 2000s, followed by some support for regional governance throughout, and then turned to support a central state revival in 2001 under the Transitional Federal Government (TFG). At the central level, Menkhaus judges these efforts largely to have failed, leaving Somalis to rely on ad hoc, local forms of governance, some of them formal, like regional governance arrangements in Puntland and Somaliland and municipal governance in towns such as Luuq, Beled Weyn, and parts of Mogadishu, and others informal such as mediation by clan elders and security from private actors.[64]

Following Douglass North's earlier discussion, Menkhaus views Somalia as a "limited access order" where elite actors jockey for position over the distribution of spoils. State-building constitutes a threat to some of them since they profit from the current state dysfunction, whereas others profit from skimming rents from foreign aid and are served well by a dependent weak state.[65] In the context of the 2011 drought, Menkhaus judges efforts to build capacity to have been found wanting:

After seven years in existence, the TFG was still unable to exercise control over most of the capital Mogadishu, had failed to advance key transitional tasks, and had almost no functional civil service. The one critical moment when it had a chance to demonstrate some basic capacity and commitment to provide lifesaving aid to its own citizens – the terrible famine of 2011, in which 260,000 Somalis died and over a million were displaced – the TFG failed utterly.[66]

Capacities have emerged belatedly. As part of a wider effort to establish state institutions, the Somalia Disaster Management Agency (SODMA) was created in 2011, the major famine year of focus in this chapter. It was

[62] Maystadt and Ecker 2014.   [63] Global Facility for Disaster Risk Reduction n.d.
[64] Menkhaus 2014, 2007.   [65] Menkhaus 2018.   [66] Menkhaus 2014, 159.

not until 2016 when a wider disaster management policy was drafted with the assistance of USAID.[67]

If we view the Al-Shabaab militia as a rival authority over much of southern Sudan at the time of the drought, then what capacities for aid delivery did it possess? Perhaps the primary capacity Al-Shabaab possessed in the lead up to the drought was the ability to limit access to territory it controlled. The militia imposed registration fees on aid groups and created bureaucratic structures of governance to try to govern and regulate the territory they came to control. This included local Humanitarian Coordination Officers who had twin purposes which were not fully compatible, one to engage with aid groups and another to deliver services of their own (or get credit for service delivery).[68] In the end, as the discussion in the section Putting the Pieces Together details, Al-Shabaab elected to shut down most access to aid groups but did not independently have the capacity to address the widespread suffering once the flow of money and resources largely stopped.

Evidence of Political Inclusion

Like state capacity, there are no perfect measures of political inclusion. One important, albeit flawed measure, captures the degree to which significant ethnic groups have representation in government. The Ethnic Power Relations (EPR) dataset identifies politically relevant groups that are discriminated against by the ruling coalition in different country contexts or are powerless.[69] Operationalizing political exclusion can be challenging. For example, Bretthauer uses the EPR dataset and identifies as a threshold for high political exclusion countries that have 20 percent of their population excluded from power-sharing in government.[70] EPR rankings are available from 1946 to 2017 and change over time if ethnic groups gain or lose political power as part of regime transitions. The data have also been geo-referenced to the range of where ethnic groups are located within countries; this is useful, as climate hazards affect specific areas within countries.

While EPR has its virtues, there are problems with this approach in relation to both Ethiopia and Somalia. Ethiopia has struggled with problems of political exclusion in recent years, notably because the government was led by the minority Tigrayans which constitute 6 percent of the

---

[67] Ali 2018. [68] Jackson and Aynte 2013.
[69] Cederman, Min, and Wimmer 2009; ETH Zurich 2018. [70] Bretthauer 2015.

population. Meanwhile, the Oromo, which account for about 30 percent of the population, felt they lacked adequate representation. Ethiopia did not technically meet the 20 percent threshold for high political exclusion because the Oromo were "junior partners" in the government, although they were dissatisfied.[71] In other countries such as Somalia, the populace is technically Somali, but there are finer forms of clan-based identity that are not captured by EPR. In a state with very weak state capacity, the degree of political exclusion may matter less because the government is barely able to function, let alone direct aid to its preferred groups. I suspect that if nothing else, an incapable administrative apparatus may be able to live off its people or foreign assistance for some time.

The EPR dataset, despite its flaws, does capture the hypothesized direction of improved political inclusion in Ethiopia between the mid-1980s and 2010. For example, EPR codes that from 1952 to 1990, more than 46 percent of the population was discriminated against and excluded from power including the Oromo, which accounted for 29 percent of the population.

After considerable regime turnover in the wake of the purge of the Derg, Ethiopia became more inclusive in the period 1996–2003 and again from 2004 to 2012, but still with high exclusion. Some 29 percent of the population was either powerless or actively discriminated against, including the "Other Southern Nations" which accounted for about 20 percent of the population. Since 2013, Ethiopia has nominally dropped below the 20 percent threshold to about 9 percent, leading one scholar to describe the regime as an "inclusive autocracy."[72] However, the Oromo have not been satisfied with the arrangement of being junior partners in a government led by a representative of the minority Tigrayan population. This is not solely based on ethnicity but because the government has carried out policies, like the expansion of Addis Ababa into ancestral Oromo territory, that the Oromo deemed contrary to their interests. Their political mobilization in the wake of the 2015 drought led the prime minister to resign in February 2018; he was succeeded by Abiy Ahmed, an ethnic Oromo, in April 2018. Ahmed has struggled to contain ethnic violence since then. Thus, for Ethiopia, we have data to suggest the country has become more inclusive over time, with formal representation for different ethnic groups, but with some residual problems with inclusiveness.

[71] Busby 2018b.   [72] Aalen 2018.

For Somalia, ethnic differences are not a meaningful basis to track inclusiveness since all are ethnic Somalis. The challenge is mapping the degree to which the state is captured by or caters to particular clans. The power-sharing agreement that was agreed in 2000 for the TFG was not a recipe for stability as parts of the country – Somaliland and Puntland in the north – have sought autonomy, and the south and central parts of the country have been embroiled in near constant warfare. Under the power sharing agreement that created the TFG, the four leading clans – the Hawiye, the Darood, the Dir, and the Rahanweyn – were each to have equal representation in government, reflected by the so-called "4.5 formula." The four clans split power equally and the remaining 0.5 shares were to be divided equally between the smaller clans, minorities, and women. The problem with this formulation was that Somalis tended to inflate the size of their clans, with nomadic groups perhaps less numerous than a number of sedentary groups, including the Rahanweyn. Bantu groups might constitute a far larger share of the population than was thought (6 percent), perhaps as much as 20 percent.[73]

There are thus reasons to believe that this formula was not inclusive and provided ample reason for groups to feel that representation in government, whatever its capacity, did not reflect true population demographics, which were likely difficult to discern in the absence of a census (the last one was completed in 1975). As one scholar noted, "The 4.5 formula ensured that all four of the main clan families enjoyed equal representation, but did not guarantee that all clans and sub-clans within those clan-families felt satisfied with their seats in Parliament."[74] In 2012, the Federal Government of Somalia replaced the TFG and inaugurated a new, slightly more consolidated era, though it appears that the 4.5 formula was retained.[75]

As Keating and Waldman argue, "Lack of inclusion in Somali society has a profound effect on the vulnerability of certain groups."[76] Marginalized groups like Bantu Somalis and the Rahanweyn clan have repeatedly been vulnerable to the effects of drought in 1991–2, 2011–12, and likely again in 2016–17.[77] This shows up in other ways beyond

---

[73] Gundel 2009; Ismail 2018.　[74] Menkhaus 2018, 19.
[75] Jama 2018; Goobjoog News 2015. Menkhaus has also discussed contemporary efforts to address clan representation; see Menkhaus 2018.
[76] Keating and Waldman 2019, 10.
[77] The United Nations Accountability Project – Somalia 2019. See also Maxwell and Majid 2016; de Waal 2007a.

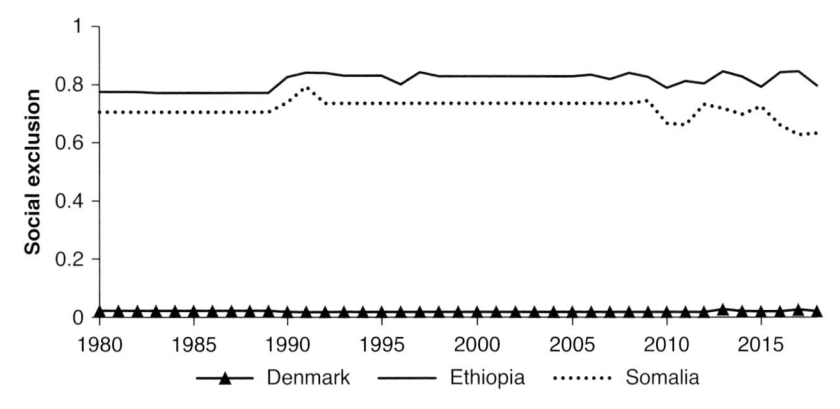

FIGURE 4.14 Social exclusion in Ethiopia, Somalia, and Denmark, 1980–2018
Source: V-Dem 2019

representation in government, also through how elite actors serve as "gatekeepers" to people's access to government services and aid.[78]

The V-Dem dataset has a couple of measures of political and social exclusion that capture related processes. Of these, the measure of equality of social group access to public services is perhaps most relevant to the impact of political exclusion on disaster preparedness and prevention.[79] Whereas political exclusion reflects the degree to which different political groups can participate in the political arena, the measure of social groups reflects the degree to which there is equality of access to public services based on other distinctions such as ethnicity, caste, language, race, and region.[80] Neither Ethiopia nor Somalia scores well on the measure throughout the entire period. Although Somalia scores marginally better than Ethiopia, it is unclear how much that matters given weak state capacity for service delivery. As a contrast to Ethiopia and Somalia, highly inclusive Denmark is also shown (see Figure 4.14).

Evidence of International Assistance

Somalia and Ethiopia provide contrast cases for the role of international assistance. Although Somalia had been heavily dependent on foreign assistance since the civil war and state collapse in the early 1990s, the period leading up to 2011 was a moment when both internal dynamics in Somalia and external preoccupations by the United States about financing terrorism

---

[78] The United Nations Accountability Project – Somalia 2019, 42.
[79] V-Dem Project 2019. [80] Ibid., 195–198.

restricted humanitarian as well as development finance. On this score, Somalia's estrangement from the international community had more in common with Ethiopia's experience in the 1980s when donors were hostile to the Derg regime.

As Menkhaus wrote, Somalia had become something of a ward of the international community after the civil war in the early 1990s, with major efforts by the United Nations post-2001 to resuscitate the state. With limited taxation capacity of its own and a very weak TFG since 2004, Somalia was overwhelmingly reliant upon international largesse.[81] Since the early 1990s and the overthrow of the Derg regime, Ethiopia became a darling of international aid donors, but as its own economic fortunes have lifted, its relative dependence on foreign aid has declined. Ethiopia's dependence on foreign aid – as captured by overseas development assistance – peaked in 2003 at 19 percent of its gross national income, declining to 6.5 percent in 2014 in the lead up to the drought. In Somalia's case, data is harder to come by, but, by 2013, foreign aid still provided about 16 percent of its gross national income in the aftermath of the drought and increased thereafter.[82]

While both countries have been heavily reliant on foreign aid, the periods leading up to their respective droughts show divergent trajectories. In terms of development finance, we can see the contrast between the two countries with respect to some select donors whose portfolios were geo-referenced throughout this period. In the lead up to Somalia's 2011 drought, we can see the number of World Bank and African Development Bank projects in 2009 and 2010 in Ethiopia compared to Somalia (Map 4.7).[83] In Somalia, there were no active World Bank projects nor any approved African Development Bank Projects during this period, while Ethiopia had many active projects, a number of them related to food security. While the politics and details of these donor choices are beyond the scope of this chapter, the absence of a functioning state in Somalia and the violence in the country during this period are among the reasons for these differences.

This is not to say that Somalia received no external assistance, but a number of major donors were not able or willing to carry out projects in the lead up to the drought. AidData's portrait of overseas development assistance to Somalia between 1999 and 2013 shows near flat development assistance but variable emergency relief. In 2010, we see a sharp drop in emergency assistance before a spike in 2011 in the wake of the drought (see Figure 4.15).[84]

---

[81] Menkhaus 2014, 158–159. [82] World Bank 2019e.
[83] Busby, Raleigh, and Salehyan 2014. [84] Tierney et al. 2011.

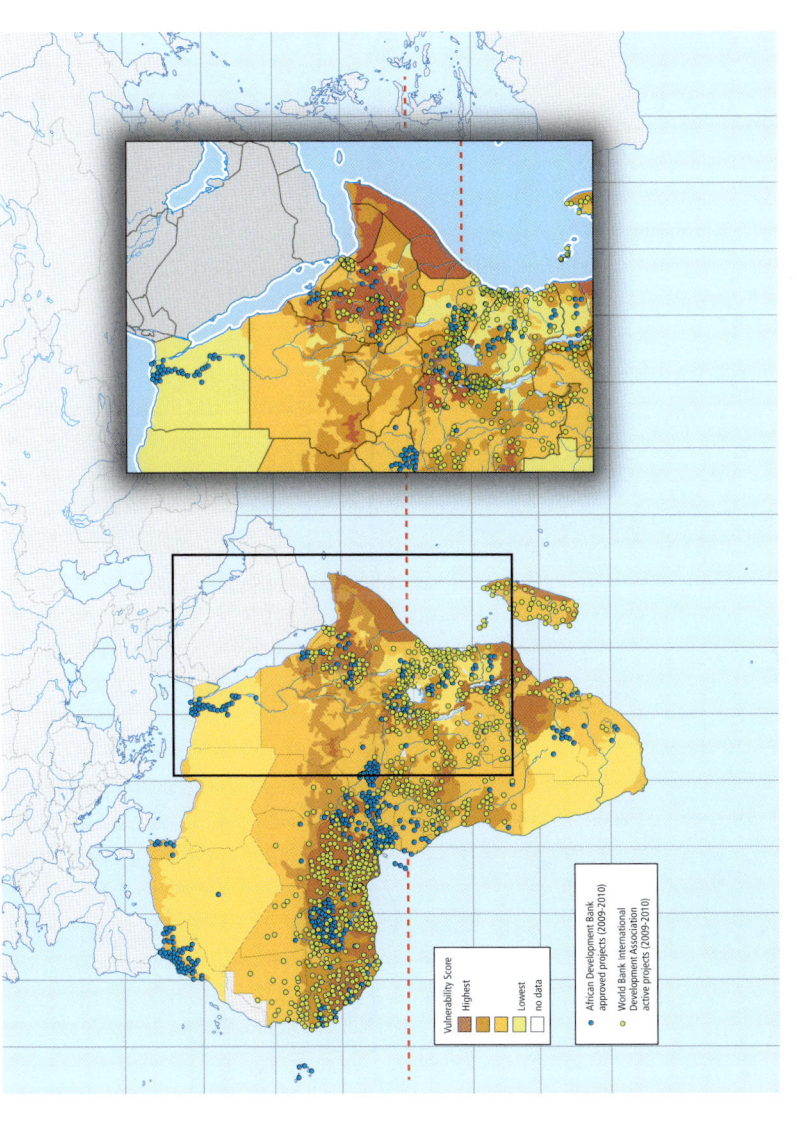

MAP 4.7 World Bank and African Development Bank projects 2009–2010 and climate security vulnerability
Source: Busby, Raleigh, and Salehyan, 2014

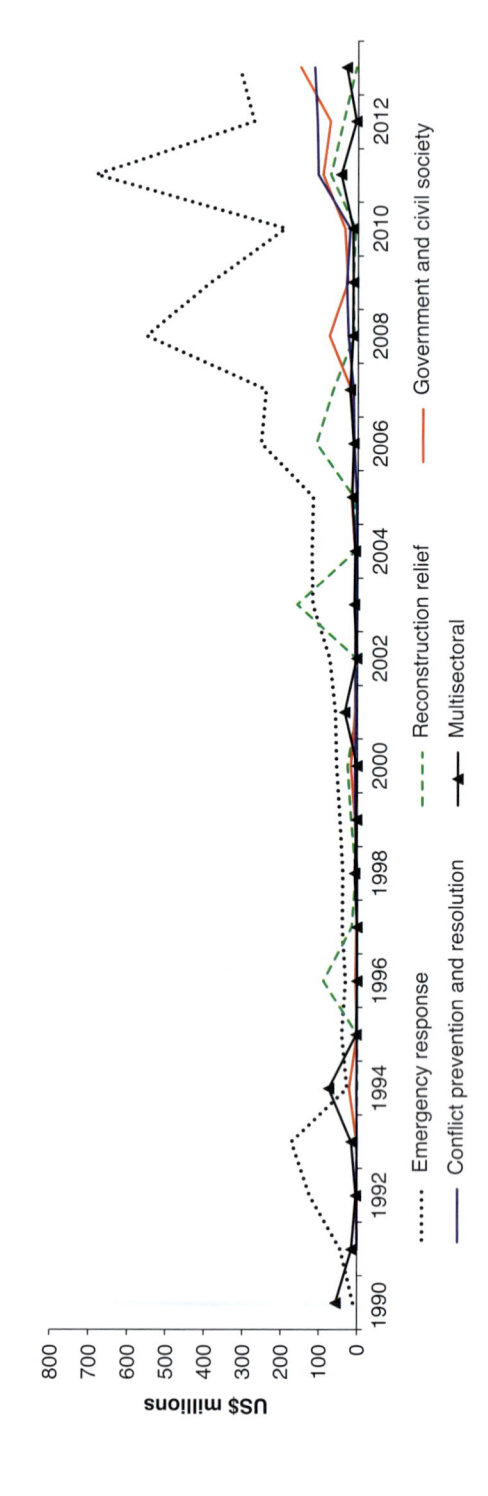

FIGURE 4.15 Foreign assistance to Somalia, 1999–2013
Source: AidData.org

Similar patterns can also be observed in another report by the government of Somalia that draws on United Nations data and other sources. This shows low but higher baseline levels of development assistance but similar patterns in humanitarian assistance to Somalia (see Figure 4.16).[85] As I discuss further in the next section, the dip in finance for humanitarian aid between 2008 and 2010 coincides with the rise of Al-Shabaab.

The situation in Ethiopia was rather different. As Map 4.7 shows, there were ample World Bank and the African Development Bank projects in 2009–10. Despite its authoritarian government, Ethiopia had been a favored aid recipient since the early 1990s. It has been among the top five African recipients of development assistance from international donors, having received more than $55 billion dollars since 1991 when the EPRDF ousted the Derg regime (see Figure 4.17).

Figure 4.17 also shows the pulse of resources that accompanied drought moments in Ethiopia's history – the belated pulse of resources to address the 1985 drought, the spike in resources in the wake of the 2010 drought, and the spike in 2015.

The patterns of limited aid to Somalia and robust aid to Ethiopia do not tell us the reasons for these patterns or how aid was distributed. In the next section, I tie the threads of capacity, inclusion, and assistance together in short narratives for both cases.

PUTTING THE PIECES TOGETHER

While difficult to discern from the trends in foreign assistance, the dip in humanitarian aid to Somalia between 2008 and 2010 observed in Figures 4.15 and 4.16 coincided with the rise of Al-Shabaab and its more extensive control over southern Somalia. In the lead-up and in the midst of the drought, Al-Shabaab was reluctant to allow aid into the country. There were reports of violence and intimidation against aid workers and efforts to impose taxes on food aid in Al-Shabaab controlled areas.[86] As Maxwell and Majid commented, southern Somalia, which faced the brunt of the crisis and was controlled by Al-Shabaab, "was largely a 'no go' area for Western humanitarian agencies."[87]

The World Food Programme (WFP) withdrew from South Central Somalia in January 2010 in the wake of a report that Al-Shabaab and other armed groups had diverted food aid. In response, Al-Shabaab

---

[85] Federal Republic of Somalia 2017.   [86] Meldrum 2011.
[87] Maxwell and Majid 2016, xiv.

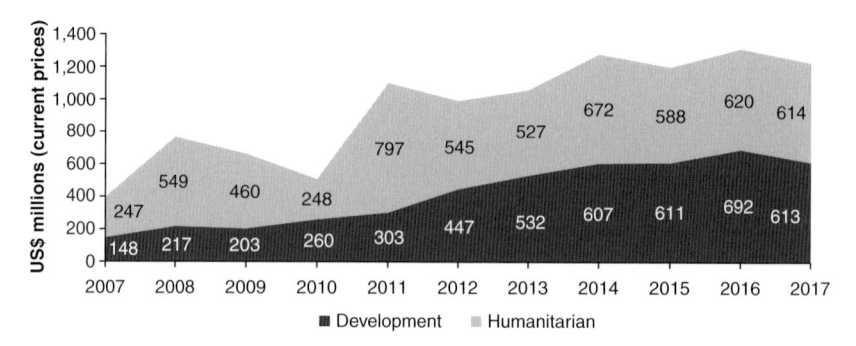

110   4 *Droughts and Famine in Somalia and Ethiopia*

FIGURE 4.16 Foreign assistance to Somalia, 2007–2017
Source: Federal Republic of Somalia 2017

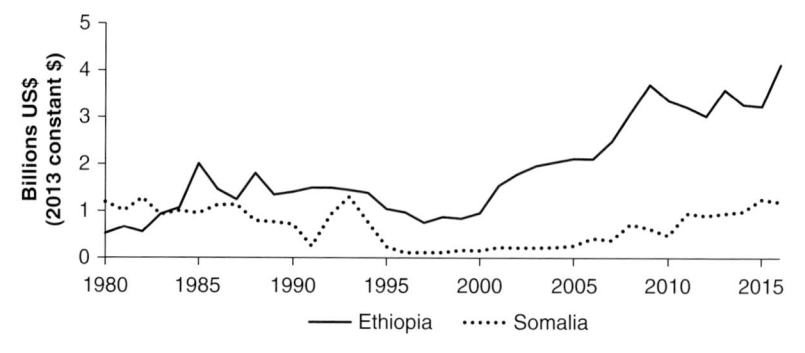

FIGURE 4.17 Official development assistance, 1980–2016
Source: World Bank via Our World in Data, https://bit.ly/3qzffSx

banned the WFP in the region. In November 2011, an additional 16 United Nations agencies and international NGOs were also banned, leaving only a handful of actors on the ground – Doctors without Borders, the Red Cross, Islamic charities, and Somalia Red Crescent Society.[88]

In addition, the US government was fearful that aid would be commandeered by Al-Shabaab and would allow it to continue its violent insurgency against a weak Somali government. After a 2010 US Supreme Court ruling, NGOs became concerned that if they carried out projects in Somalia they could be prosecuted by the US government under the Patriot Act, passed as a measure to deter finance for terrorists in the wake of the September 11, 2001 attacks.[89] In the midst of the crisis,

---

[88] Seal and Bailey 2013; MacFarquhar 2010.   [89] Atarah 2012.

USAID's Deputy Administrator David Steinberg underscored US reluctance to work directly with Al-Shabaab and the restrictions imposed by the Patriot Act: "We cannot provide anything that is interpreted as material support for a group that we consider to be a terrorist organization."[90] News reports suggest that few aid organizations received licenses from the US government to operate in Somalia even though early warning systems had anticipated the drought as far back as 2010.[91] As Maxwell and Majid concluded, "for much of late 2010 and the first half of 2011, Western donors were reluctant to provide anything in south central Somalia for fear that such funds might be diverted into the hands of al Shabaab."[92] They document that US humanitarian contributions to Somalia fell from $237 million in fiscal year 2008 to $100 million in 2009 and to less than $30 million in 2010.[93] Licenses were finally approved again in late July 2011, eight days after the famine was declared.[94]

Al-Shabaab's reticence to allow aid into the country, along with the initially chilling effect of US policy guidance, meant that international humanitarian preparedness and response measures were inadequate. Maxwell and Majid tie this delayed and failed schedule of international assistance to the excess mortality that the United Nations estimated for the country. Figure 4.18 tracks cumulative aid totals, food assistance, and excess mortality. There were three or four months of high death tolls in early 2011 before a famine was declared in July 2011 and international assistance was scaled up.

The groups most affected by the famine tended to be long-standing marginalized minority groups, the Rahanweyn and the Bantu, from the "riverine and interriverine areas of southern Somalia," the same groups that suffered in the 1991–2 famine.[95] Around 80 percent of the famine-related deaths in 1991–2 came from these groups, and they were disproportionately overrepresented in the 2011–12 famine as well. The World Bank described them as "weak agricultural communities and coastal minority groups caught in the middle of fighting."[96]

---

[90] Schmidt 2011.   [91] Halimuddin 2013.   [92] Maxwell and Majid 2016, 2.
[93] Ibid., 61.
[94] Ibid., 63. de Waal 2018, 176–177 also emphasizes that US actions prevented diaspora funds from reaching Somalia.
[95] Maxwell and Majid 2016, 6. For a similar discussion on how these groups had historically suffered from discrimination and land expropriation, see de Waal 2007a; The United Nations Accountability Project – Somalia 2019.
[96] Quoted in The United Nations Accountability Project – Somalia 2019, 43.

4 Droughts and Famine in Somalia and Ethiopia

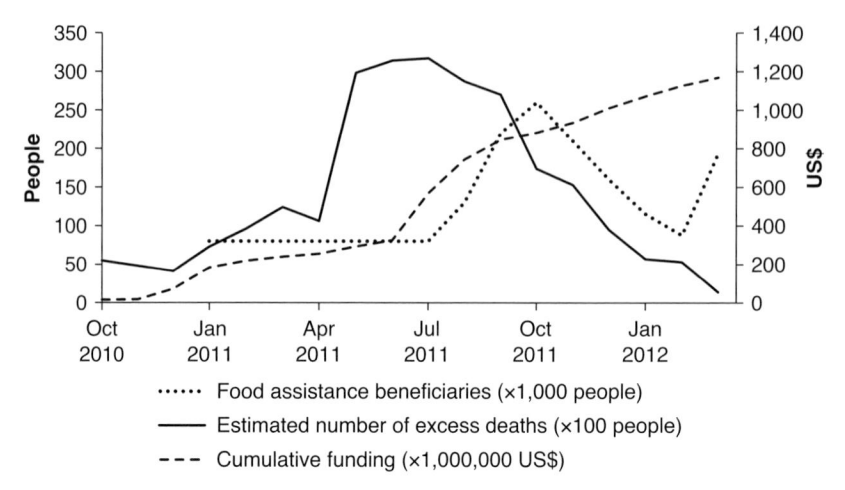

FIGURE 4.18 Mortality, funding, and aid recipients, 2010–2012
Source: Maxwell and Majid 2016

Maxwell and Majid write that the other Central Region Hawiye clans were able to self-protect through collective solidarity and mobilization of diaspora resources. These groups, they note, also negotiated with Al-Shabaab for assistance and profiteered by diverting aid from Mogadishu that was bound for displaced persons in the Lower Shabelle and Bay regions, depriving the mostly Rahanweyn and Bantu populations from resources they needed to avoid malnutrition.[97] This is consistent with the narrative that a weak Somali state had limited capacity to deliver aid, insufficient and belated international assistance was provided, and what little aid arrived initially was distributed in an exclusive manner that reinforced negative humanitarian outcomes for some groups.

The situation in Ethiopia in 2015 shows a marked contrast. Not only was there no famine, but Ethiopia had shown dramatic improvement over time. As de Waal notes, "Over half a century, Ethiopia's population has increased fourfold, and famine deaths have declined to close to zero."[98] In 1984–5, the death total from famine in Ethiopia was between 400,000 and 600,000, declining to nearly 20,000 in 1999–2000, a few thousand in 2002–3, and nearly zero by 2015–16, even as the population quadrupled.

de Waal describes the feudal rule of emperor Haile Selassie, which was largely indifferent to the suffering of its people in the 1970s, and an

---

[97] Maxwell and Majid 2016, 51. [98] de Waal 2018, 117.

equally, if not more cavalier, Derg regime that in the 1980s deliberately made the situation worse: "careful examination of how the famine unfolded – which areas were worst hit and when – shows close association with the offensives mounted by the Ethiopian army, including the latter's forced displacement of people, destruction of crops and villages and looting of livestock."[99] The Derg used food aid as an enticement to control the rebellious population, including a forced resettlement program. It denied food relief, so that Tigray, with up to one-third of the total number affected by famine in 1984–5, received just 5.6 percent of official international food assistance.[100]

de Waal also writes that the communist Derg regime was an international pariah in the lead up to the 1984 drought, which made it an unpopular recipient of foreign aid. He notes that the United States, which had been a major provider of aid in the 1970s, dramatically decreased its assistance levels after the Derg came to power. Aid averaged $15 million per year from 1977 to 1980, which was much lower than previous years. Bilateral assistance declined further to $2 million in 1981 and $1 million in 1982, before humanitarian appeals led USAID to increase spending to $6 million in 1983 and $18 million in 1984.[101] The reason for this exclusion from aid is simple in de Waal's view: "At the cynical end of the spectrum, Ethiopia was seen as a cold war enemy that was engaging in gratuitous abuse of the west, and thus deserved to be shut off from any assistance."[102]

de Waal writes that there was also concern that any aid might be diverted to support the regime rather than helping those suffering. In essence, this was a concern about exclusive institutions. With global attention on the famine, de Waal notes, Western governments dramatically increased their aid in 1985. In the case of the United States, humanitarian aid levels remained high, so that Ethiopia was the top recipient of assistance on the African continent by 1990. de Waal faults Western donors for not understanding the way the regime was using assistance: "The aim of the military policy was not to create starvation per se, but to create a population without any independent means of livelihood – i.e. to create a choice between starvation and submission for the civilian population living in areas controlled by the rebel fronts."[103] At the same time, he acknowledges this was a pragmatic choice to help people so as to placate domestic public opinion in the United States.

[99] Ibid., 119   [100] This account is consistent with Gill 2012.
[101] de Waal 1991, 358–359.   [102] Ibid., 365.   [103] Ibid., 369.

By the time a drought in 1999–2000 struck southeast Ethiopia, the country had not only experienced regime change, but was also better prepared. As de Waal notes, "The combination of an effective national disaster response and rapid provision of food aid by foreign donors, led by USAID, averted a crisis."[104] He describes the priorities of the ruling coalition that toppled the Derg as "equitable development" since 1991, which suggests more political inclusion.

de Waal portrays the government as ambivalent about its dependence on foreign donors where it sought to increase its capacity for independent action. After another near famine event in 2002, the government, in concert with aid donors, developed "A New Coalition for Food Security" with a social safety net, the Productive Safety Net Programme (PSNP). Ethiopia's efforts included early warning systems, emergency relief, and public works programs – a reflection of the government's improved capacity, backed by foreign assistance. By 2015, de Waal argues, the government was up to the task:

> By 2015, when El Niño brought the worst drought in decades, Ethiopia was better prepared than ever. In the intervening decade, the government had begun programs to help families facing food shortages with various forms of food and cash assistance. It had taken measures to mitigate the effects of droughts, rehabilitating water catchments, reforesting and building roads and clinics, especially in the countryside.[105]

Dorosh and Rashid have summarized these improvements in state capacity that occurred since the 1980s in their argument regarding why a famine was avoided in 2015:

> Public investments in agricultural technology (improved seed, fertilizer and extension) have directly raised cereal output. Expansion of road networks has connected farmers to markets, as well as improved connections between cities and regions by reducing travel time. Rapid overall economic growth has raised household incomes, increased foreign exchange earnings and enhanced government revenues. Improvements in telecommunications, better information flows and a functioning early warning system help to shorten response time for emergency relief.[106]

They draw special attention to the role of the PSNP, created in 2005, that provides food and cash transfers to poor households. That program reached nearly eight million households in 2013 and was extended in the crisis to reach a wider number.

---

[104] de Waal 2018, 120.   [105] de Waal 2016.   [106] Dorosh and Rashid 2015.

Even though Ethiopia is not a democracy, the EPRDF retains a commitment to addressing famine as a means of shoring up its legitimacy, in part because it emerged in response to the failed famine policies of the Derg era and promised to be a better steward of the government. de Waal argues: "Under the EPRDF, a strategic political commitment to preventing famine, alongside internal peace, a strong programme of economic development and generous aid donors, has meant that Ethiopia has been free of famine to a degree without precedent in the country's history."[107]

This suggests that the EPRDF was more inclusive than its predecessors, and the fact that there was no famine in 2015 suggests the government did not (unduly) play favorites with the aid that arrived. Further evidence of this is the observation that the protests that emerged in the wake of the drought were not concentrated in areas hard hit by the drought itself. Again, de Waal is clear: "The prompt and capacious relief response was politically smart, probably more so than the government realized at the time. The map of anti-government protests in 2015–16 and the map of drought relief efforts do not overlap: those in receipt of food rations evidently thought better than to riot and destroy infrastructure."[108]

That said, de Waal acknowledges that the Tigrayan heartland of the EPRDF has always received the most "organized" and "generous" responses to drought, with the southeast less well provided. He also bemoaned the resurgent authoritarian turn in the country. These observations are consistent with my argument that the Ethiopian state has improved its capacity, become more inclusive (but remains somewhat exclusive), and used international assistance to great effect to supplement its own resources and capacity where needed.

ALTERNATIVE EXPLANATIONS

One alternative explanation for the difference in outcomes between Somalia and Ethiopia in the 2010s is the presence of armed conflict in Somalia compared to its relative absence in Ethiopia. This would suggest that the main factor accounting for the difference in mortality between the countries was a function of the relative safety offered to humanitarian workers in Ethiopia compared to Somalia, where access was impeded by armed violence. In 2010 and 2011, in the lead-up to and in the wake of the drought, there were some 1,250 battles in Somalia. By contrast, in 2014

---

[107] de Waal 2018, 148–149.   [108] Ibid.

and 2015, there were just ninety-nine battles in Ethiopia leading up to and in the aftermath of that country's severe drought.[109] It could, then, be argued that the safer security situation in Ethiopia facilitated donor access and project implementation in a way that was not possible in Somalia.

Analysts have mentioned the role violence played in impeding the response in Somalia. Although he cautioned against single cause explanations, Menkhaus noted that by 2008, one-third of all of humanitarian worker casualties worldwide occurred in Somalia. Some 71 aid workers were killed between 2007 and 2009 in 139 different events.[110] By 2009, this chronic insecurity led most agencies to withdraw or suspend operations, reduce it to low levels, or subcontract to other partners.[111] Bradbury makes a similar point and notes that "Violence against aid agencies led to a total absence of international aid workers in south central Somalia in early 2010."[112]

In September 2011, Doctors Without Borders chastised other aid organizations for making it sound like money was an obstacle to addressing the crisis in Somalia when, in fact, the issue was one of limited access: "Our staff face being shot. But glossing over the man-made causes of hunger and starvation in the region and the great difficulties in addressing them will not help resolve the crisis. Aid agencies are being impeded in the area."[113] This observation makes it difficult to disentangle the violence from the problem of access itself, which as the section Evidence of International Assistance noted, was severely limited by Al-Shabaab, the Somali government, and the United States. Once the United States clarified or lifted its legal prohibitions on groups receiving aid, there was a large pulse of aid that was delivered late in 2011 (as shown in Figures 4.16, 4.17, and 4.18) suggesting the violence itself was not the major, or at least the sole, impediment to aid delivery.

To some extent, the role of violence as a major cause of the famine is similar to de Waal's argument about the 1984 Ethiopian famine. He contends that the drought itself did not cause a major disruption in food production in the years leading up to the famine, and that a single year drought in 1984 was not sufficient to contribute to famine. He suggests that the counterinsurgency campaign led by the Ethiopian government in the period 1980–5 led to food requisition by armies, blockades of food, forced relocation of populations, and rationing, all of which undermined the rural economy.[114] In the Ethiopian case, the violence was more clearly

---

[109] ACLED 2019. [110] Menkhaus 2012, 32. [111] Ibid., 33. [112] Bradbury 2010, 9.
[113] McVeigh 2011. [114] de Waal 1997, 116–117.

a function of exclusive political institutions. While the state had low infrastructural power to deliver services, it did possess some coercive capacity to forcibly resettle citizens. Such policies were both a cause of, and a response to, violence. Moreover, once international attention to the famine materialized, large volumes of aid ultimately did flow to Ethiopia, despite its pariah status in the international community.

Thus, while violence was an impediment to aid delivery in both Somalia in 2011–12 and Ethiopia in the 1980s, the blockage of aid (both internally and externally imposed in both cases) was more of a deterrent. Aid volumes ultimately increased markedly in both cases after international attention belatedly shifted to the famines and overcame opposition to aid delivery, even as some of it was captured and diverted to those perpetrating the violence.

Moreover, the violence itself in both countries was a symptom of wider governance challenges. In Somalia's case, weak state capacity meant that the country lacked even coercive capacity to have a monopoly on the use of force, let alone deliver wider services to the populace. In Ethiopia's case, in the mid-1980s the government possessed coercive capacity but lacked infrastructural power to deliver services. The combination was a state that used violence against its citizens to uproot them for putative development projects that were incompetently designed and managed. By 2015, the Ethiopian state had changed to become more inclusive and capable in its service delivery, yielding different outcomes.

CONCLUDING THOUGHTS

In the cases of Somalia and Ethiopia, the causal chain from climatic event to harvest failure and famine risk is a relatively short one. While there are other intervening factors like global food prices, local conflict, and the role of governance and donors, there is no risk of famine without a prior drought that leads to supply shocks in local food markets and impacts on livestock (though increased food insecurity may be a possibility in countries like Egypt that are more reliant on food imports even in the absence of local droughts). That makes the claims in this chapter easier to substantiate compared to other cases like the Syrian civil war, where the causal contribution of drought risk goes through more intervening variables between drought and conflict onset, including impacts on livelihoods, internal migration, and competition over resources.

In conflict cases like the Syrian civil war or the Darfur conflict, the temptation is to fall into a discussion about relative causal weight, with scholars such as Selby and de Waal dismissive of the role played by climate processes – such as rainfall anomalies. In case studies where outcomes are jointly determined by multiple factors, it is difficult to disentangle relative causal weight. The more productive question to ask is: Can you explain the outcome of these cases without mentioning the climatic/biophysical processes? In the cases in this chapter, drought is absolutely essential to the story. It is an open and difficult question to answer with confidence for cases where climate change is linked to conflict outcomes. In the next chapter, I pursue the question why drought contributed to civil war in Syria but not Lebanon, though I recognize that establishing the role played by drought in the Syrian civil war may be fraught.

Even if climate-related processes are implicated in Somalia and Ethiopia, it is less clear if the cases provide evidence of anthropogenic climate change, since the science of attribution is a young one and there are plausible natural processes like La Niña or El Niño that affected rainfall patterns. It may be that climate change is altering those natural processes as well. Moreover, the 2011 and 2015 drought episodes were not the only ones in recent memory affecting the Horn of Africa. In 2017, Ethiopia was buffeted by another drought, which tested and strained its domestic response capacity as well as the overburdened international humanitarian relief system.[115] In my 2018 study with von Uexkull, we found that both Ethiopia and Somalia were among the countries that had experienced severe water deficits in the preceding year, or were projected to through June 2019.[116] If climate change yields more frequent multiyear droughts, then we have to think a little differently about what constitutes a case and when they start and end. That gives us plenty of food for thought.

Finally, since early drafts of this chapter were written, Ethiopia has been convulsed again by ethnic violence, raising questions about the durability of improvements in both political inclusion and state capacity. As mentioned, in the wake of the 2015 drought, the Oromo protested that they were insufficiently represented in government, despite being junior partners in the ruling coalition led by the smaller Tigrayan group. In 2018, those protests led to the installation of an Oromo prime minister, Abiy Ahmed. He was able to foster reconciliation with neighboring Eritrea, which formally broke away from Ethiopia in 1993 and with whom relations have been strained. Ahmed sought further reforms, including a movement away from ethnic federalism. In 2019, he sought to dissolve

[115] Scheman 2017.   [116] Busby and von Uexkull 2018.

the EPRDF and replace it with the Prosperity Party. Tigrayans declined to join that party. In the wake of elections in August 2020, notionally delayed because of the COVID crisis, Tigrayan forces attacked government forces in the Tigrayan capital Mekele in November, leading to harsh counterstrikes by the Ethiopian government.[117] I come back to the significance of these developments in the concluding Chapter 9.

---

[117] Mules 2020; de Waal 2020; Bieber and Goshu 2020.

5

Drought in the Middle East

Contrasting Fortunes in Syria and Lebanon

We have here pointed to a connected path running from human interference with climate to severe drought to agricultural collapse and mass human migration. This path runs through a landscape of vulnerability to drought that encompasses government policies promoting unsustainable agricultural practices, and the failure of the government to address the suffering of a displaced population. Our thesis that drought contributed to the conflict in Syria draws support from recent literature establishing a statistical link between climate and conflict.
— *Colin P. Kelley et al.*[1]

Amongst other things it shows that there is no clear and reliable evidence that anthropogenic climate change was a factor in Syria's pre-civil war drought; that this drought did not cause anywhere near the scale of migration that is often alleged; and that there exists no solid evidence that drought migration pressures in Syria contributed to civil war onset.
— *Jan Selby et al.*[2]

The protest movements that swept the Middle East and North Africa in the early 2010s drew inspiration from each other. Syria was initially not thought especially vulnerable to the so-called Arab Spring, given the regime's repressive history.[3] However, encouraged by events in the region, protesters took to the streets in Syria early in 2011 demanding democratic reforms. Protests escalated after several schoolchildren were arrested and tortured in the southern Syria city of Daraa for writing anti-government graffiti.[4] In March 2011, Syria descended into a civil war that claimed

---

[1] Kelley et al. 2015, 3245.   [2] Selby et al. 2017a, 232.
[3] Daoudy 2020, 12; Hinnebusch and Imady 2018.   [4] Sterling 2012.

400,000 lives in its first five years.[5] As researchers tried to understand the circumstances that led up to Syria's civil war, a number of them surfaced a multiyear severe drought, beginning around 2006, that undermined agricultural livelihoods, led to large-scale population movements, and contributed to the protests and ensuing violence.[6] Other scholars have disputed those links, identifying water mismanagement, repression, and other government policies they believe were the primary drivers of the war.[7]

Debates about the relative importance of different factors in the lead-up to the Syrian civil war are likely unproductive, as they similarly were when Alex de Waal and Thomas Homer-Dixon debated the causal contribution of climate change to the onset of civil war in Darfur, Sudan in the mid-2000s.[8] Such debates replicate the problem Marc Levy identified in his critique of the environmental security literature of the 1990s.[9] By taking a single country and exploring whether climate drivers were associated with the civil war, scholars have not been able to explain why climate exposure contributed to conflict in that instance, but not others. Here is where comparative case studies can have value.

As indicated in Chapter 4, droughts do not lead to humanitarian emergencies in all situations, nor do they always lead to civil war. If drought did in fact contribute to the Syrian civil war, we would have to isolate the conditions that facilitated violence alongside the drought. To do that, we need a comparable case that is similar to the Syrian case but that failed to have a civil war in the wake of severe drought. Syria's neighbor Lebanon also experienced a drought around the same time but did not have a return to large-scale civil war. Syria and Lebanon share some characteristics including similar agroecological conditions and a reasonably large agricultural sector. Both states were created from the same colonial process and are complex multiethnic and multisectarian societies. Given their shared history and proximity, these cases are not

---

[5] CNN 2019.

[6] Kelley et al. 2015; Gleick 2014; Femia and Werrell 2012; Werrell and Femia 2013; Werrell, Femia, and Sternberg 2015. This sequence was captured vividly in comic strip form by Quinn and Roche 2014.

[7] Selby et al. 2017a; Fröhlich 2016; de Châtel 2014; Daoudy 2020; Thompson and Eklund 2017; Ababsa 2015. The back and forth has led to a vigorous exchange between those dismissive of the links between climate factors and conflict and those supportive of those connections; Kelley et al. 2017; Gleick 2017; Hendrix 2017b, 2017a; Werrell and Femia 2017.

[8] Homer-Dixon 2007; de Waal 2007b.   [9] Levy 1995.

entirely separable, which complicates comparative analysis, but there are no perfect comparison cases.

Lebanon has known civil war in its not too distant past – from 1975 to 1990 – and has had something of a fragile peace. Israel carried out bombing raids on Lebanon to pursue the Hezbollah militia in July 2006. Syria's fifteen-year occupation of Lebanon only came to an end in 2005 in the wake of the assassination of former prime minister Rafic Hariri. Clashes between pro- and anti-government militias in 2008 risked sending the country into civil war again. As a spillover from Syria's civil war, more than a million refugees poured into Lebanon, coming to comprise about 20 percent of the country's population.[10] Despite these pressures, Lebanon did not experience another large-scale civil war. If drought was a catalyst for, or contributing factor to, Syria's civil war, why did drought not lead to civil war in Lebanon?

The first section of this chapter provides an overview of the criteria for case selection and reviews the cases covered and how these relate to my theoretical expectations. The second section contrasts the Syrian and Lebanese experiences with each other to show why Lebanon avoided Syria's fate. This section reviews hazard exposure and presents data on capacity, inclusion, and assistance. The third section puts the pieces together and seeks to trace the causal path to violence in Syria. The fourth section explores alternative explanations for the different outcomes between Syria and Lebanon and revisits the comparability of the cases.

EXPECTATIONS AND CASE SELECTION

Before outlining the rationale for comparing Syria and Lebanon, it is worth exploring existing accounts of climate and the Syrian civil war. A number of scholars, advocates, and journalists have argued that climate change is one of the causal factors that led to the Syrian civil war. While these narratives of the genesis of the civil war highlight other factors including brutal repression by the Assad regime and mismanagement of water resources, they identify the role played by climate change. They focus on the effects of a multiyear drought that roughly started around 2006, which contributed to major declines in agricultural production, reduced rural livelihoods, and led to a significant exodus from rural to urban areas. In the telling of this climate–conflict narrative, the regime's failure to help farmers respond to the drought "eroded the social

---

[10] Government of the Netherlands 2019, 5.

contract" and undermined the government's legitimacy.[11] Gleick notes the combination of drought, crop failure, economic dislocation and migration to the cities, and social unrest.[12] Kelley et al. comment that "the migration in response to the severe and prolonged drought exacerbated a number of the factors often cited as contributing to the unrest, which include unemployment, corruption, and rampant inequality."[13] This causal story has been retold in documentaries such as *The Age of Consequences* and *Years of Living Dangerously* – the latter includes sweeping claims by journalist Tom Friedman who interviews rebel combatants who claim to be former farmers whose livelihoods were affected by the drought.[14]

The factual basis of the claims have been disputed, namely, by Selby et al. who contest whether: (1) anthropogenic climate change caused the drought, (2) the drought caused migration on the scale claimed, and (3) drought-related migration can be credibly claimed to have had an influential role in the onset of the Syrian civil war.[15] de Châtel also downplayed the importance of climate change as a causal factor in the Syrian civil war, focusing on water mismanagement and the removal of fuel subsidies as more important drivers of problems in Syria's agricultural sector.[16] Fröhlich, for her part, disputed the notion that climate migrants were heavily involved in the protests against the Syrian regime.[17]

Much of the debate rests on semantic and methodological disagreements over the meaning of cause and how to ascribe relative causal weight in case studies. Selby et al. concede that climate change could have been a factor in the civil war but then comment:

> it is worth reflecting on what Gleick and Kelley et al. mean when they insist that climate change was a contributory factor to the uprising. Does this mean that climate change-related drought was one of a small handful of factors behind Syria's descent into civil war; or that it was one amongst a thousand, or even a million, others? Is their claim that climate change was a significant factor behind the uprising; or that it was a frankly trivial one? We do not know.[18]

---

[11] Werrell and Femia 2013, 24. [12] Gleick 2014, 333. [13] Kelley et al. 2015, 3242.
[14] Friedman 2013. [15] Selby et al. 2017a.
[16] de Châtel 2014. Daoudy and Ababsa make similar claims about the importance of poor water governance and other governance issues that led to poor drought response and other political grievances that contributed to civil war; Ababsa 2015; Daoudy 2020.
[17] Fröhlich 2016. Gleick, Kelley, and others vigorously responded to the Selby et al. critique with their own claims about weaknesses in the latter's articles and the strength of their claims; Kelley et al. 2017; Gleick 2017; Hendrix 2017b; Selby et al. 2017a; Hendrix 2017a; Werrell and Femia 2017; Smith 2020a.
[18] Selby et al. 2017b, 253.

Gleick responds that they disagree over the relative importance of climate drivers:

> They [Selby et al.] try to parse the difference between whether something is a "significant" cause or a "contributory" factor and judge based on "significance." The difficulty in this approach is that "significant" is a meaningless term without quantification, but the authors do not attempt to quantify it ... At the same time, the authors regularly suggest that they agree that there was some non-zero role or link among these factors. If the only real complaint is a disagreement about the relative contribution of the many, complex factors involved, that is a far simpler and more justifiable paper.[19]

As noted above, I am reminded of the debate between Alex de Waal and Thomas Homer-Dixon over the Darfur civil war in the early 2000s. de Waal downplayed the relevance of climate in that case, suggesting that political factors were more important.[20] Homer-Dixon argued that scholars cannot ascribe relative causal weight to different causes, particularly given complex causal mechanisms with feedback loops.[21]

The problem has much to do with the slipperiness of the causal language we use. Selby et al. pointed out four different possible uses of causal analysis by scholars and commentators of the Syria case: (1) as the final spark for the conflict, (2) as the primary causal factor, (3) that climate was a significant contributory factor, and (4) that it was a causal factor of unknown or unspecified causal weight.[22] They argue the claim that climate change had no role in the onset of the conflict is itself unfalsifiable so they attempt, instead, to evaluate the robustness of the evidence connecting drought to conflict.[23] They raise some legitimate concerns about the strength of the evidence connecting each piece of the causal chain, though some of their issues can be set aside.

The Selby critique does not dispute the fact that there was a major drought that affected parts of Syria, but it faults other studies for exaggerated claims about a long-term secular decline in rainfall in the region and question whether the drought was attributable to climate change. In my view, it is not all that important to establish either a long-term drying trend or that this drought was attributable to climate change.[24] Most scholars treat Syria's drought as a short-term trigger for the civil war,

---

[19] Gleick 2017, 249.   [20] de Waal 2007b.   [21] Homer-Dixon 2007.
[22] Selby et al. 2017a, 234.   [23] Ibid.
[24] There is further evidence supporting related claims. Hoerling et al. (2012, 2146–2147) found that, since 1902, ten of the twelve driest winters in the Mediterranean occurred between 1990 and 2010. Climate change was thought to be responsible for about half of that drying.

rather than constituting a change in the long-term background trends.[25] Whether Syria's drought was attributable to anthropogenic climate change is also less important than the presence of the drought itself.[26] As I argued in Chapters 1, 2, and 3, scholars are looking to proxies for the expected effects of future climate change on security outcomes. If climate change is expected to yield more extreme future droughts, then the 2006–10 drought is a useful case, even if some dispute the clarity of the climate signal in that particular drought.[27]

Few treat the drought as the final spark for the uprising since, arguably, the precipitating event(s) happened months or years after the drought. As for whether climate change is a primary cause, a significant cause, or a cause of unknown weight, this may be irresolvable using qualitative case studies.[28] Theoretical arguments are necessary simplifications of more complex realities; they are not meant to ascribe the full causal field. While the evidence may not support all causal claims, different causal factors could, plausibly, partially explain the same outcome. Different scholars will have different judgments as to what is important to include. Some privilege sparse, parsimonious arguments while others want more causal complexity and richness. To some extent, this is a matter of taste.

Reflecting on the climate connections to the Syrian civil war, Hendrix makes some additional observations on causal pathways as part of an exchange with other authors on the Syrian civil war.

He problematizes the process of causal inference in single cases and argues that our conjectures about causality in case studies often assume certain factors are necessary conditions. The more appropriate way to think about causality, he claims, is probabilistic: "That is, climate shocks are probabilistically causal in the sense that they make something more likely. They are not deterministically causal in the sense that they are wholly responsible for the outcome." He goes further, noting: "That is, the evidence is stronger in the aggregate than is evident in any individual case."[29] As a largely quantitative social scientist, Hendrix's bet is that we

---

[25] On distinctions between trends and triggers as causes, see Hendrix and Glaser 2007; Devlin and Hendrix 2014.
[26] Again, there is evidence that the severity and duration of the Syrian drought was made twice as likely because of climate change; Kelley et al. 2015, 3241.
[27] Regional climate models project higher average temperatures, lower rainfall, days of extreme temperatures, and longer drought periods; Government of the Netherlands 2019, 4.
[28] George and Bennett 2005, 25; Bennett 2016, 36.   [29] Hendrix 2017a.

can identify the central tendencies of causation from hundreds of cases, but it is perhaps a fool's errand to try to single out the causal role in individual cases:

> When this evidence is marshalled to explain any particular event, however, it often takes on the air of a necessary condition – if but for the climate shock, the event would not have occurred. This claim is almost always impossible to substantiate and invites significant criticism – to wit, the exchange here.[30]

I agree that the lengthier and more complex causal chain between climate hazards and conflict outcomes complicates the ascription of a causal role to climate hazards. In humanitarian emergencies, it is easier to suggest, but for the cyclone or even slow-onset droughts, large numbers of people would not have been at risk of death. That said, following the work of George and Bennett, Collier, and Mahoney on process tracing, I remain convinced that we can isolate the microprocesses connecting cause and effect in individual cases, while recognizing that coming by such evidence in the Syria case is especially challenging.[31]

But even if we are able to leverage more evidence connecting cause and effect in the Syrian case, we would still lack perspective on the conditions that led to conflict in that case, which is something we can learn from paired cases.[32] In his response to Selby et al.'s original article, Hendrix argued that the case underscored the need for more contingent causal claims in the literature on climate and security: "Even if and when climate matters, it matters in a specific political, social, and economic context that must be taken into account."[33] Paired cases can help identify the scope conditions for causal claims. But what are the possible comparisons?

Feitelson and Tubi use the comparative case method to compare two river basins in the Middle East – the Euphrates (which includes Syria, Turkey, and Iraq) and the lower Jordan River (which includes Israel, Jordan, and Palestinians on the West Bank). They ultimately highlight differences in the societal response to drought between the basins to explain why outcomes in the Euphrates basin were worse. Both basins were affected by drought in the period 2007–10; however, this case selection has its own challenges. Even though drought effects transcend traditional political boundaries, the river basins themselves are not political actors so the basin as a unit of analysis is a challenging one for the evaluation of differences in societal response. Moreover, the drought

---

[30] Homer-Dixon 2007.   [31] George and Bennett 2005; Mahoney 2012; Bennett 2010.
[32] For a corrective, see Feitelson and Tubi 2017.   [33] Hendrix 2017b, 251.

effects on production do not appear all that comparable, which may be a function of the policies and societal responses of different actors. As Feitelson and Tubi note, the structure of regional economies (with Israel being a postindustrial state with desalinization capacity) made states in the region very different from each other in terms of their vulnerability to drought effects.[34]

As Selby et al. noted, northern Iraq and southeastern Turkey were also affected by the same drought.[35] However, both countries were already experiencing conflict at the time, making comparison with Syria's civil war onset problematic.[36] In any event, an examination of these three cases shows that the drought had disparate impacts on Syria and Iraq compared to Turkey, which, as an upper riparian, was able to tap upstream resources and compensate for reduced rainfall to maintain agricultural productivity.[37] So are there other cases that are potentially a better fit? In his comment on the Selby et al. articles and the wider discussion, Hendrix identifies some possible cases, "The drought that affected Syria also affected neighboring Jordan, Lebanon and Cyprus, yet widespread violence did not occur there."[38] I have used these observations as the point of departure for case selection, ultimately taking the Syria and Lebanon cases as paired cases. If the stylized narrative of Syria's civil war is correct, then we need a country with a comparable farming sector that was affected by the drought, with concomitant losses in agricultural production. Lebanon and possibly Jordan may be suitable, albeit imperfect, comparison cases for Syria.

Using countries from outside the region for comparison might be problematic because of very different agroecological conditions. Levant countries such as Syria, Lebanon, and Jordan may share a similar climate at the intersection of the humid Mediterranean and the arid Arabian desert, though there still may be significant climatic differences between them. According to the World Bank, Syria's average annual precipitation is 252 mm/year, placing the country 155th in the world. Lebanon, for its part, is wetter, reaching 661 mm/year (with a ranking of 114), while Jordan is drier with 111 mm/year (a ranking of 168).[39] However, when we compare wider fresh water resources which would include flows from

---

[34] Feitelson and Tubi 2017. [35] Selby et al. 2017a, 234.
[36] Iraq was in the throes of an ongoing civil conflict that grew out of the US invasion in 2003. Turkey has long had an insurgency in the southeast of the country, in the Kurdish region which was heavily affected by the drought.
[37] Eklund and Thompson 2017. [38] Hendrix 2017b, 251. [39] Verner et al. 2018, 38.

rivers, the differences between Lebanon and Syria are less stark. During 2007, the per capita resources available in Lebanon were 1,102 m³ per person per year compared to 855.7 m³ in Syria and only 151.3 m³ in Jordan.[40] That puts Lebanon just above and Syria just below the threshold of 1,000 cubic meters per person per year for water scarcity.[41] These observations, of course, obscure significant subnational variation, particularly in a comparably larger country like Syria, but even in Lebanon there are significant differences, with rainfall varying from 200 to 600 mm per year in the Bekaa region to as high as 1,000–1,400 mm in the mountains.[42]

Even countries that share similar climates might be different in other respects such as the size of their territory and their population, the degree of urbanization, and their dependence on agriculture. In 2007, the last point for which the World Bank provides data for Syria, the country's GDP per capita in current dollars was $2,032 compared to $2,735 in Jordan and $5,217 in Lebanon.[43] Syria is a much larger country than both Lebanon and Jordan (184,000 square km compared to 10,000 and 89,000 square km, respectively) and was far less urbanized (55.6%) than either Lebanon (87%) or Jordan (84%).[44] In 2008, Lebanon had a much smaller population than Syria – 4.1 million compared to 20.3 million – but its proportion of the population employed in agriculture was similar to Syria – 13.5% in Lebanon compared to 17.5% in Syria. Jordan, for its part, only had 3.9% of its workforce employed in agriculture.[45] In 2007, Syria's dependence on agriculture, forestry, and fishing, as a share of GDP (19.5%), was also greater than Lebanon (4.9%) or Jordan (2.5%).[46]

If we consider that the economic dislocation of bad harvests set in motion a chain of events that led to the Syrian civil war, the comparison case should have an agricultural sector sizable enough that an unhappy farming class could have stoked a protest movement, a rebellion, or, at the very least, was part of internal population movements of sufficient size to trigger domestic contestation between groups over resources. Given the small size of the agriculture sector in Jordan, it is less clear that these conditions could be met. Lebanon looks like a more plausible comparison

---

[40] Food and Agriculture Organization 2019a.   [41] Fanack Water 2019.
[42] Food and Agriculture Organization 2018, 97.   [43] World Bank 2019d.
[44] Verner et al. 2018, 38.
[45] World Bank 2019c, 2019f. Other data sources suggest Lebanon's share of agricultural employment in 2008 (2.97%) was more similar to Jordan's (3.57%) than Syria (14.53%); Roser 2013.
[46] World Bank 2019b.

TABLE 5.1 *Comparing Syria and Lebanon*

Country	Hazard events	Capacity	Institutions	International assistance	Outcomes
Syria	Droughts 2006–10	Weak capacity	Exclusive	Limited access	Civil war 2011– (Type 1)
Lebanon	Droughts 2007–10	Intermediate capacity	Somewhat inclusive	Broad-based aid delivery	Protest activity but no civil war (Types 3–6)

case, though other data sources suggest Lebanon's agricultural employment was not dissimilar to Jordan's. While no case comparisons are perfect, Syria and Lebanon, which were together carved out of the Ottoman Empire and then later cleaved by France into separate polities, are perhaps the best cases for comparison.

Here, we need to show that both countries faced severe droughts. Syria experienced a devastating civil war in 2011 while Lebanon, despite a contemporaneous drought of its own, did not. If, as I argued in Chapter 2, the project of sustaining peace in countries is largely a function of elite bargains about representation and equitable service delivery, we should find evidence to suggest that the Syrian state did not have either means or motive to service the drought-stricken communities, either in areas affected by the drought, mostly in the east of the country, or in the towns and cities many Syrians migrated to after their livelihoods as farmers became untenable.

Lebanon, by contrast, should have more state capacity and inclusion than Syria, giving the government both greater means and higher motives to respond to the drought and help farmers and consumers. This argument does not hinge on Lebanon necessarily having good governance by idealized standards, but a comparatively better response than Syria. On the margins, Lebanon may also have benefited from more international support than Syria. The case dynamics are summarized below in Table 5.1 and, drawing on Chapter 3, the possible pathways that led to that outcome are indicated.

SYRIA AND LEBANON

To recap, first, I need to show that both countries experienced drought. Second, I should be able to show some differences in state capacity between Syria and Lebanon. Third, I should be able to show that Lebanon had more political inclusion than Syria, what I take to be the key difference between the government structures. Fourth, to the extent that both countries relied on international aid in the midst of their ongoing droughts, my expectation is that Lebanon, given its more inclusive government, would ensure that resources were shared equitably.

In Syria's case, different routes could have led the country down a path to civil war. For the drought to have contributed to civil war, the declines in farmers' income could have made them more likely to participate in rebel activity in their villages. Alternatively, they may have migrated to other cities where they became open to protest, and eventually rebellion, because of low living standards and lack of opportunities. Another possible explanation is that migrant populations could have competed with urban populations in their new locations over housing, jobs, and services. If they settled in areas with coethnics, they might have swelled the number of people feeling excluded from opportunity. If they settled in more mixed areas, this competition for jobs and services could have taken on wider sectarian divisions.

This explanation does not depend on the migrants themselves having joined the protests or rebel movements in mass numbers, at least at the start. One critique by Fröhlich, based on a small number of interviews with Syrian refugees, is that there is no evidence migrants took part in the initial protests that kicked off the uprising in Syria.[47] Their own precarious standing in new urban areas could have made them less likely to participate in protests and/or violence, but their presence may have led to existing residents' dissatisfaction.[48]

Let's consider the stringency of different qualitative case methods and their application here. If we find evidence of a nontrivial number of former farmers engaged early in the protests and/or drought being a major theme of protests, this would constitute perhaps a *smoking gun test*, one that is sufficient to confirm the hypothesis that the drought was associated at least with the protests, if not the civil war. Failure to find farmers in the early cohort of protesters or the drought evoked as one of the causes is not fatal to the argument, since there are other pathways to social conflict

---

[47] Similar claims are advanced by Ababsa 2015 and Daoudy 2020. Others suggest these displaced populations were in fact part of the uprising; see Schmidt 2018; Lawson 2016; Hinnebusch 2019.
[48] Koubi 2017, 201, 2019, 374. See also Salehyan and Gleditsch 2006; Raleigh, Jordan, and Salehyan 2008; Linke et al. 2018a.

from the drought. A less stringent test is the so-called *hoop test*. If the argument passes this test, then it has at least jumped through the hoop and we cannot rule it out, but we also cannot be certain that the argument is correct. Thus, if we find former farmers well-represented among the second generation of protesters or combatants, we could argue that the drought plausibly had something to do with their availability and willingness to engage in conflict. A hoop test is stronger than a *straw-in-the-wind test* which is neither necessary nor sufficient to confirm a hypothesis, but one where at least the evidence is consistent with the hypothesis. Here, the temporal sequence of the conflict following the drought is a straw-in-the-wind test, which confirms the relevance of the hypothesis but is not in and of itself confirmation of it.[49]

Given its high level of urbanization, Lebanon might have been more at risk of domestic unrest than Syria through another mechanism, the price of food. Drought, domestically and internationally, could have had an impact on food prices, thus triggering domestic protests. Indeed, the role of increased international food prices has been mooted as one of the primary drivers of the Arab Spring and the emergence of protest politics in Egypt and other North African countries.[50] Smith has shown that countries often insulate their populations from the pass through effects of international food increases, but that where there are domestic food price increases, protests are more likely.[51] Similarly, Hendrix and Haggard demonstrated that democracies and anocracies are more likely to experience protests in the wake of food prices increases, as authoritarian governments are more likely to fear urban unrest as a threat to regime stability and seek to prevent potential protests through subsidies and other measures. With more electoral competition in democracies and anocracies, those regimes have to be more responsive to rural constituents, where food subsidies for urban consumers might come at their expense.[52] Lebanon is an interesting case with a Polity IV score of 6, the threshold between an open anocracy and a democracy, suggesting it might have been vulnerable to food-related protest activities that could have escalated given the fragility of the regime.[53]

---

[49] Collier 2011.   [50] Lagi, Bertrand, and Bar-Yam 2011.   [51] Smith 2014.
[52] Hendrix and Haggard 2015; Hendrix 2013.
[53] An anocracy is a mixed regime with democratic and authoritarian features. Syria was a –8 in the period 2008–12 before become fully authoritarian (–10) in 2013; Center for Systemic Peace 2020.

This set of explanations captures the motivations for participating in rebel movements but another – not mutually exclusive – explanation for the violence in Syria is the effects of output declines on state capacity itself, by decreasing the available tax revenue from agriculture. These effects would make it harder for the state to provide essential services to the farmers and/or to suppress violence. Such declines in state capacity may very well have predated the drought and largely be a function of other social and economic changes that made it harder for the state to afford service delivery or to suppress violence.[54]

It may be difficult to disentangle the effects of the drought itself on state capacity and a longer-term deterioration in capacity wrought by mismanagement. For example, as discussed in more detail in the section Evidence of State Capacity, some Syria scholars identify the root of the government's problems with respect to agriculture and water management with earlier decisions to pursue food self-sufficiency in water intensive crops (such as wheat) that could only be sustained with irrigation, which is itself dependent upon fuel subsidies for pumping. When the government removed these subsidies, production collapsed, hastening a further decline in the agricultural sector.[55] I return to these themes in subsequent sections.

What follows is an attempt to document evidence that supports my expectations for hazard exposure, state capacity, inclusion, and foreign assistance. As in previous chapters, I draw on area studies experts and existing scholarly narratives of the two countries' divergent trajectories.

Evidence of Hazard Exposure

The point of departure for understanding the two cases begins with a justification for the claim that the two countries experienced severe droughts in similar time frames. In the cases of Syria and Lebanon, the droughts were more directly overlapping temporally than the experience of Somalia and Ethiopia documented in Chapter 4.

As I argued in the previous chapter, there is no universal definition of what constitutes a drought. Some scholars look to rainfall measures for evidence of drought, while others see temperature as a better metric given challenges of data quality. Some indicators of drought combine information derived from rainfall, temperature, and soil conditions to identify drought periods. Given that we are mostly interested in the impacts on the

---

[54] Koubi 2017, 201, 2019, 374.   [55] de Châtel 2014.

agricultural sector, it makes sense, as in the last chapter, to focus on climatic conditions during growing seasons.[56]

For both countries, it is easy enough to establish growing season droughts in the period 2006–10 as well as a longer history of rainfall and temperature deviations. As in the previous chapter, I use rainfall and temperature data from the Climate Research Unit (CRU) at the University of East Anglia. In its climate portal, the World Bank aggregated this station-level data up to the national level to provide monthly means dating back to 1901.[57] I calculated the deviation from growing season rainfall and temperature levels for the main crop, wheat, on a rolling basis for the previous twenty years, extending the analysis back to the 1970s. For both countries, the monthly mean is based on rainfall totals for October–May based on the Food and Agriculture Organization's (FAO) crop calendar.[58]

For rainfall, this means comparing the monthly rainfall totals for October to the previous twenty Octobers and doing the same for each month between October and May. I then calculate the cumulative deviation for the entire period to show how different the total amount of rain was for that growing season compared to what farmers consider to be normal, for which a twenty-year backward time horizon seems reasonable. For temperature, I use the average difference rather than total deviation from normal temperature levels, since temperature does not cumulate in the same way as precipitation. I use both rainfall and temperature mostly to check if there are years when rainfall levels were not too far below normal levels, but we might see high temperatures creating water stress for crops that might show up in yield declines. As Kelley et al. note, "what matters for crops is soil moisture, which is influenced by temperature as well as rainfall."[59] As in the previous chapter, I complement such national level data with subnational portraits of agriculture stress for drought periods of interest using maps from the FAO.

Before presenting rainfall data, it is worth noting that the challenges of water access for agriculture and other uses become more urgent when accompanied by other changes and problems, such as rapid population growth and management of groundwater resources.[60] In Syria's case, the country's population grew dramatically from nearly nine million in 1980

---

[56] This is again inspired by the work of von Uexkull et al. 2016 on growing season droughts. I also use Smith 2014 for the methodology of deviations from rolling twenty-year monthly mean rainfall and temperature.
[57] World Bank n.d. [58] FAO 2019b, 2019a. [59] Kelley et al. 2017, 246.
[60] Null and Risi 2016, 26; Adelphi 2015.

to almost twenty million by 2007. The country was also host to some one million Iraqi refugees between 2003 and 2007.[61] Other severe droughts had buffeted Syria in the 1990s and it was also affected by upstream diversion of water by Turkey's construction of dams. Gleick shows that the average annual flows of the Euphrates at Jarabulus, just down from the Turkish–Syria border, declined from about 1,000 cubic meters per second between 1937 and 1989 to about 650 m^3 per second between 1990 and 2010.[62] Thus, the amount of water availability per capita declined from 744 cubic meters per capita in 1982 to 363 m^3.[63] Such problems are magnified when water is mismanaged, directed toward water-intensive crops such as cotton and wheat, or where irrigation techniques such as flood irrigation use more water than is needed.[64] This makes it challenging to distinguish climate and environmental stress from population and resource management. Other scholars such as Kahl bundle such dynamics under the broader label of demographic and environmental stress.[65] While these were important background conditions that shaped the wider resource envelope, there was still a marked decline in rainfall during this period that had a major impact on agricultural production.

For rainfall, the CRU data shows that 2006–9 were all below normal growing season rainfall years in Syria, with 2008 being 123 mm below normal for the growing season, the country's worst drought since 1973. While the 2006–9 drought was the most severe since then, it was not the only multiyear drought over the last forty years, as 1982–7 was also a dry period, if not by the same magnitude. Any explanation for why drought contributed to civil war in the contemporary period 2006–9 ought to be able to explain why another multiyear drought did not result in a similar outcome, arguably when even more of the country's workforce was employed in agriculture (about 30 percent in 1984) (see Figure 5.1).[66]

Even as rainfall returned to normal levels in 2010, the temperature data show that this was Syria's second hottest year since 1973, with the average growing season deviation in temperature nearly two degrees Celsius above normal (see Figure 5.2).

In Lebanon's case, the country experienced four continuous years of below normal rainfall between 2007 and 2010, with 2008 being 200 mm below the average growing season rainfall for the previous twenty years. This was the largest negative deviation in rainfall since 1973, though

---

[61] World Bank 2019f. [62] Gleick 2017, 334. [63] Ritchie and Roser 2017.
[64] The New Humanitarian 2010. [65] Kahl 2006. [66] Library of Congress 1988.

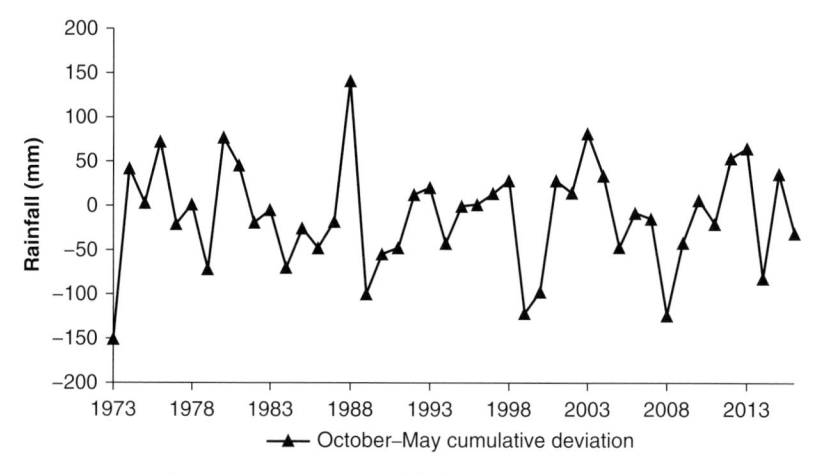

FIGURE 5.1 Syria growing season rainfall deviations, 1973–2016
Source: Author's calculations using CRU data

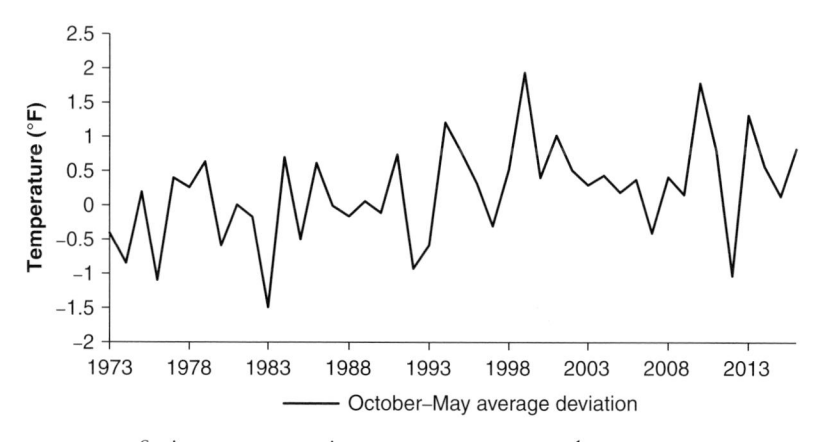

FIGURE 5.2 Syria average growing season temperature change, 1973–2016
Source: Author's calculations using CRU data

Lebanon has experienced a number of severe single-year droughts over the years including 1979, 1986, 1989, and 1999 (see Figure 5.3).

The World Bank has identified both 2010 and 2014 as significant drought years for Lebanon, though the rainfall deviation for 2010 – negative 25 mm for the growing season – is not nearly as severe as 2008. However, when we look to large temperature deviations, 2010 was an

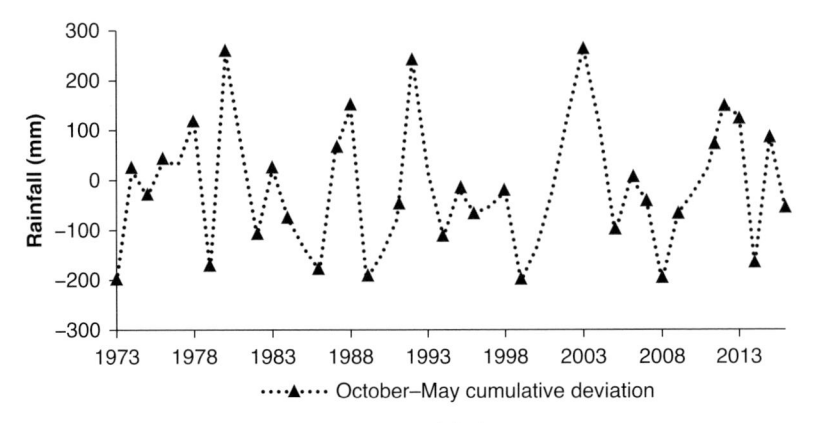

FIGURE 5.3 Lebanon growing season rainfall deviations, 1973–2016
Source: Author's calculations using CRU data

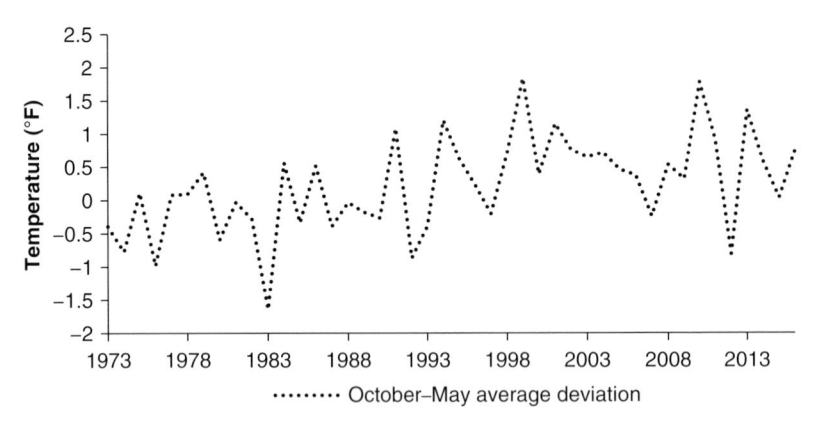

FIGURE 5.4 Lebanon average growing season temperature change, 1973–2016
Source: Author's calculations using CRU data

especially hot year when growing season temperatures averaged nearly two degrees Celsius above normal (see Figure 5.4).

Kelley et al. put the drought into context: "Before the Syrian uprising that began in 2011, the greater Fertile Crescent experienced the most severe drought in the instrumental record."[67] In a subsequent article, Kelley et al. anchor the drought in the region's wider history: "And, if instrumental data, model results and theory are not enough, an analysis of a new gridded tree

[67] Kelley et al. 2015, 3241.

ring dataset of winter/spring surface moisture availability for all of Europe, North Africa and the Middle East ... concluded that 1998–2012 was the driest 15-year period in the Levant in the last 900 years."[68] This drought, and the drying of the region, they argued, was rooted in anthropogenic climate change: "Analyses of observations and model simulations indicate that a drought of the severity and duration of the recent Syrian drought, which is implicated in the current conflict, has become more than twice as likely as a consequence of human interference in the climate system."[69]

Even if the period 2006–10 was a major drought episode in both countries, the effects were unequally distributed within Syria and Lebanon. We can see the geographic distribution of drought effects by looking at the most intense drought year of 2008 and FAO maps of agricultural stress.[70] In Syria's case, severe agriculture stress – where more than 85 percent of agricultural land experienced drought conditions – extended throughout much of the north of the country in 2008, affecting Hasaka province in the far northeast, Ar Raqqah in the central northeast, and a portion of Aleppo province in the northwest. Small pockets of drought affected other provinces, including the adjoining provinces of Hama and Homs in the center of the country as well as three southern provinces – Daraa (where early protests against the regime occurred), Damascus, and As-Sweida (see Map 5.1).[71] About 75 percent of the country's wheat production has historically been grown in the northeast of the country.[72]

The areas affected by the drought in the northeast of the country included the heavily Kurdish regions of Hasaka province and the northern part of Aleppo province bordering Turkey. The city of Al Hasakah located centrally in that province is also where a large concentration of Syrian Christians live. The drought also affected Druze populations concentrated in the south of the country in Daraa and As-Sweida provinces. In the center of the country, the drought primarily affected Sunni dominated regions in Homs province.[73] The rural areas affected by the

---

[68] Kelley et al. 2017, 246.
[69] Kelley et al. 2015, 3241. Gleick 2014 summarized additional science on the likely warming and drying of the Middle East.
[70] The FAO defines this indicator in terms of the extent of cropland affected by drought over the entire crop season: "The Annual ASI [Agricultural Stress Index] depicts the percentage of arable land, within an administrative area, that has been affected by drought conditions over the entire cropping season." It is based on the Vegetative Health Index and looks at the duration and intensity of crop cycle stress over the growing season. Food and Agriculture Organization 2019b.
[71] Food and Agriculture Organization 2019d.   [72] USDA 2008a.
[73] Locations of ethnic groups are derived from Wucherpfennig et al. 2011.

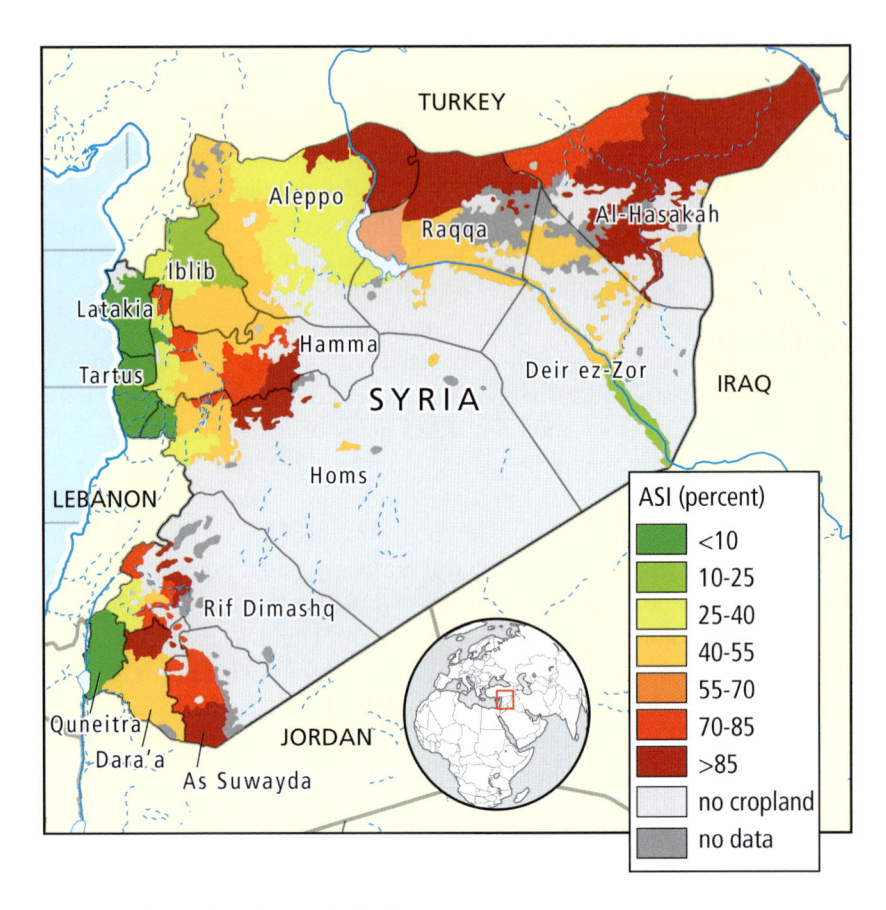

MAP 5.1 Agricultural stress in Syria, 2008
Source: Food and Agricultural Organization, https://bit.ly/3Ad9OwX

drought – particularly in the northeast of the country – tended to be the poorest parts of the country. In 2007, average annual per capita expenditures in the rural northeast were less than half – 2,051 Syrian lira (or pounds) – of their coastal urban compatriots (where the average was 4,339 Syrian lira). Between 2004 and 2007, the country experienced improved equality between regions but a deterioration in incomes in both urban and rural areas and in most regions.[74]

In Lebanon's case, the drought in 2008 mostly affected the agricultural rich lands of the Bekaa governorate in the east of the country, especially

[74] Abu-Ismail, Abdel-Gadir, and El-Laithy 2011, 24.

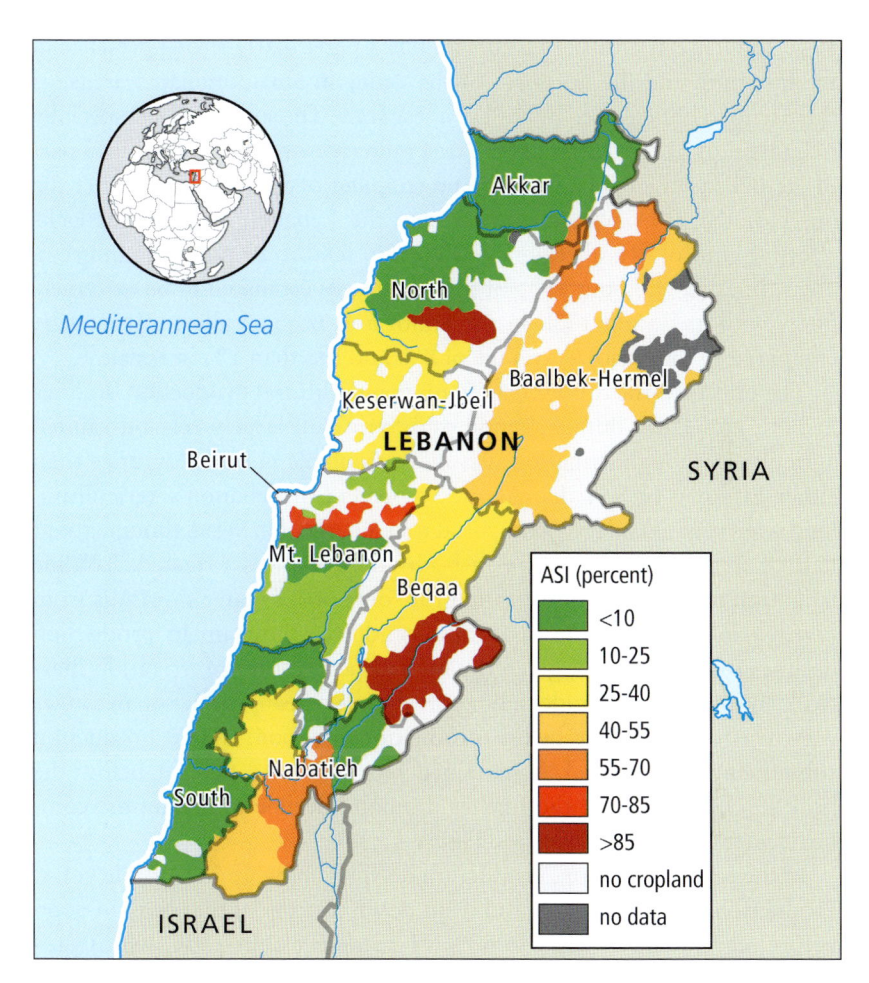

MAP 5.2 Agricultural stress in Lebanon, 2008
Source: Food and Agricultural Organization, https://bit.ly/3hn2AxS

the southern district of Rachaya, where more than 85 percent of agricultural land was affected throughout, and Nabatiye, a governorate in the far south of the country. Pockets of severe drought were also found in other governorates – Mount Lebanon and North (see Map 5.2).[75]

The Bekaa Valley, running north–south in the east of the country, is Lebanon's primary agriculture region and was greatly affected by the 2008 drought. In the southern part of the country, there are mainly small farmers, while the North and Bekaa Valley governorate has mostly

---

[75] Food and Agriculture Organization 2019c.

large commercial farms.[76] Hezbollah has a particularly strong presence in the far south and the far north of the valley in Shia-dominated areas, as these tend to be neglected by the Lebanese state. The south was an especially rich target of Israeli sorties in the 2006 military campaign. The central part of the Bekaa Valley became the main recipient of Syrian refugees after the onset of the 2011 war, an area where the Lebanese government lacks effective control. Rural regions in Lebanon tend to be poorer than urban areas. In 2004–5, about 18% of the North was estimated to be extremely poor, with 12% and 11% in the South and Bekaa Valley, respectively, compared to the national average of 8% and less than 1% in Beirut.[77]

The effects of the drought show up in agricultural production declines in both countries. Both Lebanon and Syria still largely rely on rainfed agriculture, though irrigation was more available in both countries than either Ethiopia or Somalia. About 20 percent of Lebanon's agricultural land was irrigated in 2007 compared to 9.8 percent in Syria, though about half of the wheat production in both countries was irrigated.[78] While irrigation might have compensated for some of the water shortfalls from lower rainfall totals, the drought should still show up in production declines of major crops, given the limited penetration of irrigation, particularly among smallholders. In 2008, Syria's wheat production in tonnes was nearly 60 percent below production in 2006. While production subsequently rebounded, it remained more than 20 percent below the peak level, even before the further dramatic decline following the commencement of the civil war (see Figure 5.5).

In the case of Lebanon, production declines show up in multiple years. Production peaked in 2006, and went down to 25% of that peak in 2007 before rebounding in 2008 to 93% of 2006 levels. In 2009, wheat production was 72% of peak levels before bottoming out in 2010 at 54% of peak 2006 production (see Figure 5.6). Wheat is especially sensitive to temperature spikes, with a one degree Celsius increase contributing to a 13 percent reduction in production.[79] Recall that temperatures in 2010 were nearly 1.8 degrees Celsius above normal. The World Bank notes that the 2010 drought was a significant drought year in the Lebanon context with a combination of drought, temperature, and fire leading to declines of wheat production by as much as 83 percent.[80]

---

[76] World Bank 2010. [77] UNDP 2007.
[78] World Bank 2019a. Wheat statistics for Lebanon are from Verner et al. 2018, 25; the wheat irrigation statistic of 45 percent is from USDA 2008a.
[79] Verner et al. 2018, 25. [80] Ashwill et al. 2013, 57.

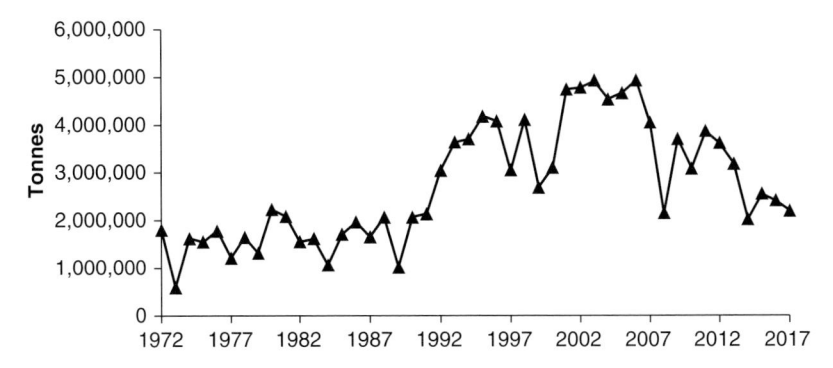

FIGURE 5.5 Wheat production in Syria, 1972–2017
Source: Food and Agriculture Organization, https://bit.ly/3h2881H

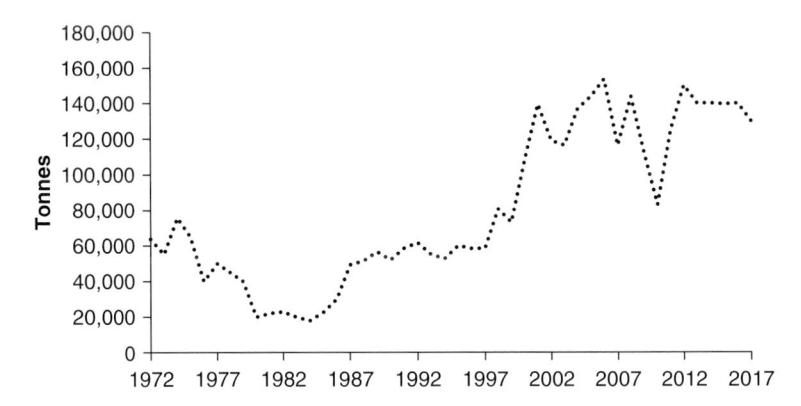

FIGURE 5.6 Wheat production in Lebanon, 1972–2017
Source: Food and Agriculture Organization, https://bit.ly/2UJAN2Q

We now have evidence of a severe drought that affected both countries, with wheat production in both experiencing a significant decline. What effect did the drought have on livelihoods? Here, a variety of policy interventions – from support for irrigation, income and food support, food for work, insurance mechanisms, assistance for asset rebuilding and recovery – could have protected people from harm. Nonetheless, there should be evidence of distress in both countries beyond production declines in terms of income declines, food insecurity, and/or livestock deaths that would have triggered demands for a policy response.

With respect to Syria, some data points have been reported even by those more skeptical of the links between climate and conflict. de Châtel notes that, "According to several UN assessments between 2008 and 2011, 1.3 million people were affected by the drought, with 800,000 people 'severely affected.'"[81] That assessment found 80% of those severely affected were living on bread and sugared tea, only good enough for about 50% of dietary needs.[82] Fröhlich cites other impacts: "Herders in the Northeast lost 85% of their livestock, affecting 1.3 million people. In 2009, according to the UN, more than 800,000 Syrians had lost their livelihoods."[83] One 2011 estimate for the United Nations International Strategy for Disaster Reduction suggested the drought was the worst in forty years, causing a tributary of the Euphrates – al-Khabour – to dry up, with wheat production in nonirrigated areas declining by 82 percent. The report suggested that nationally total livestock levels dropped from twenty-one million pre-drought to between fourteen and sixteen million afterwards.[84] A May 2008 US Embassy cable released by WikiLeaks also reported that the regime increased prices of domestically produced food in April 2008 – wheat by 40%, sugar beet by 30%, and almost a 100% increase in wheat and barley – leading to at least one food price-related protest.[85]

Lebanon's drought, which had similar effects on wheat production, is not recorded as a disaster. But, food prices did increase dramatically in the 2007–8 period, with average prices rising by 18.2 percent in 2008 alone, suggesting that the regime risked protest activity. One study suggested that the government responded by reintroducing subsidies on wheat, bread, and flour that had been in the process of being phased out.[86] The implication here is that government actions may have staunched the risk of food-related protests.

At the same time, the decline in wheat production in this period could have adversely affected farmers' incomes in Syria, potentially leading to food insecurity in rural areas, abandonment of farms, suicides, migration to urban areas, and ultimately conflict. Wheat production has been subsidized in Lebanon as a strategic priority to enhance national self-sufficiency and insulate farmers from fluctuations in global wheat prices. Lebanon still imports more than 80 percent of its foodstuffs, including

---

[81] de Châtel 2014, 525.
[82] United Nations Office for the Coordination of Humanitarian Affairs 2009.
[83] Fröhlich 2016, 40.   [84] Erian, Katlan, and Babah 2010, 15.
[85] US Embassy in Syria 2008a.
[86] UN Economic and Social Commission for Western Asia 2016, 10. This report noted that the more significant challenges to food security in Lebanon occurred later in 2014 after more than one million migrants from Syria flooded into the country, particularly in the agricultural rich Bekaa Valley in eastern Lebanon.

cereals.[87] Given that international prices can vary widely (as they did after Russia banned grain exports in August 2010 in the wake of a drought and wildfires), the decision to subsidize production of a water-intensive crop was a strategic choice.[88] Syria, as discussed in the section Evidence of State Capacity, has also historically adopted a similar, even more ambitious, commitment to food self-sufficiency.

In the case of Lebanon, through the General Directorate of Wheat and Sugar Beet Subsidy (GDCS), the government plays a critical role in subsidizing production by paying farmers a premium when international prices are low and through subsidized loans. Although production was below 2006 levels, one study suggested that robust international prices in 2007 and 2008 allowed farmers to sell all their wheat directly to the market without relying on state subsidies. The researchers noted, however, that in 2010 the impact of the drought on production made farmers even more dependent on the state for support. In a survey, less than 10 percent of farmers were willing to continue to grow wheat without state subsidy. Here, too, the implication is that were it not for state subsidy, the situation of wheat farmers would have been much worse in the midst of the drought.[89] I discuss these policies further in the next section, as we contrast the successful efforts by Lebanon to shore up consumers' access to food and farmers' income from food sales with Syria's failed efforts to protect both farmers' and consumers' interests.

The differences in governance, both state capacity and inclusion, between Syria and Lebanon loom large here. Even though Lebanon has its share of governance challenges, especially as a result of misaligned incentives from its now thirty-year old power-sharing agreement, the country managed to avoid a descent into another civil war, despite occupation from Syria and external meddling by Israel.

Evidence of State Capacity

The first dimension to evaluate is state capacity. As in the Chapter 4, I begin by examining broad indicators of governance from the World Bank and other sources before discussing in more detail agriculture-specific governance arrangements.

The World Bank governance measures capture different dimensions; a number of them – Government Effectiveness, Regulatory Quality, and Control of Corruption – may be relevant when considering the state's

[87] Ibid., 20.   [88] Parfitt 2010.   [89] Tawk et al. 2019, 199, 203.

ability to deliver services in the midst of a multiyear drought.[90] The core expectation is that Lebanon's capacity would be higher than Syria's at the onset of the drought. The drought itself could have had an effect on state capacity in terms of resources available to serve the populace, but, as suggested earlier, it is less clear how quickly a slow-onset drought could have such a negative impact on capacity.

World Bank measures typically reflect investor evaluations of country performance and are only available dating back to 1996 and for every other year until 2002. A snapshot view shows higher quality governance in Lebanon for all three dimensions in line with my basic expectations. In 2007, early in the drought period, Lebanon's Government Effectiveness was ranked 45 percentile while Syria's was 22. On Regulatory Quality, the 2007 divide between Lebanon and Syria was wider, with both countries scoring poorly on corruption (see Table 5.2).

By the advent of Syria's civil war in 2011, investors' perceptions of Syria's Government Effectiveness and Regulatory Quality actually improved. The upswing in overall effectiveness can be observed in Figure 5.7. The steep decline in perceptions of effectiveness only really crystallizes in the wake of the civil war itself.

As discussed in the Chapter 4, the PRS Group has an indicator of bureaucratic quality that dates back to 1984.[91] It, too, is based largely on investor perceptions (that indicator itself is a component of the World Bank's Government Effectiveness metric). Lebanon's bureaucratic quality reached its nadir in 1990, just as the country's long-running civil war ended. Bureaucratic quality in Lebanon improved throughout the 1990s, eclipsing Syria's in 1997 and remaining higher thereafter, although it experienced a steady decline over time. Syria's bureaucratic quality would remain lower through the 2000s with a slight improvement in 2008 (see Figure 5.8).

To some extent, these indicators confirm the expectation that Lebanon had better governance than Syria in the lead up to Syria's civil war. However, the World Bank's Government Effectiveness indicator suggests a convergence of capabilities in the 2000s to 2011, even after multiple years of drought, raising questions about the relevance of this indicator or

---

[90] World Bank 2019g. Government Effectiveness captures "perceptions of the quality of public services, the quality of the civil service and the degree of its independence from political pressures, the quality of policy formulation and implementation, and the credibility of the government's commitment to such policies." See Chapter 4, note 45 for definitions of Regulatory Quality and Control of Corruption.
[91] See Chapter 4, note 46 for a definition of bureaucratic quality.

TABLE 5.2 *Governance indicators for Syria and Lebanon*

Indicator	Country	Year	Percent rank
Government Effectiveness	Lebanon	1996	51.37
		2007	44.66
		2011	45.97
	Syria	1996	21.86
		2007	22.33
		2011	38.39
Regulatory Quality	Lebanon	1996	34.78
		2007	46.60
		2011	51.66
	Syria	1996	14.67
		2007	8.25
		2011	19.91
Control of Corruption	Lebanon	1996	31.18
		2007	19.42
		2011	19.43
	Syria	1996	19.89
		2007	12.14
		2011	13.74

Source: World Bank 2019g

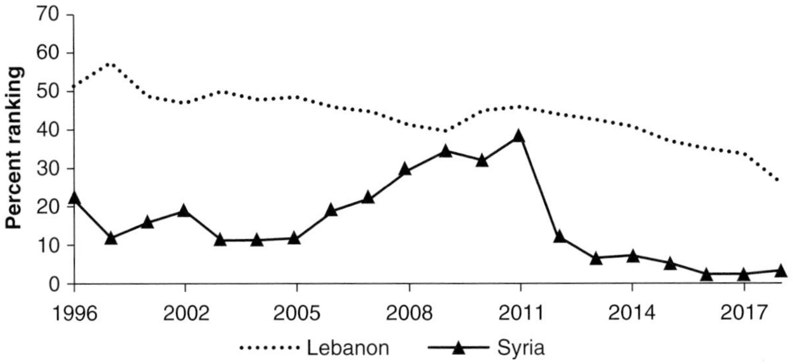

FIGURE 5.7 Government effectiveness in Lebanon and Syria, 1996–2018
Source: World Bank 2019g

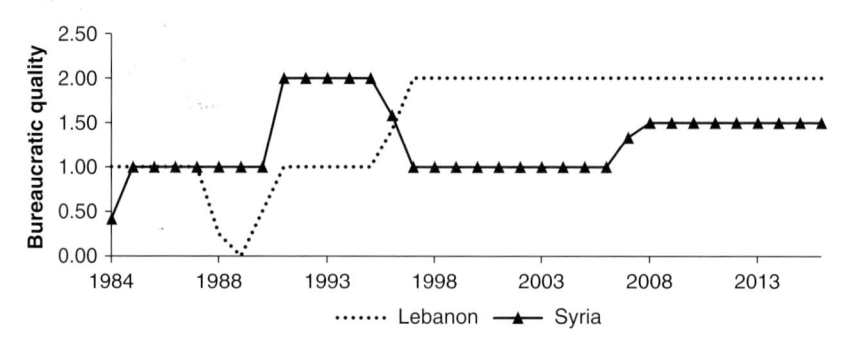

FIGURE 5.8 Bureaucratic quality in Lebanon and Syria, 1984–2017
Source: PRS Group, www.prsgroup.com

the contribution of differences in state capacity to the outcomes in both countries. It may be useful to consider the reputation that Bashar al-Assad enjoyed prior to his repressive response to the protests of March 2011. Assad came into office in 2000 after his father died. He established something of a reputation as an economic reformer throughout the 2000s as he sought to reorient the country more along the lines of a market economy, with less heavy-handed state intervention.[92] That reform agenda picked up in 2005, which corresponds to renewed investor confidence in Government Effectiveness in the World Bank data.[93]

International investors may well have judged the removal of subsidies on diesel fuel and fertilizers in 2008 to have been market-friendly and therefore a symbol of good governance. That decision on diesel fuel, however, had an impact on the irrigation pumping capacity of farmers in the midst of the later drought as well as their ability to get their goods to market.[94] Along with other policies, such decisions might have made the situation worse and less tenable for farmers, but it might not be reflected in foreign investors' perceptions of governance until the situation became mired in violence.

As in Chapter 4, we can also look for indicators of state capacity specific to drought preparedness and response. This would include prediction and early warning systems for drought, programs to prepare for droughts and insulate the population from adverse impacts including grain storage, irrigation, insurance schemes, income, and food support.

[92] Horn 2012; Haldevang 2017.   [93] Bennet 2005; Raphaeli 2006; Butter 2015.
[94] de Châtel 2014, 527.

We can also evaluate the response measures the country has established to prevent malnutrition and starvation, such as food-for-work schemes and food donations as well as programs to restart agriculture such as seed, agricultural inputs, and livestock restocking programs. Because both of the countries considered here were more reliant on irrigation than either Ethiopia or Somalia, water management institutions that regulate and apportion irrigation are also relevant. Here, I focus on policies of production subsidies in both countries, as these seem to be most relevant for thinking through differential state capacity. I discuss some further details in the next section on political inclusion.

Both countries have encouraged some measures of food production – even food self-sufficiency in the case of Syria – through systems of subsidy. These can prove costly to sustain but also problematic if removed in haste. Syria had an aggressive program to foster national self-sufficiency in food production dating back to the 1960s. In terms of water management, this was facilitated by efforts to dam rivers and extend irrigation in the northeast of the country. With respect to production, this was, as Fröhlich noted, "defined by subsidies for farm inputs and fuels, especially for strategic crops such as wheat, cotton and barley."[95]

After the introduction of diesel motor pumps in the 1960s, low-cost credit and subsidized fuel facilitated the extension of drilling wells and pumping from the 1970s to the 1990s. Groundwater levels declined significantly. Syria later initiated an annual well permitting process, ostensibly to control groundwater levels, though this became politicized and subject to corruption.[96] Even before the drought, water mismanagement and waste, alongside other factors, had led to significant declines in water availability – by as much as half according to one estimate – between 2002 and 2008.[97] Staff competence in this area was very low. According to de Châtel, "The majority of staff in the ministries of Agriculture and Irrigation has barely finished secondary school and only a small minority has a university degree."[98] Furthermore, she described the institutional structure of water management as "arcane" and "fragmented" with no less than twenty-two different ministries, councils, and directorates involved, and with little coordination between them despite similar responsibilities.[99]

Alongside this were subsidies for farmers. A 2008 World Bank assessment reviewed the country's subsidy program and noted a variety of

---

[95] Fröhlich 2016, 41. [96] de Châtel 2014, 531; see also Saleeby 2012.
[97] Fröhlich 2016, 40. [98] de Châtel 2014, 531.   [99] Ibid., 530–531.

pressures, including low oil revenues, were making it more difficult for the government to continue funding the program. Under its Annual Agricultural Production Plan, Syria created "agricultural stability zones" in 1975 to regulate land allocations for different crops with the goal of national self-sufficiency in wheat and cotton. Those licenses to operate were, in turn, tied to access to credit, inputs, and marketing services. The state regulated the sale of strategically important crops through state-controlled companies. It offered price subsidies for wheat, cotton, sugar, sugar beet, barley, and tobacco, though in some years international prices were higher than domestic prices, lessening the need for domestic subsidy.[100] The subsidies were not merely to achieve self-sufficiency, but a deliberate effort by the Assad government to build support for the regime, particularly in largely Sunni rural areas of the country.[101]

The World Bank report suggested agricultural subsidies collectively amounted to about 4% of GDP, with diesel subsidies accounting for 2.6%, fertilizer and seed up to 0.3%, and credit 0.1% (augmented by bad debts); price subsidies to cotton growers (0.9%) and beet farmers (0.1%) were other significant subsidy items.[102] The report noted that agriculture functioned as a sort of safety net where the low-skilled, rural poor were concentrated.[103] However, the subsidy program was captured by large farming interests, although it was also directed to the poorest regions of the country and included subsidies for poor cotton laborers. The World Bank's general conclusion was that Syria's statist approach to agriculture and support for "strategic crops" was leaving the country ill-equipped to take advantage of more lucrative international trade opportunities, both for selling wheat abroad but also diversifying into fruits and vegetables.[104] The Bank counseled that diesel subsidy ought to be removed alongside more support for strategic crops like wheat to dampen the social impacts of fuel subsidy removal.[105]

Syria began a process of liberalization that would scale back subsidies. This subsidy reform agenda started in the mid-1980s but picked up in 2005, when the tenth Five Year Plan (for 2006–10) was released. In 2008 and 2009, as the country faced its worst drought in decades, the government cancelled subsidies for diesel fuel and fertilizer, leading to immediate

---

[100] World Bank 2008.
[101] I thank Emily Whalen for emphasizing this point. See also Hinnebusch and Imady 2018; Hinnebusch 2019.
[102] World Bank 2008, 4, 8.   [103] Ibid., 11.   [104] Ibid., 6.   [105] Ibid., 8.

price hikes.[106] These increases in fuel prices made it harder for farmers to irrigate what little harvest they were expecting, affecting production in the final weeks before harvest and increasing the costs of transporting goods to market.[107] One reason cited for the change in policy was the size of Syria's fiscal deficit. By one account, fuel subsidies amounted to 15 percent of Syria's GDP. Repealing those subsidies apparently quadrupled fuel prices overnight in May 2008.[108] In May 2009, the government also cut subsidies for fertilizers and prices doubled.[109]

In the lead up to the drought, other choices also made it harder for the Syrian state to respond. de Châtel writes that the "the lack of social safety nets left many in the agricultural sector unable to cope."[110] In 2006, the country sold its strategic wheat reserves to capitalize on high international prices – some 1.5 million metric tons and twice as much as the year before – and thus had to turn to imports two years later as the drought undermined production.[111] Syria had been a net exporter of wheat since the 1990s and was forced to import wheat in 2008 for the first time in fifteen years.[112]

In November 2008, in a cable later released by Wikileaks, an FAO representative asked the United States Agency for International Development (USAID), the US bilateral foreign assistance program, for help with the drought. Calling the drought a "perfect storm," the official noted that the Syrian Minister of Agriculture had said "that [the] economic and social fallout from the drought was 'beyond our capacity as a country to deal with.'"[113] Once the impacts on food security were recognized, the government adopted a number of measures under the 2009 drought appeal to deliver emergency food supplies to the most affected regions as well as livestock feed, seed stock, replacement livestock, and technical assistance.[114]

As noted earlier, Lebanon subsidizes both the consumption of wheat/flour/bread as well as the production of wheat. The country has subsidized and continues to subsidize other crops such as sugar beet and tobacco; however, I focus on wheat for illustrative purposes and for which more information is available. Since the 1960s, Lebanon has diversified its agricultural production into more high value fruit trees and vegetables

---

[106] de Châtel 2014, 526.   [107] Ibid.   [108] Fröhlich 2016, 42.   [109] de Châtel 2014, 526.
[110] Ibid.
[111] Polk 2013b, 2013a. See Wikileaks released US government cable for confirmation of reserve sales; US Embassy in Syria 2008b.
[112] de Châtel 2014, 527.   [113] Quoted in Polk 2013b.   [114] US Embassy in Syria 2009.

for export (especially to Gulf countries), for which additional subsidies and incentives are provided; most are taken by large landowners.[115]

The Lebanese government estimated that wheat subsidies cost 129 billion Lebanese lira between 2007 and 2011 (roughly $85 million) on a net basis, with nearly all of the costs associated with consumer subsidies. Those subsidies were paid for with protectionist tariffs on imported cereals and other foodstuffs, foreign aid, taxes on services, and other revenue sources. There were efforts during the 2005–6 period to scale back and reduce farmer subsidies. Because farmers were unable to sell their wheat in 2006, this policy was suspended and subsidies were restored to benefit some 1,300 farmers (less than 1 percent of the country's 170,000 farmers).[116] In 2007 and 2008, international prices were higher than before and only limited subsidies were offered.

The policy of subsidy phaseout was fully reversed in 2009, with 2010 being an important year climatically as a consequence of a variety of weather extremes – high daytime temperatures, low nighttime temperatures, as well as flooding which reduced crop yields by as much as 60 percent. Because production was so low, the quantity of subsidized wheat that was bought was less than 20 percent of what the government had budgeted for. The 2010 net costs were estimated to be small, only 0.01 percent of GDP, benefiting fewer than 700 farmers.[117] While the number of beneficiaries was relatively low, per person subsidies were in the order of $6,000 to $6,500, which suggests the beneficiaries were reasonably well-off farmers, as mean per capita consumption in Bekaa Valley was only about $2,300 in 2004–5.[118] If these were local elites, price support could have served to secure their ongoing goodwill in an otherwise difficult time for production.

The more costly intervention was a cap on the price of bread that in the 2007–8 season alone had a net cost of 93 billion Lebanese lira (about $61 million), which was more than 70 percent of the subsidy total over the entire period. This constituted about 0.21% of the country's GDP in that period and nearly 1% of primary expenditures. Between April 2007 and the early months of 2008, international wheat prices increased from $200 to nearly $450 per metric ton. The government's price caps negotiated with domestic bakery syndicates had set prices at a level based on charging the bakeries about $210 per metric ton. The number of beneficiaries is

---

[115] Banfield and Stamadianou 2015, 44.   [116] Ibid., 22.
[117] Republic of Lebanon Ministry of Finance 2012, 7–9.
[118] This is based on consumption, adjusting for regional prices differences; UNDP 2007, 5.

Syria and Lebanon 151

rather unclear, but given the cost outlay and the number of tons purchased – more than 240,000 – it presumably was an order of magnitude greater than the number of farmers. In 2010–11, following additional droughts domestically and internationally, another version of the subsidy policy for consumers was reinstated with net costs of 21 billion Lebanese lira ($13 million).[119] This suggests that the Lebanese government was more concerned about the protest potential of urban consumers than that of farmers.

In terms of quality of governance, external evaluations of Lebanon's drought management capacity are decidedly mixed. A 2018 study by the FAO noted that the country had "no specific plan or strategy for drought"; there have been long-standing but unimplemented reforms of water management dating back to the 1970s.[120] Although this chapter focuses on the period 2006–10, 2013–14 is often cited as another important drought year in Lebanon which prompted some emergency actions. However, major reforms – to conserve groundwater, diminish excessive use of dams, and repair leaky infrastructure – continued to languish.[121]

In 2018, the World Bank also evaluated Lebanon's drought response and, like Syria, noted fragmentation between multiple agencies and lack of clarity over agency roles. The report cited an undated study that said the country lacked a drought emergency plan and that these problems persisted after the 2014 drought. Core conclusions were that there was no concerted drought management effort, that the country did not prioritize drought risk management, and had no drought monitoring system.[122] In 2016, a USAID project helped fund a regional drought preparedness initiative to correct these deficiencies, but this suggests that Lebanon's capacity in this area through the mid-2010s was poor.[123] Thus, while capacity may have been better in Lebanon than in Syria, Lebanon's drought-related governance was not especially good, though the state managed to maintain some subsidies to consumers and producers in the midst of the drought. It should also be noted that this state-centric account may miss the fact that the militant group/political party Hezbollah, which controls parts of southern and eastern Lebanon, is itself a potent provider of public services and welfare.[124]

---

[119] Republic of Lebanon Ministry of Finance 2012, 17.
[120] Food and Agriculture Organization 2018, 96. [121] Riachi 2014.
[122] Ashwill et al. 2013, 67–68. [123] National Drought Mitigation Center 2016.
[124] Ash 2019.

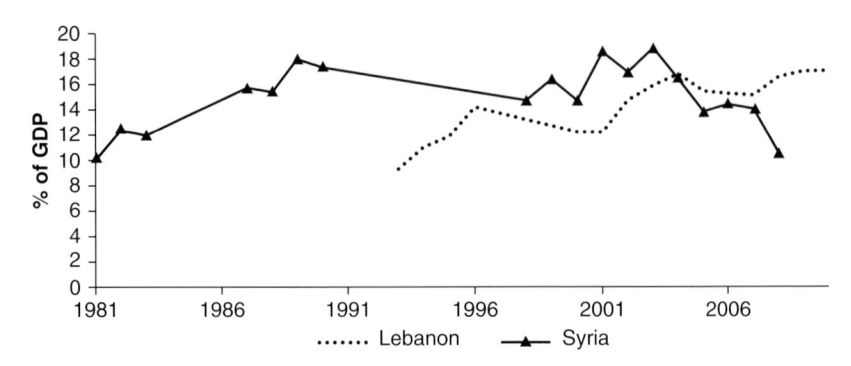

FIGURE 5.9 Tax revenue as share of GDP, Lebanon and Syria
Source: ICTD/UNU-WIDER via Our World in Data, https://bit.ly/3xa5iNL

Thus far, I have compared state capacity in Syria and Lebanon, as an ex ante difference between them, but it is also possible that the multiyear drought affected state capacity itself by reducing tax revenue from agriculture. While disaggregated sectoral tax receipts are not readily available, there are estimates of tax collection as a percentage of GDP through 2008. Neither country collects a large share of revenue in taxes. Denmark's share, for comparison, was over 40 percent. Here, we observe the two countries diverging in the midst of the drought with tax revenues as a share of GDP in Syria going down from 14% in 2005 to about 10.5% in 2008, while Lebanon's went up from 15.4% to 16.6% during the same period. While there may have been other reasons for poor tax collection in this period, this suggests that the drought could have had an effect on Syria's available resources, both for service delivery and suppression of dissent (see Figure 5.9).[125]

Evidence of Political Inclusion

The second relevant dimension is the degree of political inclusion, arguably the most important difference between these two countries with implications for equitable access to services and attentiveness of the governments to different constituencies. It can be argued that, since the 1970s, the Syrian state has exhibited a high degree of political exclusion, rewarding the minority Alawite group that controlled the government at the expense of other groups, and increasingly exclusionary under the

[125] Ortiz-Ospina and Roser 2016.

leadership of Bashar Al-Assad.[126] The Lebanese state has had elaborate power-sharing agreements in place which were reaffirmed in the wake of its fifteen-year civil war that ended in 1990. Those carefully crafted arrangements have ensured that a variety of Lebanese stakeholders have had representation in government and a means of surfacing their concerns and claims for services.

As a result of the United Nations mandate system that emerged after World War I and the partition of the Ottoman Empire, Syria and Lebanon came to be administered by France. Lebanon's boundary with Syria reflects efforts by the French to create a Maronite Christian-controlled state in the 1920s.[127] Syria and Lebanon both inherited substantial ethnic and religious cleavages upon becoming independent in the 1940s. While both have ultimately suffered from civil wars in the modern era, Syria was nominally more stable during the latter part of the Cold War when Hafez al-Assad seized power through a series of coups and concentrated power in his presidency, which lasted from 1971 until his death in 2000. Assad ruled the country largely in favor of his minority Alawite group, a sect of Shia Islam, though he did try to build support among rural Sunnis. The Alawites constitute about 13 percent of the Syrian population and have retained dominant control of the political apparatus since the 1970s, with major implications for patronage and political rewards that the Syrian state has offered to its primary supporters.

These differences are captured in the Ethnic Power Relations dataset that, as discussed in Chapters 3 and 4, charts the political representation of different ethnic groups in countries over time as governments come and go. Previous work by Bretthauer suggested that politically exclusive regimes were those where more than 20 percent of the population was excluded from power.[128] In Syria's case, the minority Alawites, as mentioned, have been the dominant partner in government since 1970. Other ethnic groups, which make up 86 percent of the population, have been excluded from political representation ever since. The majority, Sunni Arabs, which account for 65 percent of the population, moved from being junior partners in government from 1966 to 1969 to being discriminated against thereafter. Kurds, who account for 8 percent of the population, have been discriminated against throughout the country's history. Christians (10 percent of the population) have been powerless under both Hafez al-Assad and his son Bashar al-Assad who assumed power in 2000.

[126] McLauchlin 2018. [127] Sly 2013. [128] Bretthauer 2015.

Both Kurds and Druze (3 percent) have also been "powerless," according to the data.

In Lebanon's case, according to the dataset, Shias (32%), Sunnis (20%), and Maronite Christians (16%) were all senior partners in government between 1992 and 2017. Druze (6%), Greek Orthodox (5%), Armenian Orthodox (4%), and Greek Catholic (3%) have all been junior partners during this time period. Excluded populations only constituted 13% of the population; only Arab Palestinians (10%) – who lack citizenship – are formally classified as being "discriminated" against with Armenian Catholics (1%), Protestants (1%), and Alawites (1%) being "powerless." At no time since 1946 has Lebanon ever had more than 13 percent of the population excluded from political representation in government, including the lengthy period of civil war from 1975 to 1990. From 1971 to 1991, however, Shia Muslims – 32 percent of the population and historically the poorest Lebanese – were junior partners in government.[129]

As in Chapter 4, I also show the measure of social exclusion from the V-Dem dataset[130] (see Figure 5.10). Relevant for thinking about both social service provision in the lead-up and in response to drought, the measure of social groups reflects the degree to which there is equality of access to public services based on other distinctions such as ethnicity,

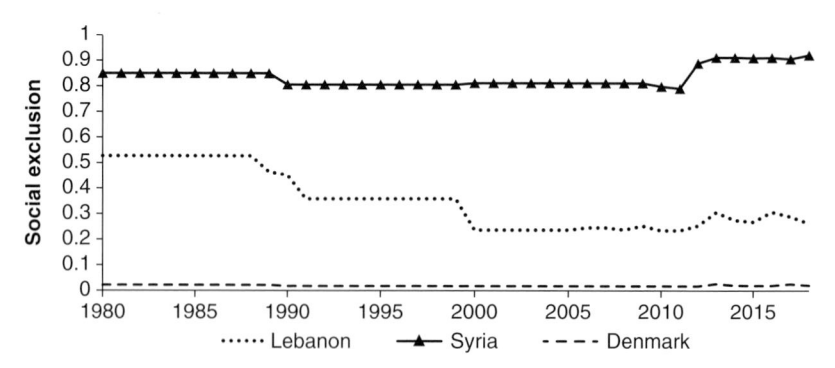

FIGURE 5.10 Social exclusion in Lebanon, Syria, and Denmark
Source: V-Dem Project 2019

---

[129] ETH Zurich 2018. Historically, rural farmers have mostly been Shia and Maronite Christians, with the Maronites somewhat better off than the Shia.
[130] V-Dem Project 2019.

caste, language, race, and region.[131] As a contrast to Lebanon, and especially Syria, highly inclusive Denmark is also shown. Consistent with the political representation figures from the Ethnic Political Relations data, V-Dem shows Syria having consistently high social exclusion with Lebanon being more inclusive throughout and improving over time.

The Assad family has ruled Syria since the 1970s with group-based favoritism to the minority Alawite sect to which they belong, which is mostly concentrated along the coast. Originally a poor group, the Assad family elevated it by appointing Alawites to key positions in the security services.[132] The Ba'ath coup that ultimately brought Hafez al-Assad to power in 1970 not only served to elevate a new elite, but also built some wider legitimacy with peasants through land reform and by buying the loyalty of an emergent middle class through public sector employment via a program of nationalization. Coupled with Arab nationalism, investments in health, education, and electrification, Assad continued to build a support base in rural areas.[133] This process of "selective goods provision" to regime supporters included the northwestern governorate of Latakia, where three-quarters of Alawites are from, as well as other regions that are part of the Ba'athist coalition including the southern region of Hawran (which includes Daraa, the locus for some of the first protests in March 2011), the rural areas around Aleppo, and the northeastern part of Deir al-Zor.[134] While the regime may have favored its Alawite base, Khaddour and Mazur note that "the regime has played a subtle game of balancing and securing the interests of other segments; the general quiescence of the Sunni business classes of Aleppo and Damascus during the uprising is clear evidence of this fact."[135] Mazur makes a similar point: "When an incumbent regime is drawn heavily from a single ethnic group, it typically favors members of that ethnic group but must strike bargains with segments of society beyond the group in order to rule."[136]

Thus, while some of the Sunni majority in Damascus were also beneficiaries of government favoritism, other Sunnis resented this arrangement, and rebellions by the Muslim Brotherhood in the early 1980s were put down with great brutality. The regime's ability to continue this model became harder over time. The national security state, which was necessary to suppress internal dissent and support the government's involvement in Lebanon was expensive. Patronage jobs and subsidized food also were

---

[131] Ibid., 195–198. [132] Sachs 2000. [133] Hinnebusch 2012, 95–96.
[134] De Juan and Bank 2015, 94. [135] Khaddour and Mazur 2013. [136] Mazur 2021.

a drain on the public pursue. As a consequence, the country's economic base deteriorated, with the state, in the 1980s, forced to embrace austerity, including massive public sector spending cuts. Partial privatization of state-owned import monopolies created a new class of beneficiaries who were dependent upon the regime for their wealth, with remaining social programs like subsidized food and jobs keeping a semblance of legitimacy with the middle and lower classes.[137]

These contradictions accelerated when Hafez al-Assad died in 2000. His son Bashar al-Assad sought to both deepen the liberalization agenda and retain some of the social protections for the masses, through a so-called "social market" economy with reforms beginning in 2005. However, he was dependent upon a narrower group of supporters in the Asad-Makhlouf family clan who were richly rewarded with patronage, which angered some old guard elites. The inexperienced technocrats that Assad recruited to pursue the liberalization agenda were not especially skilled since salaries for these new officials were low.[138] Hinnebusch describes the deterioration in the social base of the regime as it neglected its former compact with the largely Sunni peasantry: "The structural balance in the regime was shifting toward patrimonial authority at the expense of bureaucratic and inclusionary capacity."[139]

Despite oil revenue, the parlous state of the country's economic situation, made worse by Western isolation and sanctions because of Assad's support for Saddam Hussein, led the regime to abandon some of the social protections and subsidies it had maintained to retain legitimacy with the masses, starting with fuel subsidies and then agricultural inputs and price support.[140] Trade liberalization led to cheap imports coming into the country with small businesses badly affected. Hinnebusch describes the contours of this emergent system: "At the heart of the regime coalition were the 'crony capitalists' – the rent-seeking alliances of political brokers (led by Asad's mother's family) and the regime-supportive bourgeoisie."[141] This was the scene as the country entered into a multiyear drought in 2006–7.

With regard to Lebanon, managing sectarian divides has always been a challenge, as is the case in Syria. Since the 1860s, Lebanon's governance structure has consisted of different power-sharing agreements that have tried to maintain a delicate balance among the country's eighteen different religious groups or confessions, with some periods being more successful

[137] Hinnebusch 2012, 98. [138] Ibid., 99. [139] Hinnebusch 2019, 35.
[140] Hinnebusch 2012, 102. [141] Ibid., 101.

than others.[142] Because power distribution in government is based on the historic 1932 census, no census has been completed since then, even though groups have likely experienced differential population growth.[143]

In the post-independence period, the project became less tenable over time. In the 1970s, the influx of Palestinian refugees led to further cleavages between the dominant Maronite Christians and leftist groups sympathetic to the Palestinians. In 1975, disputes over efforts by President Camille Chamoun to monopolize fishing rights for his Maronite community is credited as the spark for a fisherman's strike. That strike, when suppressed by the government, was yet another driver in the escalation to what would become a long-running civil war that ultimately claimed 120,000 lives.[144]

According to one assessment, post-civil war, Lebanon was seen as carved up into a landscape of different geographies: "The civil war produced a mosaic of small territories and social spaces, in which the power of the state, and the influences of the local political elite, are relative."[145] The Taif Agreement of 1989 help put an end to the civil war through a renewed power-sharing agreement, with Syria charged with being the main power broker in the country.[146] Like other power-sharing agreements, it was not fully democratic and encumbered by inertia and the shadow of Syrian occupation. Even after the end of the Syrian occupation, the country at times lacked a president and suffered parliamentary paralysis in 2013. The evolution of pluralism and consociationalism successfully averted a return to the large-scale violence at the end of the civil war in 1990, but only barely.[147]

An interesting aspect of Lebanese politics is that it is not simply the state that engages in service provision. Sectarian parties are major service providers of health, education, and other social services and serve as intermediaries for public benefits.[148] Cammett and Issar observe that the predominantly Sunni Muslim Future Movement and the Shiite Muslim Hezbollah groups seek out-group political support through welfare

---

[142] Zahar 2005.
[143] Barshad 2019; Brown 2009. The 1932 census is widely known to have been manipulated in favor of the Christians. I thank Emily Whalen for this observation.
[144] Reilly 1982.   [145] Banfield and Stamadianou 2015, 23.
[146] The Taif Agreement shifted power from the president to a council of ministers and provided for parliamentary, cabinet, and civil service parity between Muslims and Christians regardless of demographic trends; Bahout and Bahout 2016.
[147] Kota 2012; Hartzell et al. 2016. For a more pessimistic take on the semifeudal nature of Lebanon's confessional system of parliamentary representation, see Ignatius 1983.
[148] Cammett 2015; Ash 2019.

service provision including food packages. Because the Sunni Muslim Future Movement is more deeply engaged in vote buying, it provides services more broadly than Hezbollah, which has been more ambivalent about electoral politics given its roots as a militia group.[149] Despite the many dysfunctions of the Lebanese political system, elections do provide modest incentives for somewhat more inclusive party-linked social service provision, though with considerable heterogeneity across the country.[150] In the disaster risk reduction space, this plays out, for example, in the coastal city of Tyre with multiple ambulance systems serving different sociocultural groupings.[151]

This depiction of Lebanon may oversimplify the inclusiveness of the system.[152] Notably, Shia ministers resigned from the government in November 2006, with Hezbollah and other Shia parties angling for more political power and also seeking to quell an investigation into the 2005 assassination of former prime minister Rafic Hariri.[153] Those disputes over relative influence precipitated violent clashes in May 2008. Moreover, the system increasingly seems to benefit elite groups at the expensive of the wider public.

Lebanon's power-sharing agreement may have averted a large-scale civil war in the contemporary era, though there are frequent clashes in the Bekaa and the South; moreover, the country experienced spillover conflicts with Israel in 2006 and in 2014. However, Lebanon's system has created its own problems, namely, a legacy of logrolling corruption and clientelism that has impeded service delivery and entrenched a perception that elites are serving their own interests rather than the wider public: "Hospitals, roads, schools and other projects are distributed to favored contractors according to sectarian quotas that ensure every group benefits, regardless of necessity."[154] The 2015 garbage crisis which engendered the large "You Stink!" protests and another garbage crisis in 2019 underscored the limits of this sectarian patronage-based model. A previous garbage contract was apportioned to provide valuable contracts for two

---

[149] Cammett and Issar 2010.
[150] Cammett 2014; Ash 2019. Elsewhere, Cammett examines the degree to which political parties in Lebanon are inclusive in their appeals to nonsect members and the extent to which they focus on the state or extrastate political mobilization, with Hezbollah becoming more inclusive over time but still focused on extrastate activities. The Future Movement is an example of a more inclusive political party with cross-sect appeal and focused on the state; Cammett 2014, 34.
[151] Peters, Eltinay, and Holloway 2019, 16.
[152] I thank Konstantin Ash for making this point.   [153] Slackman 2006.
[154] Yee and Saad 2019.

separate landfills (one worth $288 million and the other $142 million) – one contract went to the brother of an aide to a top Sunni politician and the other to a businessman close to a senior Christian politician. Even as these elites have gotten wealthy, much trash has been dumped in the ocean and along the coast.

Such self-dealing by elites across different groups was a major impetus for massive popular protests in 2019 that toppled the prime minister.[155] Yet the relative openness of the state to tolerate such protests provides a means by which society can relatively peacefully express its grievances and seek redress. As one observer noted in 2016 as Syrian refugees tested the regime' stability: "The Lebanese political system is definitely in need of a raft of political reforms, but the basic inclusiveness of the system remains a key bulwark at least against serious civil conflict of the kind we see in several neighboring Arab countries."[156]

This may be something of an optimistic take on Lebanon's resilience. The regime has experienced a number of challenges, even since 2019, none more so than a devastating fertilizer explosion in Beirut in August 2020 that killed some 130 people, injured 4,000, and displaced 300,000 and which led to the resignation of the government. That explosion, coupled with an economic crisis made worse by the COVID-19 pandemic, led to large-scale and increasingly violent protests.[157] I come back to the precarity of Lebanon's relative inclusiveness in the final section, Concluding Thoughts.

Evidence of International Assistance

Both Syria and Lebanon were middle-income countries, relying less on foreign aid in the contemporary era than poorer states in the international system. Figure 5.11 shows aid as a percentage of government expenditure and compares Lebanon and Syria to Ethiopia, a relatively aid-dependent state discussed in Chapter 4 (see Figure 5.11).

That said, Lebanon has relied on its diaspora community for a relatively large share of its GDP, far more than Syria (see Figure 5.12 with Ethiopia again included for comparison).

Given that both countries faced severe droughts in the late 2000s, flash appeals for humanitarian assistance could have been requested from the United Nations Office for the Coordination of Humanitarian Affairs (UNOCHA). However, there were no drought-related appeals for finance from Lebanon from 2006 to 2011, but there were from Syria.

---

[155] Ibid. [156] Salem 2016. [157] Ismay 2020; Hubbard 2020; Hubbard and Saad 2020.

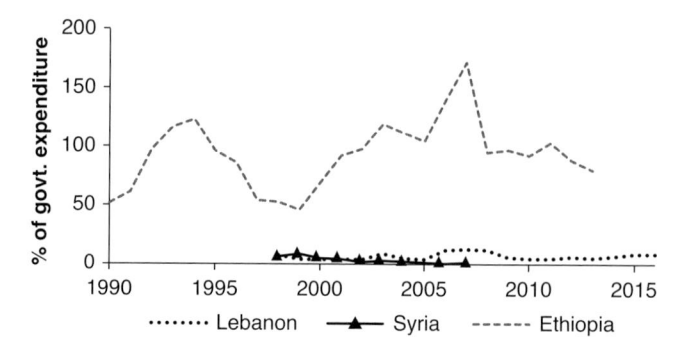

FIGURE 5.11 Development assistance as share of government expenditure
Source: World Bank via Our World in Data, https://bit.ly/3hc61IZ

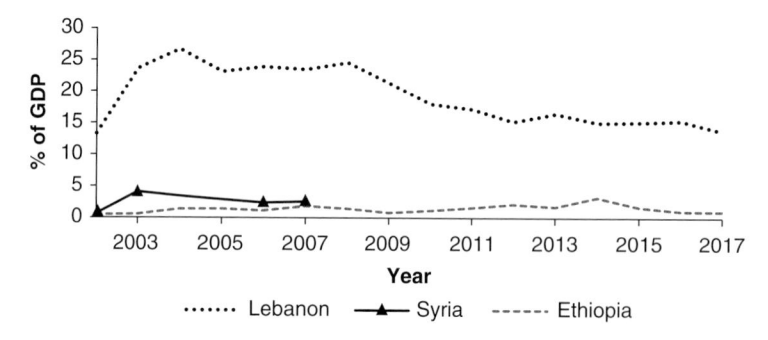

FIGURE 5.12 Personal remittances as share of GDP, 2002–2017
Source: World Bank via Our World in Data, https://bit.ly/3y6DGcl

Syria worked with the United Nations in September 2008 and August 2009 to issue emergency drought appeals, but just for the northeastern provinces of Hasaka, Raqqa, and Deir ez-Zor. No assistance was sought to aid displaced populations who had relocated to the south. Just $5.4 million of the $20.4 million requested in 2008 was provided by the international community, and only one-third of the $43 million requested in 2009 was provided.[158]

In 2006 and 2007, Lebanon did make emergency appeals for finance in the wake of Israel's thirty-four-day military operation against Hezbollah that started in July 2006 and displaced some 700,000 people in Lebanon. UNOCHA coordinated a $150 humanitarian appeal in 2006 with another smaller appeal of $20 million in 2007 for ongoing refugee support.[159] The

---

[158] Financial Tracking Service 2019b, 2019c. [159] Financial Tracking Service 2019a.

first Lebanon appeal secured more than 120 percent of the initial funding request, while the 2007 appeal only secured about 45 percent of the funding needed. While the nature of the funding appeal (refugees displaced by armed attack in Lebanon compared to drought victims in Syria) was different, the differences in the efficacy of fundraising appeals does suggest the relative isolation of Syria from the international community compared to Lebanon.

de Châtel suggests the underperformance of international community support for Syria was in part because the Syrian government was ambivalent about seeking international assistance and downplayed the drought's significance in its own media and to donors. Given the country's pride and self-image as self-sufficient in food production, fully acknowledging the drought was perceived as a bridge too far. Moreover, donors were not quite sure about the government's strategy. A drought management plan, started in 2000 and completed in 2006, was apparently not activated.[160]

Cables from the US government noted that it did not contribute to the appeal in 2008, which was limited to providing aid to just 10,000 families. Bad relations between the US and Syria seemed to be at the heart of the Obama administration's reluctance to support aid, though the challenges of actually delivering assistance through the World Food Programme were also noted.[161] In 2009, the United Nations asked the United States to make a $10 million contribution, which might have helped to persuade other donors to make contributions of their own.[162] However, the United States did not contribute to that effort either.[163] A 2010 cable released by Wikileaks showed that the World Food Programme continued to have difficulty getting its funding appeals supported in that year. Of the more than $22 million requested in November 2010, little more than $5 million had been mobilized by February, limiting the number of beneficiaries to 240,000 which was 60,000 less than intended. The Obama administration had wanted to call the crisis an emergency while the Syrian government was reluctant to label it as such.[164] Syria's experience as donor outcast is akin to Ethiopia in the 1980s under the Derg and Somalia in the lead-up to the 2011 famine. When countries are led or substantially controlled by groups or individuals deemed untrustworthy by the international community, this decision can be as significant as efforts by recipient countries to block aid themselves.

---

[160] de Châtel 2014, 527–528.   [161] US Embassy in Syria 2009.
[162] US Embassy in Syria 2010a.   [163] Financial Tracking Service 2019c.
[164] US Embassy in Syria 2010b.

PUTTING THE PIECES TOGETHER

Thus far, I have shown that Syria and Lebanon both experienced severe droughts that led to major declines in agricultural production. I have shown that these countries differed in state capacity and political inclusion and traced the differences in subsidy regimes and drought response in both. What is the sequence of events that leads from drought in both countries but to civil war only in Syria?

Here, we need to show that the drought had something to do with the protests either in terms of the claims people made or the people who made them. Do we see drought among the reasons cited by protesters? Are people displaced by drought among the protesters? If neither of these conditions holds, is there another pathway by which the drought could have affected the outcome?

Fröhlich disputes that northeasterners displaced by drought were in a position to protest, given their tenuous standing as migrants in the south, but this view is not shared by other scholars. Drought could have played an important role in swelling the ranks of underemployed young men who took part in the protests and, later, violence. They might not have been the first to participate but could have eventually joined in. Even if migrants were not involved in dissent against the regime, that is not fatal to the argument connecting climate migration to conflict. As I have argued, the presence of large migrant populations competing for housing, jobs, and services under the Syrian regime might have triggered dissatisfaction by longtime residents, even if migrants themselves shunned participation in the protests. While drought might have surfaced as a reason for the protests themselves, it is quite possible that other drivers were the sparks for the protests and ultimate violence that took place.

In this section, I focus on Syria and discuss Lebanon more fully in the Alternative Explanations section that follows. Ide surveys the Syrian evidence and concludes that: (1) the links between climate change and the drought are plausible but not proven; (2) there is strong evidence of the drought leading to massive loss of agricultural livelihoods – but that evidence is contested; (3) the evidence for massive rural to urban migration is also contested; and (4) the role of migration in intensifying grievances is plausible, but limited information exists.[165]

As Ide notes, there is good evidence to suggest that the drought did displace large numbers of people from the north of Syria to southern and

---

[165] Ide 2018.

western cities, but there is disagreement among scholars about the size of the displacement and whether migrants themselves participated in protests.[166] In a November 2008 cable released by Wikileaks, the US Embassy in Syria reported the fears of the FAO representative Abdullah bin Yehia: "Without direct assistance, Yehia predicts that most of these 15,000 small-holding farmers would be forced to depart Al Hasakah Province to seek work in larger cities in western Syria."[167] Yehia worried that "15,000 unskilled laborers would add to the social and economic pressures presently at play in major Syrian cities," already burdened by Iraqi refugees, inflation, middle class dissatisfaction, and "a perceived weakening of the social fabric and security structures."[168] In a 2009 report for the emergency aid appeal, the United Nations estimated the size of internal migration: "Migration figures range from 40,000 to 60,000 families. 36,000 families have reportedly migrated from Hassakeh Governorate alone."[169] In 2009, reportedly about 60–70 percent of villages in the governorates of Hassakeh and Deir ez-Zor were deserted.[170] In a June 2009 cable released by Wikileaks, the US Embassy in Syria reported that some 250,000 to 300,000 had migrated out of the region according to FAO, mostly to seek casual labor in major cities like Damascus, Aleppo, and Homs as well as casual farm labor near the Jordanian border in Daraa and As-Suwaida.[171] A February 2010 cable from the US Embassy noted that there was some reverse migration back to the northeast as the rains had returned, but there was still immense human suffering in the region. Moreover, the cable noted it was still "taboo" to acknowledge the scale of the migration publicly.[172]

Some researchers report even higher numbers of displaced Syrians. Femia and Werrell, Gleick, and Kelley report as many as 1.5 or even 2 million displaced persons as a result of the drought.[173] Selby et al. critique this estimate as wildly out of proportion to most estimates of between 40,000 and 60,000 families, or about 300,000 people. They go on to dispute whether or not the drought was all that critical in driving even those numbers, given existing seasonal migration from the region

---

[166] Nasser, Mehchy, and Abu Ismail 2013, 26; Erian, Katlan, and Babah 2010, 32; Wodon et al. 2014, 55; Abu-Ismail, Abdel-Gadir, and El-Laithy 2011, 24.
[167] US Embassy in Syria 2008b, 4.   [168] Ibid.
[169] United Nations Office for the Coordination of Humanitarian Affairs 2009.
[170] de Châtel 2014, 527.   [171] US Embassy in Syria 2009.
[172] US Embassy in Syria 2010b.
[173] Kelley et al. 2017; Gleick 2014; Femia and Werrell 2012; Werrell, Femia, and Sternberg 2015; Werrell and Femia 2017.

and other drivers such as economic liberalization.[174] Ide's conclusion is that authors from competing camps are potentially exaggerating their differences, and the discipline would be better served by acknowledging the challenges of estimating precise numbers in data-poor Syria:

> In sum, the figure of up to 1.5 million refuges used by Kelley et al., but also by Femia and Werrell, Feitelson and Tubi, Gleick, and Werrell et al. is almost certainly overstated. But it is still very likely that several hundreds of thousands of additional people migrated from the drought-affected areas to the outskirts of urban centres. This number is more significant than implied by the estimates provided by Selby et al. and could have had considerable negative impacts on social service provision and resource availability.[175]

The precise size of drought-related migration may not be knowable, given the challenges of understanding the baseline levels of seasonal migration and other disruptions that were occurring in Syria at this time, including Iraqi refugees and the return home of Syrians after the country's occupation of Lebanon ended.[176] It is difficult to disentangle the physical effects of the drought from the governance failures of response. As de Châtel concludes: "Similarly, climate change per se – to the extent that its predicted effects would already be visible – did not drive Syrians into the street in protest; it was the Syrian government's failure to adapt to changing environmental, economic and social realities."[177] Those, like me, who make conditional claims about the circumstances under which climate effects lead to negative security outcomes acknowledge this point.

The specific magnitude of the migration may be less important than the social consequences. Were the numbers large enough to potentially trigger the kind of problems the FAO feared, namely, local contestation over resources? What happened to the people who stayed but whose livelihoods were adversely affected by the drought? According to de Châtel, although there had been seasonal migration of young men in the past, this drought was so severe that whole families migrated to southern governorates, with a number settling in tent camps outside the southern city of Daraa,[178] the site of early protests against the regime in March 2011. Hinnebusch also sees the drought playing a part in Daraa: "In Dera, formerly a base of the Ba'ath, where it [the protests] began, the loss of work opportunities in Lebanon, corruption and drought had encouraged Salafism among unemployed youth."[179]

---

[174] Selby et al. 2017a, 238.   [175] Ide 2018, 351.
[176] Selby et al. 2017a, 239; Ash and Obradovich 2020, 6.   [177] de Châtel 2014, 522.
[178] Ibid., 526.   [179] Hinnebusch 2012, 107.

Elsewhere, Hinnebusch notes, "Different from Egypt but somewhat similar to Libya, the uprising was geographically dispersed and away from the capital, beginning in the rural peripheries, then spreading to small towns, suburbs, and medium-sized cities, where its foot soldiers were unemployed youth, refugees from drought, and others among the 'losers' of a decade of post-populist neo-liberalism."[180] This would make a potentially strong case for those dislocated by the drought to be among the early protesters against the regime. Hinnebusch goes on to suggest that those resettled rural groups ended up in suburbs where they got to see some of the regime's favoritism to Alawites first-hand – which became a source of grievance in and of itself. He notes that "Alawis were given preferences for scarce jobs and Alawi crony capitalists [were] enriching themselves on their insider connections. All of this was making the rural areas, small towns, and suburbs hotbeds of discontent that would be mobilised in the uprising."[181] He elaborates on this, writing that the discontented areas were precisely those where different groups came into contact with each other: "Initial centres of grievances were mixed areas where Alawis and Sunni lived together as in Latakia, Banias, and Homs. The uprising then spread to Hama and Deir Ez-Zor, traditional bastions of Sunni piety resentful of the secular regime."[182] A related source of grievance in Hinnebusch's view was competition for housing, made worse by the influx of displaced populations from the drought, refugees from Iraq, and rising real estate prices as a result of inflows of money from the Gulf.[183]

This picture of drought-driven migration and mobilization is echoed by other scholars. Schmidt suggests the drought-ravaged northeast led to mass migration to crowded slums in the north, especially Idlib and Aleppo: "[I]t was also especially from those segments and places, that the opposition militias recruited its greatest number and most dedicated activists."[184] Lawson makes a similarly sweeping argument about the combined effects of drought and agricultural mismanagement: "Unable to survive any longer in rural areas, tens of thousands of impoverished laborers flooded to the fringes of the country's major cities, where they clustered in shanty towns that were almost entirely devoid of public services. It was in such marginal districts that the popular Uprising broke out in the early spring of 2011."[185] Goldsmith echoes these claims: "Drought and government neglect of rural Syria since the mid-2000s

---

[180] Hinnebusch 2019, 41. [181] Ibid., 36. [182] Ibid.
[183] Hinnebusch and Imady 2018, 4. [184] Schmidt 2018, 38. [185] Lawson 2016, 86.

meant that masses of unemployed rural Sunnis became easy targets for jihadist recruiters from 2012 as the regime escalated its brutal repression and the Uprising turned towards armed resistance."[186]

Dukhan makes a similar claim: "However, the collapse of the rural economy of tribal communities in the south and east of Syria during Bashar al-Assad's regime due to drought, lack of development projects and the mismanagement of al-Badia resources ignited the Syrian uprising to start in tribal regions."[187] He cites examples of tribal belts around Damascus, Homs, and Palmyra that swelled with migrants after the drought and saw the Alawites getting better jobs and living in better conditions.[188] Kilcullen and Rosenblatt argue that drought led villages to empty out and recreate themselves as "vast, insulated neighborhoods of urban poor" who armed themselves for protection.[189]

Other scholars, however, such as Ababsa have a different view. She acknowledges the size of the drought and displacement, suggesting that as many as 300,000 families were driven to Aleppo, Damascus, and other cities. However, she concludes that political factors as much as climatic factors drove that exodus, citing historic depletion of the water table, failures of previous irrigation schemes, and removal of fuel subsidies in the midst of the drought that made it too expensive to pump water, drive tractors, or drive water out to farmers' herds. Moreover, Ababsa argues, displaced populations who ended up working in greenhouses in Daraa and in coastal cities like Latakia and Tartous did not participate in the uprising: "They did not take part in the revolts in Der'a and on the coast, being too poor to get politicized according to many observers. Nevertheless, the presence of so many displaced peasants, proof of the failure of the Jezira development, was an important context element in which the revolts occurred."[190]

Daoudy goes further and notes that the "revolts did not take place in the rural areas struck by drought."[191] While she acknowledges dislocation to suburbs of Damasacus and Der'a, she claims "there is no evidence that these migrant farmers contributed to the uprising," pointing to interviews she carried out with civil society members and a few migrant farmers. She then points to the demands made by protesters and notes that neither food nor water were among their demands.[192] To some extent, this may miss the role played by drought, to the extent that it had one, in the uprising, as I explore further below.

---

[186] Goldsmith 2018, 147; see also Saleeby 2012. [187] Dukhan 2014, 21.
[188] Dukhan 2018, 108. [189] Kilcullen and Rosenblatt 2014, 33.
[190] Ababsa 2015, 210. [191] Daoudy 2020, 12. [192] Ibid., 14.

Ash and Obradovich try to connect the migration within Syria through a unique data strategy. They use lights at night as a proxy for population density and record changes in light penetration between 2005 and 2010 to suggest population growth and decline. They then correlate such changes with the likelihood of protest, finding that areas in the northeast that experienced a decline in light intensity were associated with lower protest risks, while Sunni Arab areas that experienced higher light intensity (and hence were recipients of displaced populations) were more likely to experience protests.[193]

While this study has some bearing on the connections between drought-related migration movements and protest, Ash and Obradovich note that in their qualitative review of demands by protesters at the early stage of the uprising in March and April 2011, the drought was not mentioned, nor were many economic demands in general, the emphasis being on freedom and calls for reduced corruption. They see this as providing some support for one of their hypotheses that people migrated to areas with similar kin and religious identities; so, rather than foster intergroup cooperation, migration served to accentuate the claims of the now larger identity group.[194] Ash and Obradovich note the presence of shared ties between sending and receiving areas in their technical appendix:

> Specifically, there is evidence Syria's migrants and locals shared tribal, in addition to sectarian, connections. In particular, Syria's Sunni Arabs have a complex patchwork of kinship networks and these sometimes transcend regional boundaries . . . . Several tribes and tribal confederations have populations in both areas stricken by drought and those that received migrants: there are Baggara in both Aleppo and Deir ez-Zour, Fadan in Raqqa and Allepo's Ain al-Arab, Al-Abda in Hasakeh and Hama and Al-Harb in Hasakeh, Aleppo, Damascus and Homs.[195]

Ash and Obradovich's argument provides a provocative new line of potential research inquiry but also underscores the evidentiary challenges of tracing these links definitively. For example, another study by De Juan and Bank, also using nighttime lights, found both selective distribution of electricity to regions preferred by the regime and lower levels of violence in those areas compared to less favored areas. They also argue that the violence they observed does not appear to be correlated with Sunni

---

[193] Ash and Obradovich 2020.   [194] Ibid.
[195] See ibid., appendix C2. For discussion of social networks and labor migration in Daraa, see Leenders 2012.

settlement areas, which they suggest is because the early uprising did not have a strong sectarian character. At first blush, this finding that regime favoritism led to more violence is consistent with my argument about Syria's political exclusion, and coupled with Ash and Obradovich's research, we have some support for the connection to drought.

However, Ash and Mazur suggest De Juan and Bank's findings are spurious given problems in using uncorrected nighttime lights data.[196] Ash and Mazur argue convincingly that electricity and other utilities were not selectively provided to regime supporters, but other sorts of services were, including public employment and newly privatized government services like the mobile phone market. They also make the case that De Juan and Bank's use of overly aggregated data at the subdistrict level obscures the fact that nearly all of the early fatalities in the uprising were from Sunni Arab areas. When disaggregated to the town level, 99.5 percent of the fatalities through to November 2012 were in Sunni-majority areas.[197]

It is, perhaps, not surprising that two to three years after the beginning of the drought, there is no explicit mention of it in the protest activity in 2011, although the relative absence of economic claims is somewhat surprising. Moreover, as Selby et al. note, while Daraa is often identified as the first case of mass post–Arab Spring protest in Syria, there were at least two other protests in Damascus as well as ongoing protests in the Kurdish region dating back to February 2011.[198] As noted earlier, the US Embassy cables of May 2008 also signalled that there was at least one food-related protest by farmers in the coastal city of Tartous, but that it was quickly dispatched with water hoses: "A minority-run police state with heavy-handed internal security services, the SARG [Syrian Arab Republic Government] keeps a close watch on any civil unrest that could pose a threat to the regime."[199] Nearly three years later, such demands for redress for food-related concerns may have morphed into broader calls for political reform.

If the government's drought policy was not an explicit source of criticism, what about the protesters: Do we know anything more about who they were? It is still unclear which social groups were the main protagonists in anti-government activity. For those interested in whether the drought played a role in the onset of the Syrian uprising, the findings

---

[196] De Juan and Bank 2015; Ash and Mazur 2020.
[197] Ash and Mazur 2020, see page 34 in the supplementary material.
[198] Selby et al. 2017a, 240.   [199] US Embassy in Syria 2008a.

discussed thus far suggest the need for more fine-grained micro-level work to understand how the uprising started and escalated in different cities across Syria. Efforts to reconstruct these foundations are potentially subject to sweeping generalizations based on limited data. For example, the journalist Tom Friedman traveled to Syria for the documentary *Years of Living Dangerously* and interviewed some rebels who claimed they had previously been farmers, which led him to conclude that the drought fueled the civil war.[200] Even the more methodologically rigorous work by Fröhlich is based on about thirty semi-structured interviews from two refugee camps and two cities in Jordan.[201]

Mazur has carried out a number of city-level studies to identify the dynamics of mobilization in the uprising. He argues that minority-led governments often make clientelistic linkages with leaders of ethnic majority groups to stay in power. When those regimes antagonize majority groups, however, mass protests can result with regime-allied ethnic intermediaries unable to contain the unrest. While Syria's early nonviolent uprising was characterized by cross-ethnic participation, the violent phase of the uprising was almost wholly Sunni.

In a case study of Homs, Mazur makes the case that rural village networks remained intact as the city grew, with largely unregulated settlement on its periphery.[202] Rural to urban migration to Homs has a long history, with the city's population swelling from about 140,000 in 1960 to around 750,000 in 2010. While Alawis were among the migrants to Homs, in 2010 most were Sunnis: about half the city had a Sunni tribal background from the tribal steppes in the east.[203]

Mazur suggests that regime violence against Sunnis ultimately made it impossible for Sunni intermediaries with ties to the regime to contain the protests. While the uprising originated with urban elites, their poorer Sunni compatriots joined when it became clear there were sufficient numbers for sustained protests. But the explicit call was for freedom, not food. One popular chant was, "We don't want bread, we want our freedom and dignity."[204] In another study of the origins of the uprising in the eastern governorate of Dayr-al-Zur, Mazur finds a similar dynamic of poorer community members from peripheral towns slowly joining the protests: "[V]iolence against local community members constituted the

---

[200] Friedman 2014.
[201] Fröhlich 2016. Pearlman 2016 interviewed some 200 refugees in 2012 and 2013, overwhelmingly men and a narrow majority from urban areas.
[202] Mazur 2020b, 483–484.   [203] Ibid., 498.   [204] Ibid., 507.

'spark' that activated solidarities in a way that material deprivation alone had not."[205]

What role then for the drought in the conflict? To some extent, Mazur's research supports Selby et al.'s argument that there were already plenty of rural migrants in the cities from previous waves of migration, so adding in several hundred thousand more (or whatever the ultimate number was) would not have been enough to change the structural situation appreciably. Selby et al. estimate that with population growth, Iraqi refugees, general rural-to-urban migration, the return of migrants from Lebanon, and some outmigration from Syria, drought-related migration accounted for at most between 4 and 12 percent of urban growth between 2003 and 2010.[206] Did a pulse of 200,000–500,000 people in 2008–2009 – if you accept their numbers – make much of a difference to the uprising?

Unless we can find out know more about who ultimately joined the uprising, we may never know if Selby et al.'s assessment is correct. Even if former farmers were not among the first wave of urban upper-class protesters, did they join later? One way to get at this would be to have more information about the demographic profile of the uprising (How represented were recent Sunni migrants from rural areas?). Nearly ten years into the conflict and so much loss of life, it may be hard to find out other than through more extensive interviews with refugees.

This discussion perhaps misses an important aspect about how the drought fundamentally weakened tribal authority in the east, which perhaps facilitated groups' further radicalization. Khaddour and Mazur note that one of the biggest effects of the conflict was the weakening of traditional tribal authority in the east where previously one leader could effectively contract with outside actors on behalf of the group.[207] Fragmentation of tribal authority wrought by the conflict provided an opening for radical groups like the Islamic State and Jabhat al-Nusra. In his study of Dayr-al-Zur, Mazur writes that "leaders of revolutionary demonstrations and the armed rebellion were overwhelmingly young and largely came from outside the historical bayt al-ʿashīra lineages; many came from poor and marginalized parts of their tribes."[208] Left unmentioned in these accounts are the effects of the drought that, just before the conflict, severely disrupted the population and social networks, perhaps accentuating the challenges for traditional tribal intermediaries.

---

[205] Mazur 2020a, 182.   [206] Selby et al. 2017a, 243.   [207] Khaddour and Mazur 2017.
[208] Mazur 2020a.

Alternative Explanations 171

In the language of the stringency of qualitative tests, my argument connecting the drought to migration and the conflict has survived a hoop test but there is not yet a smoking gun. We do not have drought explicitly among the demands for redress by protesters, but we have evidence of large-scale movements of people from drought-affected areas and protests being more likely in recipient regions. Broad claims by a number of scholars link the droughts explicitly to the protests, but we still lack extensive micro-level confirmation that former farmers themselves eventually participated in the uprising in large numbers or of how the disruption of agriculture in the east made radicalization and conflict resolution less likely.

ALTERNATIVE EXPLANATIONS

I have argued that Lebanon did not experience a civil war of its own during the period 2006–10 because it had a somewhat more capable government response, flawed but more inclusive governance, and was somewhat better able to tap into international support systems, particularly from its large diaspora population. Setting aside for the moment the strength of the evidence connecting climate to conflict in the Syrian case, what explains the differences in outcomes between the two countries if not the factors I have identified?

Perhaps the strongest alternative explanation for the discrepant outcomes between the two countries is that Syria and Lebanon are different in too many important respects. One distinct difference in the mid-2000s was Lebanon's memory of a civil war, which was catastrophic for the country. The fifteen-year war resulted in more than 120,000 casualties; between 600,000 and 900,000 people left the country, which only had a population (in 1975) of about 2.5 million.[209] The memory of the civil war perhaps served as a deterrent for groups in Lebanese society, for all its problems, from pursuing large-scale violence to secure their political objectives.[210]

Other differences between the countries may also make them less than perfect comparisons. Lebanon was richer, more urbanized, and less dependent on agriculture than Syria. Indeed, the effects of the drought, though significant for wheat production, do not show up in terms of large-scale human suffering. Ash suggested Lebanese farmers in the Bekaa Valley, for example, could more readily make their way to Beirut, given

---

[209] Murphy 2006.   [210] I thank Emily Whalen among others for raising this point.

its close proximity (roughly a couple of hours), even if seasonally, to make ends meet. Syrians, more fundamentally, had to uproot themselves in the wake of the drought to move from the northeast to more far-flung places like Daraa.[211] Travel time from Al Hasakah to Daraa, for example, is more than ten hours. This is an important concern, though there is ample evidence of fairly significant seasonal migration from the east to the south prior to the civil war in Syria.[212]

This line of argument suggests a couple of different possibilities: (1) that the drought was not as severe in Lebanon and (2) that the countries are different from each other in fundamental ways in terms of socioeconomic development. With respect to the former, the logic of that critique would only reinforce the causal significance of physical exposure to drought. If the drought had been more severe, so the logic goes, Lebanon too would have been at risk of conflict and a breakdown in state institutions. The second explanation suggests that the political differences in governance in terms of capacity and inclusion are less central to the outcome than in Syria and also that Lebanon has divergent levels of economic development. If Lebanon had been poorer and more agriculturally dependent, would it too have succumbed to violence during this period?

As suggested in the section Evidence of State Capacity, the regime is vulnerable to food price shocks in urban areas, and the government undertook efforts in 2008, at great expense, to insulate the populace from these problems. More broadly, Lebanon's situation has been precarious, both in the lead-up to the drought and during its aftermath. The two countries are intertwined, with the Syrian state functioning as an occupying force until the assassination of Lebanon's prime minister in 2005 forced Syria's withdrawal. Lebanon weathered difficult protests and violence between 2006 and 2008, including an Israeli air campaign against Hezbollah in 2006 as well as violent protests in 2008 in response to the government's efforts to shut down a Hezbollah communications network. In the aftermath of Syria's civil war, Lebanon took in more than one million Syrian refugees. In recent years, Lebanon has experienced a variety of, sometimes violent, protest movements related to multiple garbage crises, a spate of wildfires, and the aftermath of the huge Beirut fertilizer explosion. Despite all these developments, Lebanon has, at the

---

[211] Konstantin Ash, personal communication, April 15, 2020.
[212] Daoudy 2020, 160; de Châtel 2014, 526–527.

time of writing in late 2020, not lapsed again into large-scale violence (though it might).[213]

In the earlier discussion of Political Inclusion, the focus was on different elite groups having formal representation in government. But, if we think of inclusion as fundamentally linked to legitimacy, ordinary citizens feeling they have voice opportunities can be an essential safety valve for citizens to express grievances peacefully without risk of repression and deepening anti-regime sentiment. The protests that started in Daraa in Syria in 2011 were brutally repressed and this repression ignited broader Sunni solidarity across the country that then escalated into cycles of violence that the state and local intermediaries could not control.[214]

Lebanon's protest movement, which started in 2019, has been multi-sectarian and youthful, and the sectarian system that has enabled elites to profit at the expense of the people is subject to collective enmity.[215] While persistent protests suggest Lebanon has some expressive mechanisms for the public to press for change, what happens if protests never lead to change? Amnesty International has noted the government's increased use of repressive tactics and excessive use of force, including rubber bullets, tear gas, and beatings.[216] Although freedom of speech is protected in the Lebanese constitution, Human Rights Watch has documented an uptick in attacks on free speech since 2015, and, in July 2020, a coalition of organizations announced a joint campaign to defend free expression in Lebanon.[217] So Lebanon's relative good fortune compared to Syria may not last.

Would the Syrian civil war have started were it not for the drought leading up to it? Would early interventions supported by international partners to help the agricultural sector, particularly in the eastern part of the country, have helped avoid the onset of the later conflict or diminished its potential to escalate? These are still difficult, if not impossible, questions to answer. Those critical of assigning a causal role to climate change in conflict onset fear governments will instrumentally seize upon the effects of climate to "absolve" them of responsibility for the consequences that follow.[218] The comparative method deployed here highlights the importance of governance factors and why climate change leads to security impacts in some instances and not others.

---

[213] Hubbard 2020. [214] See Mazur 2021, chapter 4. [215] Al-Khalidi and Knecht 2019.
[216] Amnesty International 2020. [217] Human Rights Watch 2020.
[218] Daoudy 2020, 157.

Another set of alternative explanations focuses on the reasons for Syria's conflict other than drought. As discussed in the section Expectations and Case Selection, scholars of Syria's experience have identified a host of other explanations that they believe better explain Syria's civil war – economic and water mismanagement, a youth bulge, a lack of job creation, corruption, and lack of political freedom.[219] As I have argued, differences between scholars on the role of climate in Syria's civil war often come down to different assessments of relative causal weight, for which process tracing is not all that well suited. We can rule out certain explanations; if, for example, the timing was wrong – a drought that occurs after a civil war starts cannot be the cause of it. However, we are less well-placed to make definitive judgments about the importance of different causal factors when they all plausibly help explain a case.

CONCLUDING THOUGHTS

In reading and writing about Syria and Lebanon's contrasting fortunes, I worry that they have more in common with each other, which bodes ill for both. Diverse multiethnic states with considerable religious diversity and cleavages are difficult countries to govern, particularly if their postcolonial national identity has never been especially strong.[220] In the era of Hafez Assad, Syria relied on authoritarian brute force to crush dissent. By the 2010s, the regime's relative coercive capacity had declined. We can see this both in terms of putative territorial control and the size of the military relative to the population, both of which decreased after Bashar al-Assad became president.[221] As the Syrian state became modestly more anocratic

---

[219] de Châtel 2014; Daoudy 2020.
[220] Roessler and Ohls 2018; Wantchekon 2000. While the Syrian state's administrative capacity may have improved by the 2000s, its relative coercive capacity over societal groups declined, perhaps facilitated by groups' access to external resources to support them (including access to weapons). This to some extent corresponds to a move from what Roessler and Oehls call a strong state, weak rival, combination to a weak state, strong rival situation. While the former is a conflict-prone ethnocracy, the latter is a repressive minority-rule regime that is unstable. The stable situations are those where groups have relatively equitable power and can deter each other.
[221] Following Hanson and Sigman, I use V-Dem's metric of the central government's territorial control (v2svstterr) which shows a modest decline from 93.75% of the country's territory in 1999 to 85.5% in 2000, declining further to 84.25% in 2010, 81.25% in 2011, and then dropping steeply to 61% in 2012 as the civil war takes off. In terms of the size of the military as a share of the population, Syria's military was 4.16% in 1983 at the height of the government's repressive powers, steadily declining thereafter to 1.39% in

Concluding Thoughts 175

in the first years of al-Assad, the Sunni majority was always going to be less likely to accept the status quo. Government mismanagement of water resources and rural-to-urban migration were long-running trends that affected the country's trajectory. In this context, the drought accentuated contradictions that were already in Syrian society.

In Lebanon's case, an external occupying force – the Syrian military – functioned as an enforcer after the end of the country's civil war, approving key appointments, keeping the various groups in check, and limiting the potential for spiraling conflict.[222] While this was anti-democratic, the withdrawal of the enforcer may pose a risk to stability if different groups become dissatisfied with the status quo. There will be new elites who could mobilize among the disaffected youth to radicalize the population along sectarian lines. Any of the various triggers over the last few years – drought, the influx of a million refugees, wildfires, multiple garbage crises, the fertilizer explosion – could, individually, have pushed Lebanon over the edge into more significant armed conflict.

The fact that all of these have occurred and Lebanon has still not fallen into the abyss may merely be a matter of luck. The self-dealing by Lebanese elites has alienated them from mass publics and created an opening for repeated protest movements. This is similar to the way the Syrian protests initially evolved, as Hinnebusch has argued: "The shock troops of rebellion were young, unemployed, deprived people with little stake in the status quo, widely dispersed and unknown to the government, hence quickly producing new leaders to replace those arrested or killed."[223]

The role of leaderless mass protests is somewhat at odds with the arguments I have made about elite pacts in fostering and/or undermining stability, and suggests a number of important considerations as we turn, in Chapter 6, to why international actors might care about climate security concerns outside their own borders and what the policy agenda ought to be in a world where more regimes will be tested, as has been the case in Syria and Lebanon.

2009 and, after a brief uptick, bottoming out at 1.35% in 2012; Hanson and Sigman 2013; V-Dem 2019; Roser and Nagdy 2013.
[222] Wantchekon 2000, 349.   [223] Hinnebusch 2012, 107.

6

Cyclones in South Asia

The Experiences of Myanmar, Bangladesh, and India

> I am both saddened and frustrated to know that we have been in a position to help ease the suffering of hundreds of thousands of people and help mitigate further loss of life, but have been unable to do so because of the unrelenting position of the Burma military junta.
> — *US Admiral Timothy Keating, June 2008 in the wake of cyclone Nargis*[1]

> It was a huge, meticulously organized lifesaving operation and, it seems, a success story for the government's early-warning system.
> — *Hari Kumar, Jeffrey Gettleman, and Sameer Yasir on India's preparations for Cyclone Fani, 2019*[2]

In May 2008, a major cyclone devastated the Ayeyarwady Delta in Myanmar, leaving 700,000 homeless. Three-quarters of the delta's livestock were killed, half of the fishing fleet sank, and a million acres of rice paddies were inundated with saltwater.[3] Myanmar's authoritarian regime did not request nor permit significant foreign aid. The US Navy, having made fifteen unsuccessful attempts to receive authorization to deliver aid, ultimately ordered its ships to depart in early June.[4] In the end, some 140,000 people died.[5] Contrast that with the more recent experiences of India and Bangladesh.

In April 2019, a cyclone bore down on the northeast coast of India, yet again exposing millions in the poor state of Odisha to high winds and flooding. The storm crossed the Bay of Bengal and also lashed neighboring

---

[1] Associated Press 2008. [2] Kumar, Gettleman, and Yasir 2019b.
[3] *The New York Times* 2009. [4] *The New York Times* 2008.
[5] Zarni and Taneja 2015.

Bangladesh. In India, the authorities issued millions of text messages, mobilized thousands of emergency workers, and broadcast evacuation warnings across a variety of media – television, radio, loudspeakers, sirens, and buses.[6] India's actions involved more than immediate evacuation and reflected efforts to direct people to the many purpose-built emergency shelters that had been constructed over many years.[7] Mass casualties were avoided. In the end, fewer than sixty people died in India as a result of Cyclone Fani, which compares favorably to a storm that struck the area in 1999 and claimed more than 10,000 lives.

Nearby Bangladesh was also affected by a much-diminished Fani and, like India, carried out similar emergency evacuation procedures and was able to avoid mass casualties. Bangladesh has had its share of severe cyclones. In 1970, the catastrophic Bhola cyclone set the stage for the country's breakaway from Pakistan after as many as half a million people died. In 1991, nearly 140,000 died after a massive cyclone struck a still under-prepared Bangladesh. Since then, the country has invested in early warning and preparedness and experienced fewer and fewer deaths over time.[8] India and Bangladesh's successful efforts to reduce the death toll from storms are a stark contrast to their own respective pasts, but also to those of Myanmar.

This sets the stage for the subject of this chapter. Turning to the cyclone prone Bay of Bengal, how do we account for the differential success of Myanmar, Bangladesh, and India in preventing large-scale loss of life from exposure to cyclones, compared to each other and to themselves over time?

In this chapter, I provide more support for each country case in terms of how capacity, inclusiveness, and aid intersected at different points in time to contribute to the outcomes in question. The first section reviews the cases covered and how these relate to my theoretical expectations. The second section contrasts the Myanmar case with more recent cyclone events in Bangladesh and India, and uses within case variation in Bangladesh and India to show their improved preparedness and response over time.

EXPECTATIONS AND CASE SELECTION

I identified cyclones of similar intensity that made landfall in populated areas with the capacity to lead to large-scale loss of life. In some cases, those storms did indeed lead to mass casualties, while in others they did not. The

---

[6] Kumar, Gettleman, and Yasir 2019a.   [7] Kumar, Gettleman, and Yasir 2019b.
[8] Cash et al. 2013.

task is then to explain the differences in outcomes, comparing country experiences to each other and using within-case variation over time.

When it faced Cyclone Nargis in 2008, Myanmar had weak state capacity, exclusive political institutions, and restricted entry of external aid delivery. The preparations and response to the storm led to catastrophic loss of life. In recent years, Bangladesh and India have faced exposure to severe cyclones but largely escaped major loss of life. However, that has not always been the case.

In previous decades, Bangladesh and, to a lesser extent, India had low state capacity. Before its founding, Bangladesh suffered from high political exclusion when it was appended to Pakistan. Bangladesh and India both suffered large casualties from cyclones (Bangladesh in 1970 and 1991 and India in 1999). Bangladesh and India have developed improved state capacity over time, and both (especially Bangladesh) have become more inclusive. Where Bangladesh has relied on foreign aid to buttress its state capacity to a great extent and distributed aid in a broad-based manner, India has been more self-reliant but used aid strategically to good effect. Cyclones in recent decades have been much less deadly (India in 2013 and 2019 and Bangladesh in 2007 and 2019). The case dynamics are summarized below in Table 6.1 as well as the predicted pathway that led to that outcome.

Myanmar, Bangladesh, and India

Before substantiating the argument, I provide an overview of cyclones and climate change and a thumbnail sketch of the key episodes in each country, starting with the catastrophic outcomes in Myanmar, followed by a discussion of Bangladesh and then India.

Tropical cyclones are extreme weather events that occur when low-pressure areas form rotating winds. They occur in warm ocean waters in the Atlantic and Pacific Oceans and are characterized by high winds and potentially catastrophic storm surges. As warm, moist ocean water evaporates, it rises, cools, and forms spinning clouds. The most severe storms can reach winds speeds of 157 miles per hour with storm surges in excess of 20 feet.[9] They are given different names – hurricanes, typhoons, and cyclones – in different parts of the world. In the Indian and South Pacific oceans, they are known as cyclones.

The evidence linking climate change to the increased incidence and severity of cyclones is not conclusive. Scientists are more confident that climate change will increase the *severity* of cyclones rather than their

---

[9] Emanuel 2005, 21.; US Department of Commerce 2018.

TABLE 6.1 *Comparing Myanmar, Bangladesh, and India*

Country	Hazard events	Capacity	Institutions	International assistance	Outcomes
Myanmar	Category 4 cyclone – 2008	Weak capacity	Exclusive	Limited aid delivery	Significant deaths 2008 (Type 1)
Bangladesh	Category 4 and 5 – cyclones 1970 1991 2007	Weak capacity→ Improved capacity	Exclusive→ Becoming more inclusive	Broad-based aid delivery by 2000s	Mass deaths 1970, 1991 (Type 1, Type 2) Limited deaths 2007 (Type 4, Type 6)
India	Category 4 and 5 – cyclones 1999 2013 2019	Intermediate and increasing capacity	Somewhat inclusive	Limited reliance on aid	Significant deaths 1999 (Type 4) Small death totals 2013, 2019 (Type 6)

number. Both wind speed and rainfall associated with cyclones will likely increase, but the number may even decline.[10] Even if scientists do not fully understand how climate change will alter cyclone strength or frequency, cyclones will intersect with climate change in other ways. With climate change induced sea-level rise, the potential for increasingly catastrophic storm surges will make even existing cyclones potentially more damaging.[11]

In 2018, the World Meteorological Organization's Task Team on Tropical Cyclones summarized their understanding of the links. In addition to higher storm surge levels associated with sea-level rise, they wrote that higher moisture content in the atmosphere will likely lead to higher rainfall rates associated with cyclones. They also concluded that storm intensities will likely increase with climate change, leading to a higher proportion of storms that reach the most intense Category 4 and 5 storms. They reiterated that the science is not yet settled on how climate change will affect the number of cyclones.[12]

Asia is home to some 90 percent of the cyclone victims worldwide, including those killed or otherwise affected.[13] The Bay of Bengal, especially the mouth of the Ganges River, is one of the most cyclone prone regions in the world. The extended delta at the base of the Himalayas is subject to monsoon flooding from the drainage of major rivers including the Ganges, the Brahmaputra, and the Meghna which bring sediment downstream and make coastal areas attractive for farming.[14] Those low-lying marshy areas of the coastal plain and the many islands offshore, however, are flood prone. Even a small rise in sea level will produce extensive flooding because of the relative flatness of much of the terrain. The shape of the bay and the shallowness of the waters offshore also favor large storm surges, particularly when accompanied by high-tide events.[15]

Cyclones often sweep up the funnel shaped bay to bring heavy winds and rain along with large storm surges that can inundate whole villages.[16] Mortality in such swift-onset disasters usually occurs fairly swiftly in Phase 1, what disaster experts refer to as "the immediate post-impact phase," the first forty-eight hours after a storm. The next 3–10 days, Phase 2, of secondary post-impact consists of body retrieval, medical services,

---

[10] IPCC 2014b, 7, 2011, 113.   [11] Dasgupta et al. 2011, 2014.   [12] Knutson 2018.
[13] Guha-Sapir and Vogt 2009, 273, examined the period 1960–2008 but this trend for storms (which includes extra-tropical storms, convective storms, and tropical cyclones) continued through 2019 with more than 90 percent of fatalities and those affected occurring in Asia. See also CRED 2019 and Busby and Krishnan 2015.
[14] Penna and Rivers 2013, 236.   [15] Emanuel 2005, 221.
[16] Penna and Rivers 2013, 236–237; Frank and Husain 1971.

removal of debris, and opening of roads, with some but a declining chance of death. Phase 3, from ten days to three months, is the period of post-emergency relief when emergency funds have to be used with deaths uncommon, though possible if there are infectious disease outbreaks. Phase 4 is the period of post-emergency recovery and reconstruction with death rates more or less returning to what they were prior to the cyclone.[17]

Throughout the eighteenth and nineteenth centuries, cyclones in this region killed more than 10,000 on at least six occasions, and several storms killed hundreds of thousands.[18] Bangladesh, perhaps the most densely populated agrarian country in the world, has been especially vulnerable, having experienced some 58 percent of total fatalities from storms between 1970 and mid-2019.[19] Three of the deadliest cyclones in history – the 1970 Bhola cyclone, the 1991 cyclone, and the 2008 cyclone Nargis – are among those examined in this chapter.[20]

Myanmar is the anchor case in this chapter as it serves as a relatively contemporary reminder of the role played by weak state capacity, exclusive political institutions, and limited or denied foreign assistance. Cyclone Nargis buffeted Myanmar in 2008, just as the military regime was about to hold a constitutional referendum intended to address political pressure from longtime opposition leader Aung San Suu Kyi to democratize (and potentially bar her from assuming political office). The government's poor preparations and response (which included rebuffing attempts by the international community to help) meant that 140,000 people died in coastal communities in the Ayeyarwady Delta.

The story of Bangladesh's origins is even more troubling but exemplifies similar dynamics. In 1970, Cyclone Bhola hit Bangladesh (then East Pakistan) and India's West Bengal. Between 300,000 and 500,000 died, making it the most deadly cyclone ever recorded. Poor response to the storm contributed to the move for independence from Pakistan, which the latter sought to suppress through violent reprisals that claimed hundreds of thousands of additional lives. The large-scale loss of life from Bhola was a function of Pakistan's limited government capacity to exercise sovereign control over East Pakistan nearly 1,000 miles away, but it also reflected exclusive political institutions that catered to West Pakistan's Punjabi-speaking majority. East Pakistan's majority Bengali population was considered inferior – in the words of Amir

---

[17] Guha-Sapir and Vogt 2009, 273–274. [18] Emanuel 2005, 221.
[19] CRED 2019. The 1970 Bhola cyclone alone constituted 36 percent of the fatalities in this dataset.
[20] Weather Underground n.d.

Abdullah Khan Niazi, head of the Pakistani Forces in East Pakistan in 1971, as a "low-lying land of low-lying people."[21]

Despite some improvements in early warning systems and cyclone shelters in the two decades that followed, a 1991 cyclone had catastrophic effects, claiming some 138,000 lives. It would have been worse had it not been for some external aid from the US military in the immediate aftermath of the storm, in what was called Operation Sea Angel.[22] This storm set in motion a much more vigorous early warning system and expansion of cyclone shelters. By 2007, Cyclone Sidr, which made landfall along the same path as the 1970 cyclone, would still kill around 3,000 people, but the death toll was a much lower order of magnitude. In 2019, Bangladesh largely escaped from major casualties when Cyclone Fani buffeted the region.

Like Bangladesh, India has also seen improvements over time in its capacity to reduce deaths from cyclones. In 1999, a devastating Category 5 cyclone smashed into Odisha state in eastern India on the Bay of Bengal. Around 10,000 people were killed. In 2013, another Category 4 cyclone struck the same state. In this instance, 50 people died, as the country evacuated more than 500,000 people from low-lying areas, the largest such evacuation in more than twenty-three years.[23] While donors such as the United States Agency for International Development (USAID) worked with India on early warning systems and disaster preparedness, India hardly relied on disaster aid for preparedness or recovery.[24] In 2019, Cyclone Fani exacted about the same level of mortality, and the government's response was praised for preventing large-scale loss of life.

For each country, I provide evidence of each indicator (hazard, capacity, inclusion, and aid) and then the outcome for the case or cases where there is within-case variation. Substantiating the argument requires several pieces of confirmatory evidence. First, we need to demonstrate that the countries all faced similar cyclone exposure from the storm events cited above. Second, we need to show differences in state capacity between Bangladesh, India, and Myanmar in the 2000s and 2010s, as well as differences between Bangladesh and India compared to themselves in earlier periods. Third, we need to be able to show that Bangladesh had more exclusive political institutions in the 1970s with improvement by the 1990s and 2000s. India should have had reasonably inclusive political institutions throughout the entire period, perhaps becoming marginally more inclusive over time. Myanmar's political institutions in 2008 should have been highly discriminatory. Fourth, we

---

[21] Jones 2010, 226–228.   [22] Berke 1991.   [23] Press Trust of India 2013.
[24] Konyndyk 2013.

should be able to show that Myanmar experienced an aid blockage in 2008 while Bangladesh should have had large volumes of aid distributed fairly equitably in 2007. Finally, while showing that the indicators are consistent with my expectations, I also compare my interpretations to those of regional experts and practitioners to ensure my argument conforms to the work of scholars with more ground-level experience.

Evidence of Hazard Exposure

I consider several cyclone events to generate comparisons between countries and within countries over time – Myanmar in 2008, Bangladesh in 1970, 1991, and 2007, and India in 1999, 2013, and 2019. It is important to establish that these storms had similar physical potential to kill large numbers of people. We can evaluate the potential deadliness of storms through two measures of cyclone strength: the Saffir–Simpson 5-point scale (SSHWS) and the Indian Meteorological Department's (IMD) 7-point scale of tropical cyclone intensity. Both are based on wind speed, and all these storms were in the upper categories for cyclone strength. Launched in 1973, the Saffir–Simpson scale is the metric for measuring cyclone severity that most are familiar with, if not the name. It has five categories from 1 to 5, with 5 being the most severe. The Saffir–Simpson scale has been revised over the years and, prior to Hurricane Katrina in 2005, included both a wind speed measure and a storm surge measure in its overall categorization. For Hurricane Katrina, the overall storm score was Category 3, though the storm surge measure itself was ranked as Category 5. Because of potential confusion between conflicting measures, the revised scale released in 2010 is based solely on wind speed.[25]

Category 5 storms have their highest 1-minute sustained wind speeds at 157 miles per hour or higher (greater than 252 km/h) and can create catastrophic damage. The National Oceanic and Atmospheric Administration's (NOAA) website describes the potential damage: "A high percentage of framed homes will be destroyed, with total roof failure and wall collapse. Fallen trees and power poles will isolate residential areas. Power outages will last for weeks to possibly months. Most of the area will be uninhabitable for weeks or months." Category 4 storms, with wind speeds between 130 and 156 mph (209–251 km/h), can also produce catastrophic damage, though sturdier homes can survive. Tree fall, power outages, and a period of uninhabitability are also associated with Category

[25] Walker et al. 2018.

4 storms. Even Category 3 storms, with wind speeds between 111 and 129 mph (178–208 km/h), are considered major hurricanes because they too can lead to significant loss of life and property damage.[26] As mentioned, Hurricane Katrina was a Category 3 storm based on wind speed.

Three of the tropical cyclones discussed in this chapter are among the most deadly in recorded history; the 1970 Bhola cyclone was the most deadly and the 1991 Bangladesh cyclone and the 2008 cyclone Nargis were also among the top 10.[27] All of the storms under consideration in this chapter were Category 3, 4, or 5 storms under the 2010 version of the Saffir–Simpson scale.[28] The two most damaging storms (the 1970 Bhola cyclone and the 2008 cyclone Nargis) were either Category 3 or Category 4 storms. Of the Category 5 storms, two were less damaging (1999 cyclone, 2013 Phailin) while the 1991 cyclone, which some describe as Category 5, led to large-scale mortality. These mortality figures are notional, since death counts, particularly from decades ago, are likely imprecise. Some reports suggest that the Bhola storm mortality in 1970 was closer to 500,000. Moreover, they may only include direct deaths from storms and not the follow-on effects of disease and starvation that often occur when the response in inadequate. In the case of the Bhola cyclone, they certainly do not include the deaths from violence that ensued in the civil war that cleaved East from West Pakistan to produce independent Bangladesh.

Looking solely at the Saffir–Simpson or IMD scales to compare cyclone strength can also be misleading. Because the Saffir–Simpson scale is now solely based on wind speed, other attributes that can contribute to damage such as storm surge are not captured.[29] Storm surges, particularly when they accompany high-tide events, can lead to extensive flooding, particularly in coastal plains or if storms originate in bays which can funnel storm surge overland.[30] Inland flooding may be responsible for as much as 90 percent of the damage from cyclones.[31] Capturing storm surge is not entirely straightforward since countries may not possess tide gauges and other measurement capacities at all locations where a storm comes ashore. Equipment can also be damaged during a storm. Moreover, where storm surges accompany high-tide events, the magnitude of storm surge is

---

[26] National Hurricane Center, National Oceanic and Atmospheric Administration n.d.
[27] Weather Underground n.d.
[28] The IMD scale is based on three-minute average of sustained winds. The two most severe categories include extremely severe cyclonic storm (wind speed 167–221 km/h) and super cyclonic storm (wind speed in excess of 222 km/h); World Meteorological Organization 2015.
[29] Seo and Bakkensen 2016.   [30] Kumar et al. 2015; Phys.org 2019.
[31] Kumar et al. 2015, 347.

sometimes the combined total of both the surge and the tidal event. In other instances, the reported surge is simply the additional increment of the storm alone. Table 6.2 shows cyclone severity, wind speed/storm surge attributes, and the death total from each cyclone.

Authors are not always clear about how they refer to storm surge. In some cases, they may mean the combined total of surge and tides, while others use storm surge in a more limited way to capture the additional increment in addition to the tidal flow. The precise storm surge is less important than the general magnitude. Storms with higher surge are likely to lead to greater loss of life due to drowning. Of these, Phailin (2013) and Fani (2019) had the least storm surge and the least potential for large-scale mortality.

These indicators only provide preliminary evidence that these cyclones are worthy comparisons. At first blush, it appears that storms with the highest storm surge are the ones associated with the largest number of fatalities, suggesting that the physical force of the storms could be responsible for differences in fatalities. On the other hand, even among storms with high storm surge (Nargis [2008]; Bhola [1970], 1991, 1999; Sidr [2007]), there is considerable variation in mortality with some severe storms such as 1999 and Sidr producing fewer deaths by large orders of magnitude. This suggests that policy is an important parameter that affects whether cyclones produce large-scale loss of life beyond the physical attributes of the storms themselves.

That said, another important difference could be that the areas affected might not be comparable in terms of human population. A storm that primarily affected a scarcely populated area could hardly exact the same sort of damage as a cyclone that barreled through a densely populated part of the world. As Table 6.2 shows, all the areas the storms hit affected millions of people. These are notional numbers, likely an extrapolation of population in affected areas at the locations where storms passed over. For the 1970 Bhola cyclone, when reporting and population size estimates were less available and less sophisticated, they should be taken with some caution.

For countries like India and Bangladesh where we are tracking within-case variation, we also have to recognize that fatalities as a share of the population go down as the population gets larger. According to the World Bank, Bangladesh's population in 1970 was slightly more than sixty-four million. By 1991, it was more than 105 million, by 2007 more than 142 million, and by 2018 it exceeded 160 million.[32] The time frame of study for India is more truncated, and the events of interest all occurred in Odisha, a poor state on the northeast coast of India (its capital city is

[32] World Bank 2019f.

TABLE 6.2 *Notable cyclones in Myanmar, Bangladesh, and India*[a]

Cyclone	Saffir–Simpson scale and IMD rankings	Maximum sustained winds/ storm surge	Deaths[b]	Affected
Myanmar April–May 2008 (Nargis)	Extremely severe cyclonic storm (IMD scale) Category 3 tropical cyclone (SSHWS)	167 km/h (104 mph) 16 ft[c]	138,366	2,420,000
Bangladesh November 1970 (Bhola)	Extremely severe cyclonic storm (IMD scale) Category 4 tropical cyclone (SSHWS)	205 km/h (130 mph) 35 ft[d] 26 ft[e]	300,000	3,648,000
April 1991 (Marian or Gorky)	Super cyclonic storm (IMD scale) Category 4 tropical cyclone (SSHWS)	235 km/h (146 mph) 29 ft[f] 29 ft[g]	138,866	15,438,849
November 2007 (Sidr)	Extremely severe cyclonic storm (IMD scale) Category 4 tropical cyclone (SSHWS)	213 km/h (130 mph) 13–20 ft[h] 28 ft[i]	4,275	8,978,766
India October–November 1999	Super cyclonic storm (IMD scale) Category 5 tropical cyclone (SSHWS)	260 km/h (160 mph) 20 ft[j]	10,378	13,870,008
October 2013 (Phailin)	Extremely severe cyclonic storm (IMD scale) Category 4 tropical cyclone (SSHWS)	213 km/h (130 mph) 8 ft[k]	65	13,230,004

| April 2019 (Fani) | Extremely severe cyclonic storm (IMD scale) Category 4 tropical cyclone (SSHWS) | 215 km/h (130 mph) 5 ft[l] | 64 | 10,000,000 to 16,000,000[m] |

Notes

a All wind speed and IMD scale categorizations are from IMD, except for Cyclone Bhola; India Meteorological Department 2018a. Cyclone Bhola details are from NOAA Hurricane Research Division 2015. Saffir–Simpson scale classifications are based on maximum sustained wind speed.
b Deaths and affected totals come from the EM-DAT International Disasters Database, except for Cyclone Fani.
c Fritz et al. 2011.
d NOAA Hurricane Research Division 2015.
e Penna and Rivers 2013.
f Khalil 1993.
g Penna and Rivers 2013.
h Tasnim, Esteban, and Shibayama 2015; Ikeuchi et al. 2017.
i Penna and Rivers 2013.
j Kalsi 2006.
k India Meteorological Department 2013.
l Floodlist News 2019.
m European Commission's Directorate-General for European Civil Protection and Humanitarian Aid Operations 2019; Indian Red Cross Society 2019.

Bhubaneswar). In 2001, shortly after the deadly 1999 storm, the Odisha census showed a population of nearly thirty-seven million.[33] In 2011, shortly before the much less deadly 2013 storm, the census showed a population of nearly forty-two million.[34] By the time cyclone Fani hit in 2019, its population was estimated to be about forty-seven million.[35]

A slightly different way to get at the potential population affected is by examining the storm tracks of individual storms and comparing those to the population totals in the areas affected.[36] This is challenging since census estimates, particularly for earlier periods, tended to undercount poor households and enumerating the population in highly mobile, island communities was, in any case, difficult.[37] That is one reason why estimates of the death total from Bhola range from 250,000 to as high as 500,000. It was both unclear how many people lived in the area to start with, and there was not an easy way to tally the deaths, especially with fisherman washed out to sea. The last census that was conducted prior to independence was completed in 1961, after which time there had been both outmigration to West Pakistan and population growth.

Still, it is helpful to get a sense of the trajectory of the storms to see the extent to which they affected populated areas. Myanmar anchors the chapter as the contemporary negative outcome case so I start there. In 2008, Cyclone Nargis started south in the Bay of Bengal but instead of heading north into Bangladesh made a somewhat unfamiliar eastward turn and crossed into Myanmar in the Ayeyarwady delta. This flat delta extends inland at sea level for nearly 200 km.[38] With much of the mangrove tree cover removed, the cyclone barreled in from the sea and caught the population largely unawares on the evening of May 2. Nearly 140,000 died, most of them by drowning (see Map 6.1).

Map 6.1 also shows how the cyclone tracks overlap with the map of population density for Myanmar according to the estimate by Columbia University's 2010 Gridded Population of the World.[39] The cyclone impact area is among the most densely populated regions of the country, which otherwise is much less densely populated than neighboring Bangladesh (pictured upper left). Bangladesh has, in fact, long been one of the most densely populated regions of the world. In 1970, it already had a population density of 493 people per square kilometer, making it the

---

[33] Government of India 2016. [34] Government of India 2019.
[35] Indiaonlinepages.com 2019. [36] India Meteorological Department 2018b.
[37] Penna and Rivers 2013, 238. [38] Webster 2008.
[39] CIESIN, Columbia University 2018.

Expectations and Case Selection   189

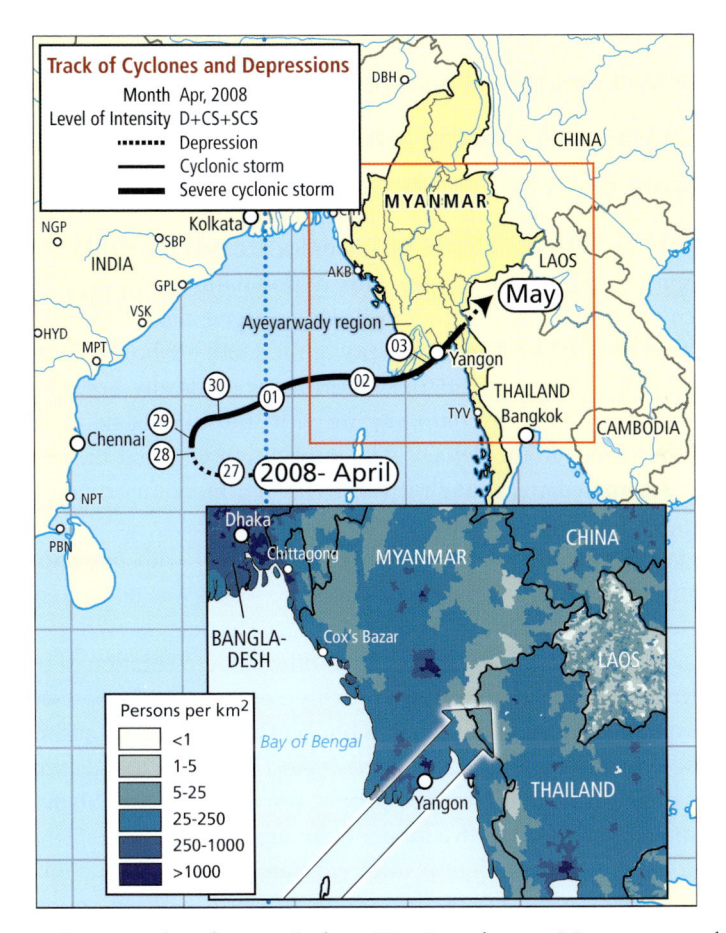

MAP 6.1 Storm tracks of 2008 Cyclone Nargis and 2010 Myanmar population density
Sources: India Meteorological Department 2018b and Gridded Population of the World, v.4.11, https://bit.ly/369Dz3R

tenth most densely populated country in the world behind island countries and territories such as Barbados, Malta, Hong Kong, and Singapore.[40]

The Bhola cyclone track in 1970 shows the storm trajectory was west to east, heading south of Dhaka but north of Chittagong as it passed over the delta. The cyclone started as a tropical depression in the Bay of Bengal on November 8 and made landfall on the afternoon of

[40] World Bank 2019f.

November 12, flooding the densely populated villages on the Ganges Delta and inundating a number of *chars* (islands) with a large storm surge that accompanied the lunar tide (see Map 6.2).[41]

Some 100,000 of Bhola Island's 900,000 estimated residents died. Around 2.5 million were estimated to have become homeless.[42] One contemporaneous study in *The Lancet* found that 46 percent of the population of Tazumuddin, a small subdistrict (known then as *chanas*, now as *upazilas*) on the coast with a population of 104,000, died. Children and women were the most likely to have died. The study concluded that all of the residents of many of the small offshore islands likely died.[43] Emanuel suggests the 1970 storm was so deadly because at the time "there was no way to communicate with" those living along the coast and on the islands in the delta.[44] The implication was that the death toll was almost immediate, though it is unclear if a more vigorous response would have prevented a significant number of deaths.

The last census in Pakistan had been carried out with populations by district in 1961, including then East Pakistan, which was undergoing both an outflow of population from East to West Pakistan as well as overall population growth. Map 6.3 shows an unofficial representation of preindependence district boundaries; I have roughly overlaid a crude representation of the overland trajectory of the 1970 Bhola cyclone.[45] Barisal and Comilla districts were the most affected. In the 1961 census, these districts had populations of 4.2 and 4.4 million people, respectively. Only Mymensingh (seven million) and Dacca (five million) had higher populations.[46]

The 1991 storm had a slightly different trajectory, coming up the Bay of Bengal but making landfall on the evening of April 29 about 50 km south of Chittagong along the southeastern coast. Cyclones were not named in the region until 2004, though the US military dubbed this storm Marian and others sometimes refer to it as Gorky.[47] The death toll overall neared 140,000, mostly attributed to the storm surge, with an additional 2,000 deaths from lack of access to clean water and sanitation in the weeks after the storm. The island of Kutubdia, off the coast near Chakaria, Cox's Bazar, sustained a large number of casualties, with more than 20,000 of the island's population of 110,000 dying and 80–90 percent of structures

---

[41] ESRI n.d. [42] Penna and Rivers 2013, 243. [43] Sommer and Mosley 1972.
[44] Emanuel 2005, 223. [45] Mahmood 2019.
[46] This source reports seventeen districts, and Map 6.3 shows twenty so boundaries may have changed after the 1961 census, but this gives a rough idea; The Economic Weekly 1961.
[47] India Meteorological Department 2010.

Expectations and Case Selection 191

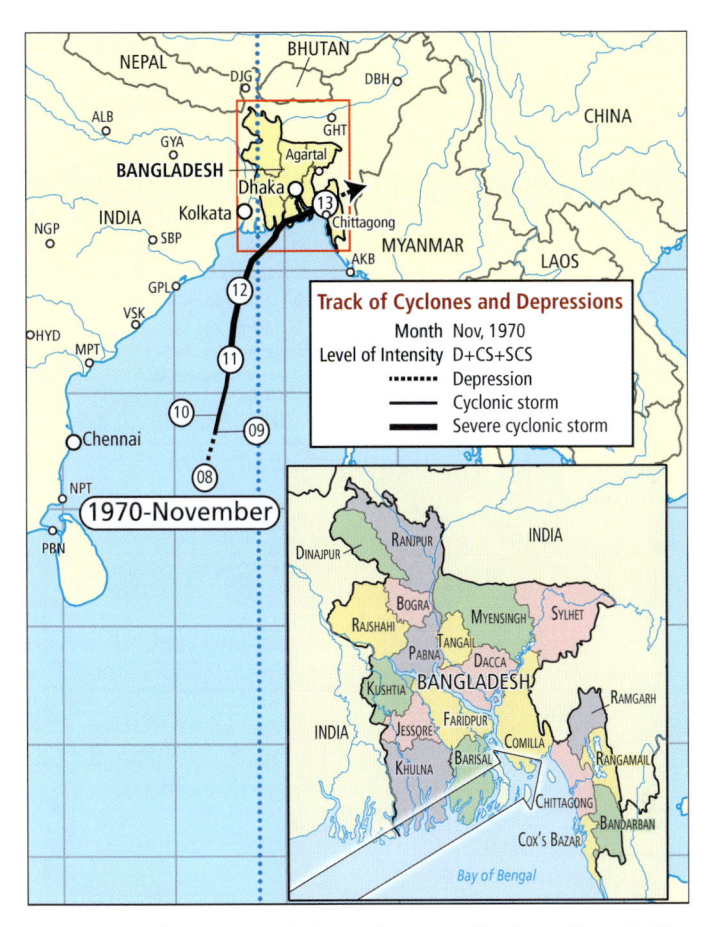

MAP 6.2 Storm tracks of 1970 Bhola Cyclone and districts of East Pakistan
Source: India Meteorological Department 2018b

destroyed.[48] The port city of Chittagong sustained extensive damage, although more limited casualties. In Map 6.3, I have overlaid where the cyclone tracked on a map of population density for 1991, which shows the cyclone made landfall in the densely populated Chittagong district.[49]

In 2007, Cyclone Sidr had a trajectory that came in from the Bay of Bengal and passed over the Sundarbans mangrove forests toward the densely populated capital city of Dhaka. The storm made landfall on November 15, causing more than $240 million in damages, including extensive damage to the rice crop and fishing boats (see Map 6.4).

[48] USAID n.d. [49] Government of Bangladesh 1991.

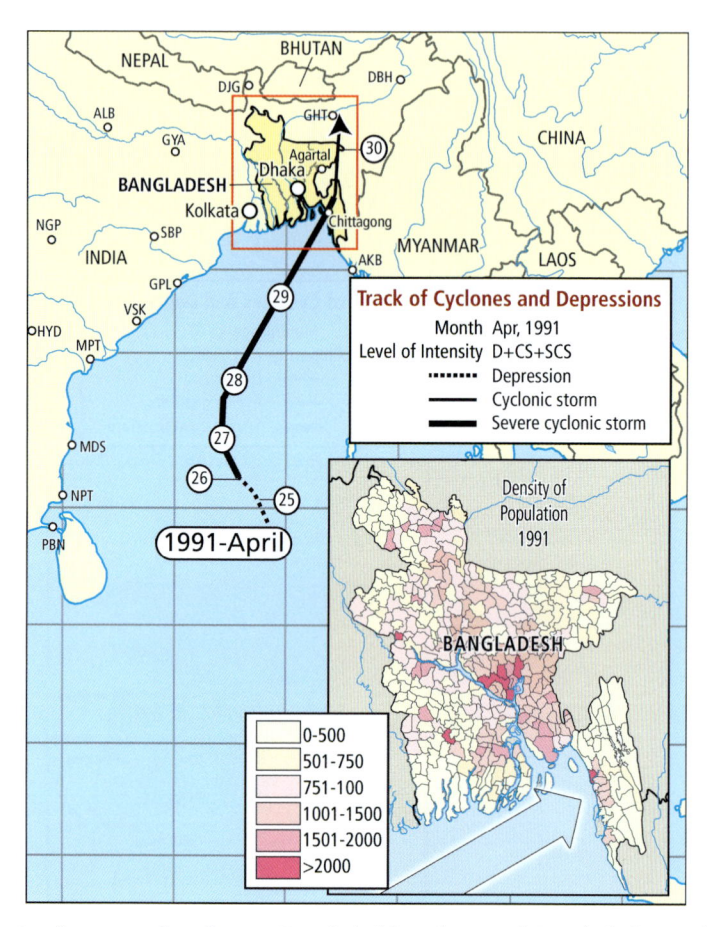

MAP 6.3 Storm tracks of 1991 Bangladeshi cyclone and Bangladesh population density
Sources: India Meteorological Department 2018b and Government of Bangladesh 1991

Columbia University mapped the storm's trajectory over land as it passed over the Sundarbans, triggering a twenty-foot storm surge, and then moved east toward more populated areas across the >extensive low elevation coastal zone.[50] While the overall death toll – 4,200+ deaths – was modest by Bangladesh's historic standards, a number of towns such as Barguna and Patuakhali lost several hundred lives. Although loss of life in Dhaka was limited, the city sustained severe damage from winds and flooding, with widespread power and water outages.

[50] Masters 2007.

Expectations and Case Selection 193

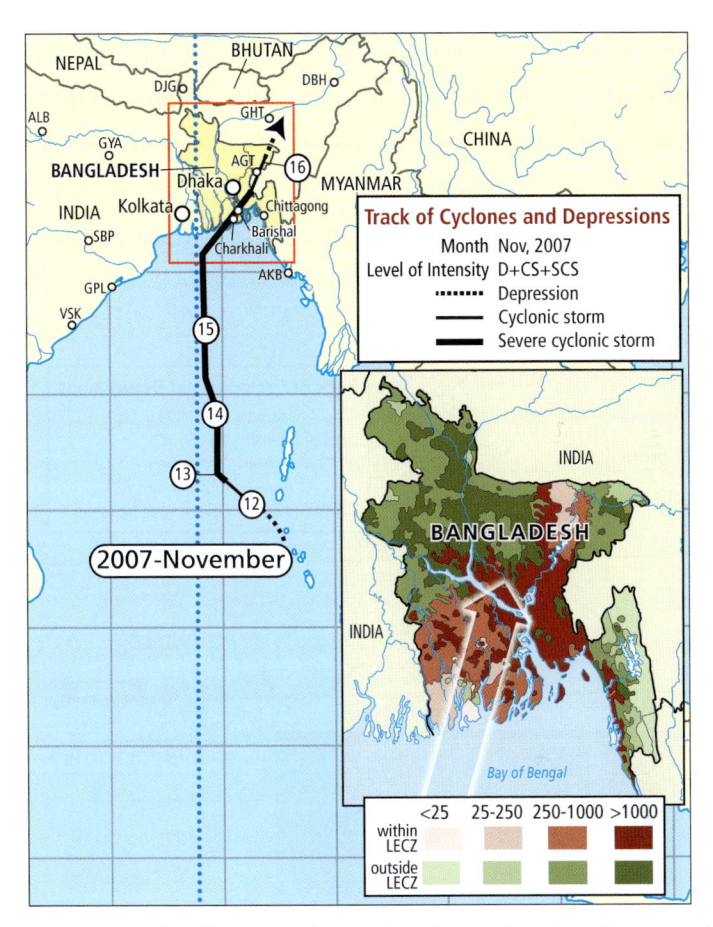

MAP 6.4 Storm tracks of 2007 Cyclone Sidr and 2000 low elevation coastal zone population density
Source: India Meteorological Department 2018b and Center for International Earth Science Information Network (CIESIN), Columbia University, https://bit.ly/3wjwDM5

In terms of the cyclones of interest that affected India, Odisha state has long been vulnerable to cyclones, with reports dating back to antiquity. In the post-independence era, Odisha has been buffeted by cyclones in at least twelve years, including 2019.[51] Perhaps as many as 320 million people in India are vulnerable to cyclones.[52] The 1999 storm

---
[51] Pal, Ghosh, and Ghosh 2017, 351 include the following years: 1971, 1973, 1977, 1981, 1983, 1984, 1985, 1987, 1989, 1999, and 2013.
[52] Yadav and Barve 2017, 387.

that affected Odisha state shows the storm coming up the Bay of Bengal and then turning overland, making landfall on October 29 near the seaport town of Paradip and heading south along the coast. Despite preparations and warnings, India sustained more than 10,000 deaths, most of them in the Jagatsinghpur district where Paradip is located.[53] At the time, this was the strongest storm in Odisha state in 114 years (see Map 6.5).[54] While this is not the most densely populated area in India,

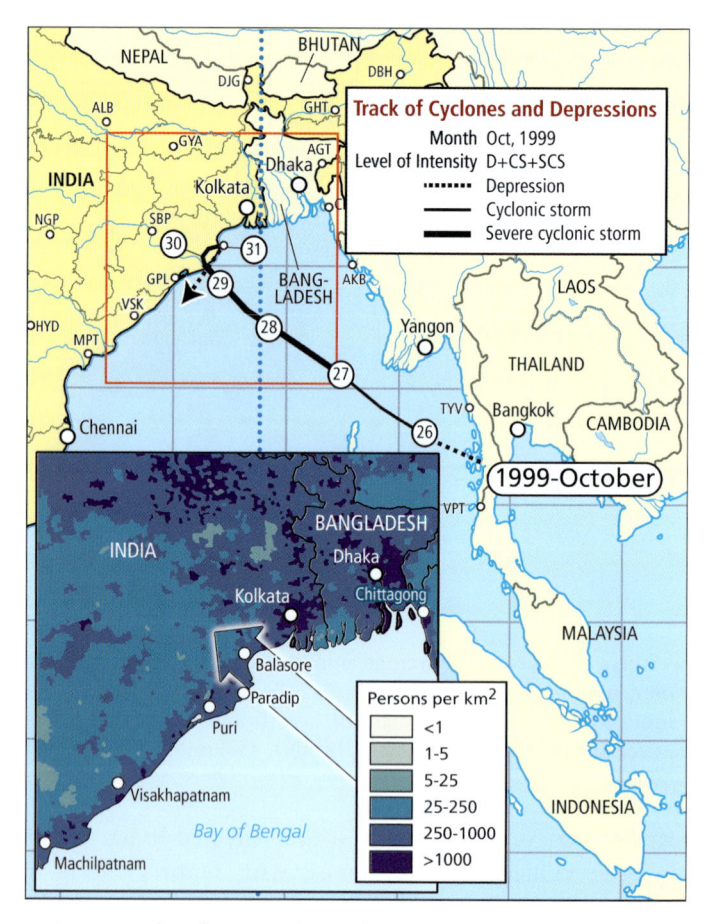

MAP 6.5 Storm tracks of 1999 India cyclone and 2000 population density
Sources: India Meteorological Department 2018b and Gridded Population of the World, v.4.11, https://bit.ly/369Dz3R

---

[53] UN Disaster Management Team 1999. [54] India Meteorological Department 2000.

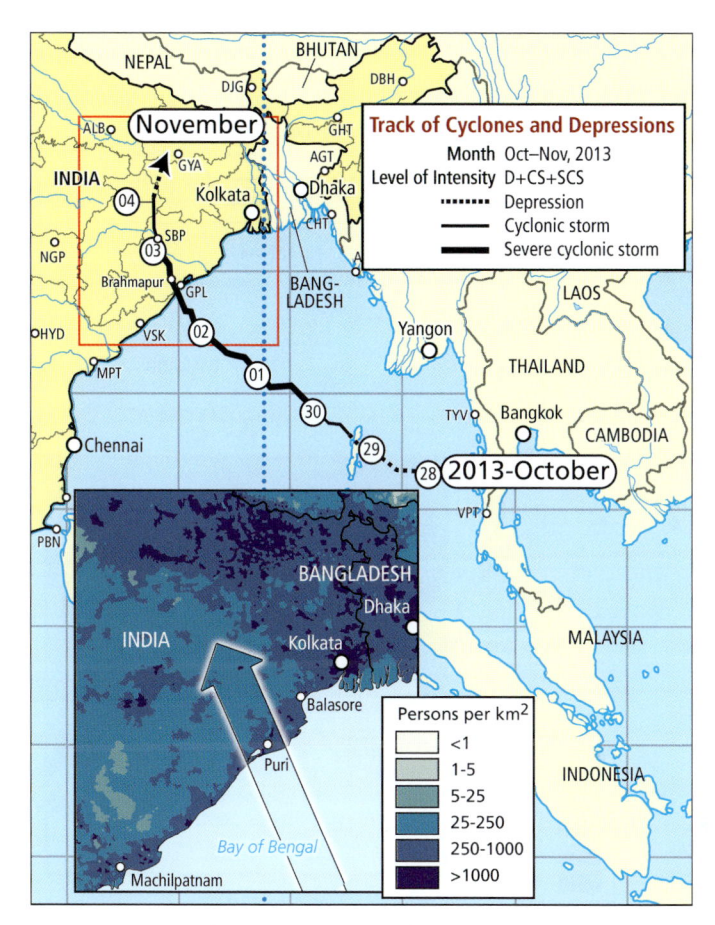

MAP 6.6 Storm tracks of 2013 Cyclone Phailin and 2015 population density
Source: India Meteorological Department 2018b and Gridded Population of the World, v.4.11, https://bit.ly/369Dz3R

the country is generally densely populated. The most affected area of the coast still had between 250 and 1,000 people per square kilometer.

In 2013, Odisha state was again affected by a severe cyclone, but here, the death total was much reduced, some sixty-five, despite affecting more than thirteen million people, roughly as many as the 1999 storm. Cyclone Phailin made landfall on October 12, further south than the 1999 storm, with significant damage affecting the city of Brahmapur, and then headed further inland (see Map 6.6). Like the 1999 storm, Phailin did not affect the most densely populated areas of India, but the region is fairly densely

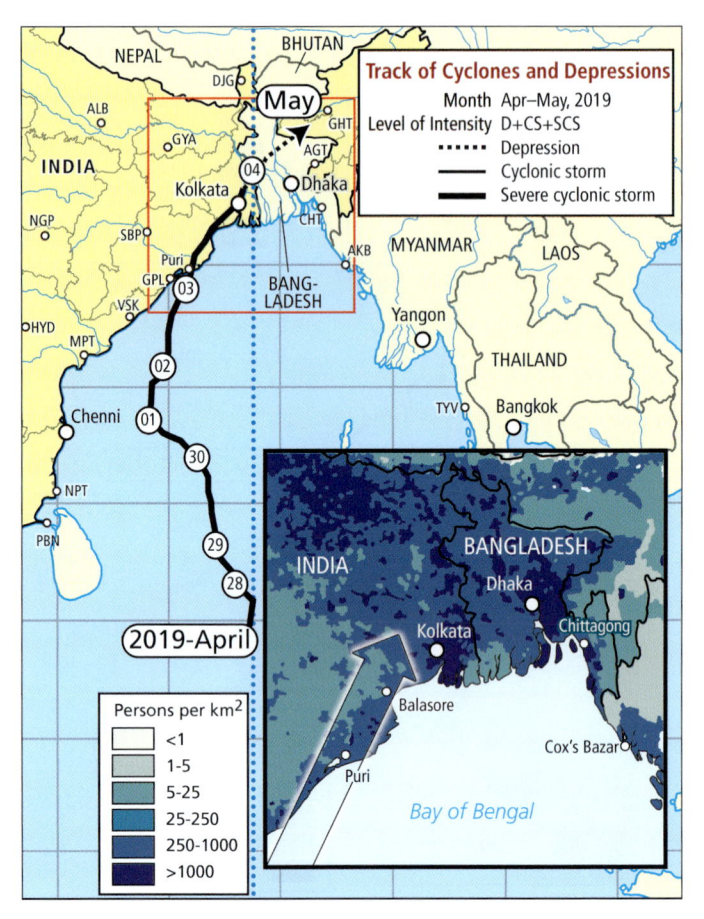

MAP 6.7 Storms tracks of Cyclone Fani and projected 2020 population density
Source: India Meteorological Department 2018b and Gridded Population of the World, v.4.11, https://bit.ly/369Dz3R

populated, with some 250 to 1,000 people per square kilometer in the most affected zone.

Finally, cyclone Fani battered the coast of India in May 2019, making landfall as a Category 4 storm with wind speeds of 115 miles per hour as it touched down near the city of Puri in Odisha state with a population of 200,000 (see Map 6.7). The storm's trajectory was similar to the earlier Odisha state cyclones, making landfall in an area that was relatively densely populated.

BACKGROUND ON GOVERNANCE IN MYANMAR, INDIA, AND BANGLADESH

Having shown that the cyclones were of sufficient strength to endanger large populations, we next explore two dimensions of governance – state capacity and political inclusion.

Evidence of State Capacity

To assess state capacity, I evaluate how these three countries compare to each other and, for India and Bangladesh, how they compare to themselves over time. When the Bhola cyclone hit in 1970, Bangladesh was not yet an independent country. Indeed, Pakistan's failure to respond adequately to the storm in East Pakistan precipitated the push for independence. Thus, the first cyclone of interest dates back to the 1970s, prior to the availability of World Bank data on governance. World Bank governance data is not even available for the 1991 cyclone that buffeted Bangladesh.

To capture state capacity, an index of political risk that predated the World Bank indicators – and, indeed, is one of the data sources that has informed the Bank's assessments – is available. Once again I use the PRS Group's indicator of "bureaucratic quality" from the International Country Risk Guide (ICRG). Like World Bank indicators, the PRS Group's indicators are based on expert assessments. Although they do not extend back to the 1970s, they are available from 1984 to the present. Perhaps the most relevant indicator to capture from the ICRG data is what they call "bureaucratic quality." The ICRG describes bureaucratic quality as a "shock absorber" for changes in government characterized by some autonomy and established processes to recruit and train workers.[55] To the extent that this captures the government's ability to implement policy, both to anticipate potential harms and respond to them, this is likely a decent proxy for the government's ability to prepare for and respond to disasters.

The scale is 0 to 4 with Western countries like Denmark occupying the upper bound. As Figure 6.1 shows, India's bureaucratic quality throughout is high for the entire period, with the country mostly having a score of 3 on the index (in 2017) along with countries such as Argentina, Croatia, Malaysia, Poland, and South Korea. Through 1992, both Myanmar and

[55] PRS Group 2012.

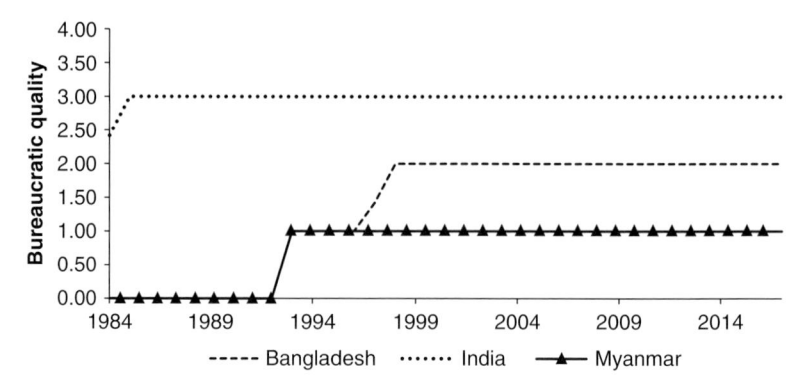

FIGURE 6.1 Bureaucratic quality in Myanmar, Bangladesh, and India
Source: PRS Group 2012

Bangladesh had no bureaucratic capacity. In 1993, Bangladesh moved up modestly to 1 where it remained until 1997, and then stepped up again in 1997 and in 1998, reaching the midpoint of 2 where it has remained ever since. This put Bangladesh in company with Botswana, China, Colombia, Ecuador, Iran, Indonesia, and other lower middle-income countries in 2017. Thus, at the time of the 1991 cyclone, Bangladesh had no bureaucratic capacity, and this was presumably true at the time of the 1970 cyclone as well given the trajectory since 1984. By 2007 and Cyclone Sidr, Bangladesh's capacity had significantly improved.

Myanmar, for its part, did increase in capacity from 0 to 1 in 1993 alongside Bangladesh but never improved further. By 2017, Myanmar was in the company of low-capacity countries like Azerbaijan, Congo, Mozambique, Nicaragua, Nigeria, Russia, and Yemen. Only the Democratic Republic of the Congo, North Korea, Somalia, and Sierra Leone were thought to have worse bureaucratic quality.

The indicator of bureaucratic quality is likely an imperfect one to capture government preparedness for climate-related natural disasters. The 2018 INFORM index on risk management assesses different dimensions of coping capacity including a measure of institutional capacity that includes a self-assessment of disaster risk reduction and two metrics of governance, one based on corruption perceptions from Transparency International and the other the World Bank measure of Government Effectiveness. On the self-assessment of disaster risk reduction under the Hyogo Framework, India scored 1.8, Bangladesh 3, and Myanmar 7.1. Lower scores are better, reflecting self-assessments of better disaster risk reduction policies. These self-assessments were completed in 2015 for both Bangladesh and India and in 2009 for Myanmar, one year after Cyclone Nargis.[56]

None of the countries score all that well on either corruption or Government Effectiveness. Bangladesh was regarded as having marginally more corruption (7.4) than Myanmar (7.2), with India ranked as 6 (for comparison, Denmark, for example, was ranked as 1). On Government Effectiveness, India was 4.8, Bangladesh 6.5, and Myanmar 7.5. This put Myanmar in company with countries such as Iraq and Venezuela, while India shared rankings with Vietnam and the Philippines. Bangladesh was in line with Egypt and Cameroon.[57] While only a static snapshot from near the end of the period for all three countries, these data points correspond to my theoretical expectations that India had better governance throughout followed by Bangladesh and Myanmar, a distant third.

We can also assess disaster capacity by thinking about what essential services are required to prepare for and respond to cyclones, including meteorological capacity, dedicated personnel, and preparedness and response plans.[58] Capacity should also show up in terms of some programmatic and physical infrastructure such as early warning systems (mass media broadcasts, sirens, loudspeakers), evacuation supplies (vehicles), evacuation facilities (such as purpose-built shelters), and emergency response infrastructure (helicopters, boats, pre-packaged food and water, emergency medical equipment), as well as natural barriers to prevent flooding such as embankments and mangroves.

For India and Bangladesh, where we wish to observe within-case variation over time, improvements in state capacity would be observed in different metrics. In terms of hazard forecasting, in addition to the presence of forecasting units we should see increased methodological sophistication with the ability to forecast impending storms with further anticipation. For evacuation and response options, we should see not only the presence of emergency evacuation and response measures but quantitative increases, for example, in the number of volunteers and purpose-built shelters. For the three countries, I review the forecasting, planning, and evacuation capabilities. I discuss the responses in the sections on Political Inclusion and International Assistance.

---

[56] Marin-Ferrer, Vernaccini, and Poljansek 2017.
[57] These scores were for 2016 with respect to corruption and 2015 for Government Effectiveness.
[58] Cash et al. 2013, 2095 distinguish between preparedness, readiness, and response.

Forecasting

For cyclone forecasting, there are global and regional models that local meteorological agencies can rely on, though countries often have their own capacity.[59] Myanmar lacked forecasting capability at the time of Cyclone Nargis but would go on to develop better equipment. By 2018, it had three radar facilities and thirty new weather observation systems.[60]

By contrast, both India and Bangladesh have better forecasting capacities that have improved over time. Since 1973, India has been the designated regional specialized meteorological center (RSMC) for tropical cyclones for the World Meteorological Organization.[61] The Bangladesh Meteorological Department (BMD) is responsible for forecasting cyclones, but its capacity to give much advance warning is limited compared to their peers in India. Bangladesh's forecasting capacity, as of 2018, was three days or seventy-two hours lead time.[62] The Indian Meteorological Department (IMD) has been able to forecast cyclones since 2003, with twenty-four hours advance notice. That was expanded to 72 hours maximum advanced lead time in 2009 and to 120 hours in 2013.[63]

While the RSMC provides information for forecasts, local actors may have additional data that can inform country-level forecasts. The BMD has a Storm Warning Center and five radars, including three Doppler radars, to inform its cyclone warnings, though no specialized staff dedicated solely to cyclone warnings.[64]

Early Warning, Evacuation, and Planning

Myanmar's capacity at the time of Cyclone Nargis was poor. Not only did the country lack forecasting capacity, but early warning systems were virtually absent as well in 2008. Most of those who died were delta farm families asleep in their beds. As one villager from the town of Labutta said: "Villagers were totally unaware. We knew the cyclone was coming but only because the wind was very strong. No local authorities ever came to us with information about how serious the storm was."[65]

In Myanmar's case, the country was spectacularly unprepared. As one observer commented: "[F]ew countries have been less prepared and less willing to respond to a major disaster than Burma." He expanded on this: "The Burmese government failed to implement these essential measures in any meaningful way during the critical days and weeks following Cyclone Nargis. This failure resulted from the government's lack of logistical

---

[59] Paul 2008.   [60] Gottlieb 2018.   [61] Bandyopadhyay 2012.
[62] World Bank 2018, xv.   [63] Ibid., xiv.   [64] Ibid., 37.   [65] Casey 2008.

capacity to respond effectively to a disaster of such magnitude and its distrust of the intentions of mainly Western governments and aid organizations."[66] In its 2009 self-assessment under the Hyogo Framework, the Myanmar government admitted its lack of institutional capacity and resources to carry out the job: "The first is a lack of an overarching disaster management law and secondly, there is a lack of clarity on how to integrate DRR [disaster risk reduction] into each ministry as well as the focal ministry to support other ministries."[67]

For its part, Bangladesh's efforts have improved over time. As Cash et al. note: "Mortality and morbidity from these events have fallen substantially in the past 50 years, partly because of improvements in disaster management."[68] Even before independence, there were rudiments of a cyclone warning system developed by civil society, with the Red Crescent Society taking the lead in 1966.[69] In 1972, Bangladesh established a Ministry of Relief and Rehabilitation and also created the Cyclone Preparedness Program (CPP) in the same year. By 2007, the CPP was able to mobilize some 43,000 volunteers, up from 20,000, to help provide early warning and assist in the evacuation.[70] However, after independence, the focus was mostly on response and rehabilitation rather than early warning.

The 1991 cyclone spurred a move toward more holistic efforts, including preparedness. In 1993, a Disaster Management Bureau was created to succeed the Disaster Coordination and Monitoring Unit. In the 1990s, the Ministry was renamed the Ministry of Disaster Management and Relief.[71] Bangladesh also has a national coordinating body – the National Disaster Management Council – to join together various agencies and levels of government that have disaster-related portfolios.[72] Under the Hyogo Framework, Bangladesh submitted four separate triennial communications between 2007 and 2015. The country now has extensive bureaucratic systems at all levels of government to address disaster preparedness and response.

Some of the first reports from the pre-Hyogo period in the late 1990s acknowledge lack of capacity, which, to some extent, would prove critical in the 2001 cyclone. In its 1998 submission to the International Decade for Natural Disaster Reduction, the Bangladeshi government listed the obstacles they faced, which included: (1) no risk map, (2) not using modern

---

[66] Stover and Vinck 2008.   [67] Government of Myanmar 2010, 5.
[68] Cash et al. 2013, 2094.   [69] Tasnim, Esteban, and Shibayama 2015, 48.   [70] Ibid., 49.
[71] Habib, Shahidullah, and Ahmed 2012, 31.   [72] Diya and Bussell 2017.

technology, (3) bad warning system, (4) public education not up to the mark, (5) lack of coordination, (6) telecommunication and road communication not well connected, and (7) weak management and administration.[73] By 2009, not long after Cyclone Sidr, the government of Bangladesh submitted its first self-assessment under the Hyogo Framework. It recognized that much had been achieved to instantiate an institutional commitment to risk reduction, though more remained to be done: "Substantial achievement [has been] attained but with recognized limitations in key aspects, such as financial resources and/or operational capacities."[74]

In India's case, the country experienced a sea change in its approach to disasters in the wake of the 1999 cyclone with more prioritization of preparedness over response, though the 2004 tsunami which claimed the lives of 10,000 Indian citizens may have been the final catalyst for the passage of legislation in 2005.[75] The 2005 Disaster Management Act set in motion a series of institutional changes including a State Disaster Management Authority in Odisha as well as a District Management Authority in Ganjam which would serve to coordinate government and civil society.[76]

After the failure to respond to the cyclone of 1999 as well as an earthquake in 2001 in Gujarat, disaster functions were transferred in 2002 from the Ministry of Agriculture to the Ministry of Home Affairs.[77] The 2005 act sought to clarify the responsibilities at federal, state, and district levels. The district level lead responsible for coordination, supervision, and monitoring is the District Magistrate/Collector, though the District Management Authority is nested in a wider hierarchical structure at state and federal levels.[78] To facilitate "unity of command," the District Magistrate has legal authority to demand cooperation from all other local government agencies responsible for infrastructure and utilities, with noncompliance risking imprisonment.[79]

The 2005 act not only created a tripartite structure of disaster management agencies but also led to the creation of a National Disaster Management Institute (NDMI), responsible for the training and capacity of state officials and responders.[80] The eleventh and twelfth five-year plans (2007–12 and 2012–17, respectively) were both dedicated to disaster preparedness. Odisha state actually created its state level Disaster

---

[73] Government of Bangladesh 1998, 2. [74] Government of Bangladesh 2009.
[75] Venkataraman, Raj, and Abi-Habib 2018. [76] Pal, Ghosh, and Ghosh 2017, 352.
[77] Ibid., 353. [78] Ibid., 352. [79] Ibid., 354. [80] Ray-Bennett 2018, 34.

Management Authority in 1999, not long after the cyclone in that same year, which predated the national-level 2005 act.[81] That agency set in motion a number of changes including the construction of some 200 additional cyclone shelters, discussed further in the section on Emergency Shelters below.

By 2014, India, like Bangladesh, had made four self-assessments under the Hyogo Framework, and the government described the variety of institutional changes and resources mobilized since 2005 to address disaster risk reduction and emergency response, with some $265 million for hazard proofing infrastructure, $88 million for capacity building, and $5.6 billion for state disaster response, among other investments.[82] We can compare this with some of the self-reported funds Bangladesh had made available for similar purposes. By 2014, the Bangladeshi government had moved further to address its limitations but still faced challenges of mobilizing resources. In fiscal year 2013–14, the government set aside some $30 million for disaster response, with additional contributions of nearly $3 million from the United Nations, $10 million from NGOs, and $3 million from the Red Cross/Red Crescent.[83] That $46 million set aside for disaster response is less than $0.30 cents per person for disaster response in a country of 159.4 million people (in 2014). India, by contrast, had set aside more than $4 per capita for disaster response.[84] As I discuss in the next section, Bangladesh had also invested in emergency shelters.

Emergency Shelters

Purpose built cyclone shelters – typically cement structures on stilts – can prevent large-scale loss of life from flooding if they are available in sufficient quantity, accessible to populations, remain in good condition, and if people are willing to use them. Haque argues that vulnerable coastal areas should have cyclone shelters within a 2 km walking distance of villages.[85]

With regard to Myanmar, a USAID official who had worked closely on the response said the country had no shelters in place. And its lack of preparedness went further: "Not only did the country lack a weather radar network that could predict cyclones, it also had no early warning system, storm shelters or evacuation plans." By 2018, the official reported that the country finally had plans to build fifty shelters in Rakhine state.[86]

---

[81] Pal, Ghosh, and Ghosh 2017, 357.   [82] Government of India 2014.
[83] Government of Bangladesh 2015.
[84] India's population in 2014 was nearly 1.3 billion.   [85] Haque et al. 2012.
[86] Gottlieb 2018.

In 1970, Bangladesh was much like Myanmar in 2008. Little progress had been made by Pakistan to build shelters in East Pakistan before the Bhola Cyclone. Hossain claims that despite a government program that been in place since 1960 to prioritize cyclone protection, early warning, and shelters, "Cyclone shelters built after 1960 were few, distant from the coasts and islands where the most vulnerable lived, and unsuited to people's needs."[87]

Tasnim et al. note that purpose-built shelters only began to be constructed in earnest in 1985.[88] In 1991, Bangladesh had 311 cyclone shelters with capacity for 350,000 people.[89] However, when the 1991 cyclone struck, only 1 in 5 cyclone shelters was usable as a result of flooding.[90] That said, where people had access to shelters, the death rate was 3.4 percent compared to 40 percent on islands that lacked access to shelters.[91] Prior to 2007, Bangladesh had 1,500 purpose-built shelters, each able to accommodate 5,000 people. In the wake of Cyclone Sidr, the state sought to build an additional 2,000.[92]

Tasnim et al. provide different numbers, suggesting that the country had as many as 512 shelters in 1992, rising to 3,976 in 2007, though each with the capacity of only 1,500 to 2,000 people (however, more than 1,500 had been damaged by river erosion and were in disrepair).[93] Whatever the precise figure, Bangladesh had clearly made strides by 2007 to protect its population, although it still had some way to go.[94] By 2011, the country had invested more than $10 billion in embankments and shelters.[95] As Tasnim et al. comment: "During Cyclone Sidr in 2007 it was clear that Bangladesh had made remarkable progress in its disaster preparation and prevention skills."[96] Mitchell et al. report that some two million were evacuated to those shelters during Cyclone Sidr.[97]

Odisha state in India only had 23 cyclone shelters before the 1999 cyclone, enough for 42,000 to seek shelter. In its wake, the state decided to construct a number of shelters.[98] By 2019, some 203 shelters had been constructed in vulnerable coastal areas of the state, with another 52 under construction. The goal was to ensure that everyone had access to a shelter within 2.25 km, accessible without having to cross a natural barrier such as a stream.[99]

---

[87] Hossain 2018, 193. [88] Tasnim, Esteban, and Shibayama 2015.
[89] Bern et al. 1993, 73. [90] Haque et al. 2012. [91] Siddique and Eusof 1987, 3.
[92] Haque et al. 2012. [93] Tasnim, Esteban, and Shibayama 2015, 49, 50. [94] Ibid., 37.
[95] Dasgupta et al. 2011, 2014. [96] Tasnim, Esteban, and Shibayama 2015, 37.
[97] Mitchell et al. 2014, 25–26. [98] Iwasaki 2016. [99] OSDMA 2019.

Across these various indicators of state capacity broadly understood and more specific to cyclones, I find support for my theoretical expectations that Myanmar's capacity was limited at the time of the 2008 cyclone, that Bangladesh has experienced marked improvement over time since its secession from Pakistan, and that India has had a higher level of capacity than both countries throughout the period, but also improved its disaster preparedness and response capability over time.

Evidence of Political Inclusion

To capture political inclusion, we again begin by exploring the Ethnic Power Relations (EPR) dataset to understand the differences between the three countries and how India and Bangladesh have evolved over time.[100]

From 1990 to 2011, Myanmar was run by a military regime which meant that a number of small groups were either actively discriminated against (Chinese, Muslim Arakanese, and Karenni) or powerless (Wa). Together, these groups constituted about 8 percent of the population. However, other larger groups excluded themselves from political participation – the Shan (8.5%), Kayin (7%), Zomis (2.1%), Buddhist Arakenese (2%), Mons (2%), and Kachins (1.5%). If we combine these ethnic groups, more than 30 percent of the population was excluded from power at the time of the 2008 cyclone, which suggests a highly exclusionary regime on ethnic grounds.

In some ways, this may both overstate and understate the degree of inclusion in Myanmar. A clique of military leaders was running the country at the time of the cyclone; this was both anti-democratic but also an attempt to contain the risk of the Bamar majority further marginalizing ethnic minorities. As David Steinberg argues, "Effectively, the issue for the central government is how to ensure that an ethnic (Burman) majority government has legitimacy among the various minority groups."[101] Myanmar has struggled since independence with forging a cohesive national identity in a country with 135 recognized ethnic groups. The dominant Bamar ethnic group accounts for 68 percent of the population, and the state has suffered from insurgencies by minority ethnic groups throughout its history, dating back to its founding. So, perhaps rather paradoxically, to prevent majoritarian discrimination against minority groups, the military did not permit a transition to

---

[100] Cederman, Min, and Wimmer 2009; ETH Zurich 2018.
[101] Steinberg 2012, 223; see also Taylor 2005.

democracy, excluding Bamar-led groups like Aung San Suu Kyi's National League for Democracy from coming to power.

In Bangladesh's case, from 1947 until 1971, the primary ethnic group in then East Pakistan, Bengalis, were discriminated against by the government based in West Pakistan even though they constituted about 55 percent of the total population in the country. Therefore, at the time of the Bhola cyclone in 1970, the population was marginalized from power and access to government services; this helps us understand the inadequacy of the response and the political uprising in the wake of the cyclone. After independence, discriminated groups like Bengali Hindus never constituted more than about 11 percent of the population. Thus, at independence, the government became much more inclusive in terms of political representation.

In India, throughout the period of interest, from 1999 until the present, there are no discriminated minority groups in the EPR dataset. There are small groups like the Manipuri, Naga, Bodo, and Kashmiri Muslims who are powerless, but they constitute less than 1 percent of the population. In the period 2015 to 2017, other Muslims – some 11 percent – were regarded as powerless, but the country remained broadly inclusive in terms of representation and the EPR dataset. This has, however, started to change since the election of Prime Minister Narendra Modi, particularly after his reelection in 2019.

As in Chapter 5, we can also examine V-Dem's indicator for social exclusion, which captures the extent to which access to services is equitably available to all groups based on religion, language, caste, ethnicity, region, or migration status. Lower scores indicate more social inclusion. This measure shows that India was more inclusive than either Myanmar or Bangladesh. Bangladesh was marginally more inclusive than Myanmar throughout the period and experienced modest improvements in inclusiveness before deteriorating and converging with Myanmar in the mid-2010s as that country's quasi-democratic transition led it to become slightly more inclusive (though Myanmar's subsequent treatment of the Rohingya Muslim minority in 2016–17 suggests an escalation of ethno-religious majoritarian domination). These results are largely in keeping with my theoretical expectations.

As in Chapter 4, I include Denmark as a point of contrast to the other countries studied. It is interesting to note that India still has relatively high social exclusion, likely a legacy of its caste system and other sources of inequality that continue to bedevil the country despite its democratic status.[102] This lingering inequality is thus at odds with Sen's depiction of

[102] Chhotray 2014, 218.

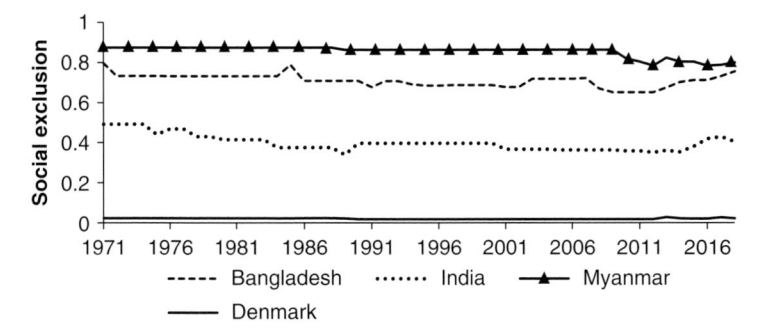

FIGURE 6.2 Social exclusion in Myanmar, Bangladesh, India, and Denmark, 1970–2018
Source: V-Dem 2019

the democratic Indian state as responsive to the needs of its citizens – politicians have to be reelected and free media surface information that politicians ignore at their peril.[103] The reality is that even in democratic countries, some degree of social exclusion can leave communities vulnerable.

Bangladesh's social exclusion scores are somewhat higher than my theoretical expectations. This may be a function of lingering caste and religious diversity. Still, these high scores are puzzling, particularly since Pakistan, prior to independence, was rated as much more inclusive – and even more inclusive than India – with a score of 0.317 compared to India's 0.491 in 1970. Pakistan's scores have remained slightly more inclusive than India since the 1970s. One of the primary reasons Bangladesh fought for independence from Pakistan was the Pakistani government's desultory response to the Bhola cyclone, with the violence suggesting a degree of social exclusion that went well beyond the cyclone itself (see Figure 6.2).

Evidence of International Assistance

Having shown variation in state capacity and inclusion, largely in accord with my theoretical expectations, what does the evidence show for foreign assistance? States with limited capacity can compensate for weakness in disaster preparedness and response through international assistance broadly defined, which can facilitate improvements in both. A country that receives little or no international assistance preceding or in the wake

[103] Sen 1981.

of a hazard exposure cannot compensate for weak state capacity. Even if a country receives substantial foreign assistance, much also depends on subsequent distribution of that aid. If the aid is distributed to allies or government partisans rather than on the basis of need, this should result in differential outcomes in terms of suffering as well as a potential source of grievance and subsequent political mobilization.

The trajectory since 1960 is one of significant but dwindling foreign aid to India over time, with the country becoming much less reliant on foreign assistance by the 1990s. In the case of Bangladesh, Pakistan – its parent country prior to independence – was a significant beneficiary of foreign aid. When it gained independence, Bangladesh received rising levels of foreign aid and continued to receive significant aid, with a dip in the 1990s before surging again in the 2010s. Myanmar, for its part, received little foreign assistance in absolute terms until 2008 in the wake of Cyclone Nargis; there was a bigger spike in 2013 after it made a more earnest effort to open up politically, before later backsliding (see Figure 6.3).

Given differential population and economic size, this trajectory is not entirely revealing. In terms of aid per capita, from 2000 onwards Bangladesh received more than $10 per capita, higher than Myanmar and India, with India's aid per capita only negligible. Aid per capita to Myanmar did not exceed that for Bangladesh until 2013 (see Figure 6.4).

Foreign aid, however, may not capture all forms of international assistance. For example, emergency assistance in the form of humanitarian aid may not be captured as official development assistance (ODA). The UN Office for the Coordination of Humanitarian Affairs (OCHA) collects data

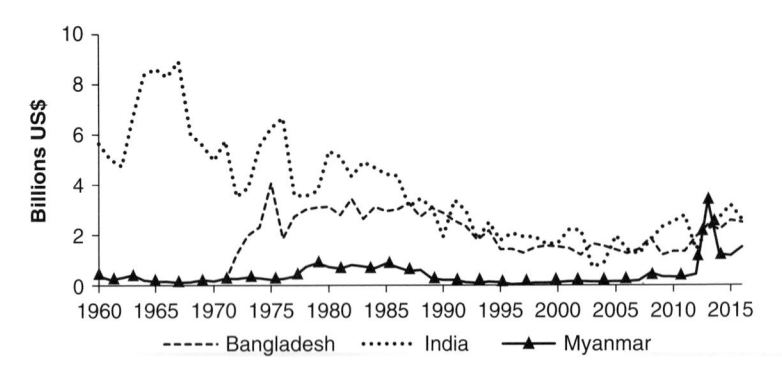

FIGURE 6.3 Net official development assistance in Myanmar, Bangladesh, and India
Source: World Bank via Our World in Data, https://bit.ly/3jwqNnU

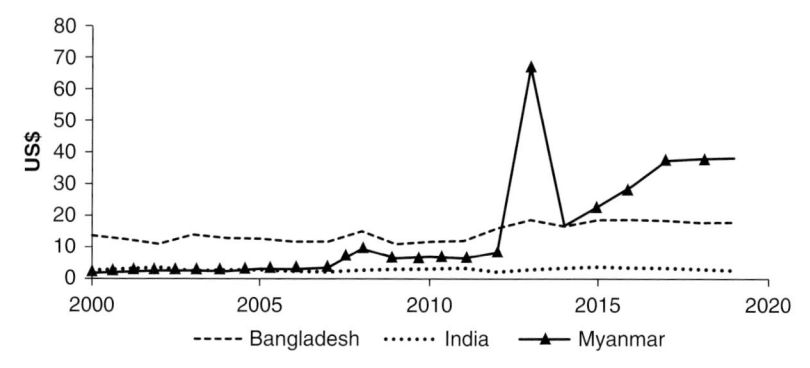

FIGURE 6.4 Aid per capita in Myanmar, Bangladesh, and India, 2000–2019 (constant 2016 US$)
Source: OECD via Our World in Data, https://bit.ly/2TqqYqu

on contributions to humanitarian appeals for disaster relief, refugee crises, populations displaced by civil war, and so on, through the Financial Tracking Service (FTS). Some governments include humanitarian contributions as ODA while others do not, but OCHA FTS explicitly says they do not track development finance.[104] Data are only available dating back to the early 2000s. Although individual project data are available, I aggregated the data for the destination country and year. Assistance volumes are smaller than ODA but show important spikes corresponding to some of the key events of interest; notably, Bangladesh received a spike of donations in 2007, in excess of $300 million, in the wake of Cyclone Sidr, while Myanmar received more than $600 million after Cyclone Nargis. India, throughout, rarely received large sources of humanitarian funds, relying mostly on internal resources (see Figure 6.5).[105]

Taken together, the data on ODA and humanitarian funding suggests that Bangladesh and Myanmar were far more reliant than India on external resources to compensate for weak state capacity – both generally and in the midst of emergent humanitarian crises. This suggests that the distribution of aid is worthy of further exploration in both Bangladesh and Myanmar. For Bangladesh, we should expect to see that the country did a better job after independence (notably in 1991 and 2007) in distributing external assistance equitably and quickly, compared to 1970 when it was still part of Pakistan, and compared to Myanmar in 2008, where aid

[104] OCHA Financial Tracking Service n.d.a.
[105] OCHA Financial Tracking Service n.d.b.

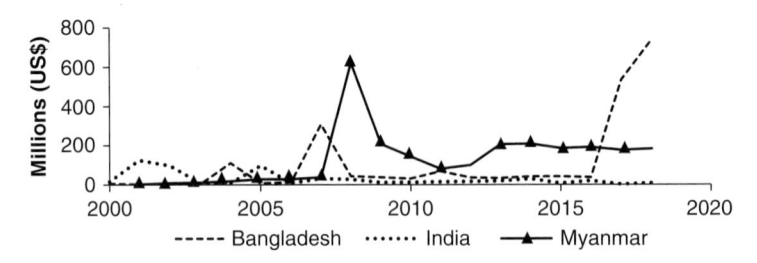

FIGURE 6.5 Humanitarian funding, Myanmar, Bangladesh, and India, 2000–2018
Source: OCHA Financial Tracking Service, https://fts.unocha.org/

supplies would be expected to have been thwarted or directed to serve some communities and not others. The funds for such programming are not solely dedicated to response but have also been important for preparedness. In Bangladesh's case, international donors helped support the country's improved early warning systems, scale-up of construction of emergency shelters, and efforts to mobilize community volunteers.[106]

PUTTING THE PIECES TOGETHER

To recap, in Myanmar's case, we have a single instance of a disaster in which the government's response should be inadequate, slow, and, where provided, biased toward allies of the regime. While the lack, or slow pace, of a response could also be explained by low state capacity, a response that provides aid to some groups and not others but diverts aid or resources to certain groups is indicative of exclusionary politics. In Bangladesh's case, we should expect the response to the Bhola cyclone to reflect East Pakistan's then political exclusion, with the response from the Pakistani government being inadequate, slow, and biased toward local favorites. After independence, we should expect successive responses, such as they were, to be distributed relatively equitably to societal groups, and the response to improve over time as capacity improved. In India's case, we should expect relatively equitable distribution between groups throughout the entire study period, with improved warning and service delivery over time as state capacity improved.

With respect to Myanmar, Guha-Sapir and Vogt claim that one of the primary reasons for the high death toll was the lack of early warning: "[S]tate authorities did not warn the inhabitants, despite repeated

---

[106] Diya and Bussell 2017, 3.

bulletins from the Indian Meteorological Department. The last warning came 48 hours prior and specified the cyclone force and the area of landfall."[107] Moreover, they note that external emergency assistance was delayed until "well after the event."[108]

Larkin interviewed Burmese who had ventured as private citizens to deliver aid in the wake of the storm. They found widespread devastation, with most of the bamboo and wooden houses destroyed. Even the few concrete buildings were now rubble. In one village, 80 percent of the community had died or were missing. Survivors had clung to branches to survive the surge. A Burmese businessman delivered food and medical supplies, and he reported that no help at all had been provided by the regime: "During their journey, Chit Swe [the businessman] and his colleagues did not see any other assistance being delivered. There were no soldiers or Navy boats on the water and no aid workers in the villages. Many villagers said that the help the businessmen gave them was the first they had received."[109] Another shopkeeper who had ventured to the delta town of Kunyangon to deliver food and cooking oil and had taken photos which he showed to Larkin:

[He] clicked through an alarming number of dead-body photographs before coming to a picture that showed crowds of people squatting down on either side of a dirt road, holding their hands up toward the camera. "They have nothing. They have no money. They have no shirts. No shoes. Nothing. And there is no help for them. I saw no officials there to assist them."[110]

In Myanmar, the regime impeded the disaster response by refusing visas to development workers and rebuffing the offers of foreign militaries – notably, the United States – to provide disaster relief. Fearful that the Americans and other foreigners would use the disaster as an excuse to invade or force political reform, the regime was reluctant to allow in outsiders, especially with the important referendum looming. Mirante damned the regime for its response:

The world then watched as Burma's military regime swung into action – moving swiftly to reject disaster relief teams, not only from the United States but from Qatar and Indonesia. Experts from the United Nations and international relief organizations were made to wait for permission to enter a country whose people desperately needed their help.[111]

---

[107] Guha-Sapir and Vogt 2009, 275.   [108] Ibid., 276.   [109] Larkin 2011, 24.
[110] Ibid., 41.   [111] Mirante 2008a.

Larkin noted that when word got out internationally that tens of thousands in remote and inaccessible villages needed food and water, international actors were ready to help but were turned away:

> In neighboring Thailand, the U.S. government had loaded a C-130 cargo plan with lifesaving relief supplies that would have taken under an hour to reach Burma, but the craft was not given clearance to land at Rangoon's airport. The United Nations World Food Programme had three planes ready to fly in from Bangladesh, Thailand, and Dubai in the United Arab Emirates. The planes were loaded with vitamin-fortified biscuits for hungry survivors who may not have been able to eat for some days and would be in need of instant nourishment. These biscuit-laden planes were also denied clearance. A flight from Qatar carrying relief materials and aid workers managed to land at Rangoon airport but was immediately forced to take off again without unloading any of its contents.[112]

She reports that foreign aid workers had their movement restricted to Rangoon, preventing those with expertise from traveling to the delta. Police were posted at bridges and jetties along the Rangoon River to check ferry passengers for permits to travel.[113]

A week after the cyclone, despite offers from the international community, the government released a statement on May 9, suggesting that it was "not yet ready to receive search-and-rescue teams as well as media teams from foreign countries."[114] The government pledged to carry out its own response. Bilateral aid would be welcome, but only if entrusted to the regime to distribute, which was, as Larkin argued, "unacceptable for most Western donors, who require accountability, transparent procedures, and the ability to track the delivery of the goods they donate."[115]

Mirante provides several examples of the regime using local or foreign resources for purposes other than benefiting those affected by Cyclone Nargis. She notes that the country went ahead with a 7,000-ton export shipment of rice to Sri Lanka in the wake of the storm, as part of a larger order of 50,000 tons, at a time when the World Food Programme was seeking permission to import 15,000 tons of rice. She also describes how, when the regime did allow foreign aid, this benefited the military or local functionaries:

> The aid the regime did not refuse was often delivered in "dump and leave" fashion, unloaded and warehoused by the Burmese army – after all, its nearly half million soldiers need biscuits and blankets too. Some of the relief goods were outright stolen and sold in the markets. An unknown amount made it to the Delta, much of it only to be delivered into the hands of the junta's local government flunkies.[116]

---

[112] Larkin 2011, 8.   [113] Ibid., 9.   [114] Xinhua 2008.   [115] Ibid., 10.
[116] Mirante 2008b.

As Mirante and others noted, the regime was determined to go through with a scheduled constitutional referendum that would legitimate it. The referendum was scheduled to take place a week after the cyclone, and it did not want anything to deter its preparations.[117]

While some restricted amounts of aid were allowed in under those conditions, Larkin notes that it was nothing compared to the speed and volume of assistance that the US military provided within forty-eight hours of the 2004 tsunami which affected much of the region. In that case, the US response ultimately totaled 18,000 soldiers and 430 sorties per day a couple of weeks after the tsunami.[118] Larkin describes a handful of flights that the authorities allowed to land in Rangoon a week after the storm, provided the regime was able to receive and distribute the aid themselves. The authorities would inspect cargo and confiscate communication and IT equipment. A US Embassy staffer said the contents of five C-130 planes which should have been taken to the delta by Americans to be distributed were instead commandeered by the regime: "Instead they are being loaded onto Burmese military vehicles, and we have no idea where they're going. No one is telling us anything. We're bringing in all this stuff and it's all going into a big black hole."[119] Rumors had it that it was being resold in markets, that soldiers were distributing the aid so that the military would get credit for the operations, or even stolen by the generals' wives.

A year after the tragedy, a Human Rights Watch official reflected on how the ruling State Peace and Development Council (SPDC) impeded the international response: "The SPDC denied visas to foreign disaster response experts and refused to allow US, British, and French warships waiting off the coast to offload their supplies."[120] It was only after the referendum had taken place, the Human Rights Watch official noted, that the international response was blessed by the regime after United Nations Secretary-General Ban Ki-moon intervened: "Three weeks after the cyclone, the Tripartite Core Group, a relief and reconstruction mechanism devised and jointly directed by the UN, the Association of Southeast Asian Nations (Asean) and the Burmese government, was established."[121]

The response to the 1970 Bhola cyclone in East Pakistan is eerily similar to the response to Cyclone Nargis. The earlier vignette about the Pakistani general who referred to those in East Pakistan as "low-lying people"

[117] Mirante 2008a. [118] Larkin 2011, 47. [119] Quoted in Ibid., 48. [120] Adams 2009.
[121] Ibid.

suggested Pakistan's rulers in the west of the country were prejudiced toward the population in the east. Other accounts reinforce this observation. Reilly comments that "the central government in West Pakistan seemed to turn a deaf ear to East Pakistan's suffering" and noted that the first aid given to the East came from Pakistan's enemy, India, and other international donors. Pakistan's president reportedly flew over the delta and said the damage was exaggerated and then delegated response efforts to another official rather than handling it himself. Images of "bodies dangling in tree tops and hollow-eyed, traumatized children" ultimately spurred the government to action but the government in West Pakistan had already incurred intense reputational damage, both internationally and in East Pakistan. On November 23, East Pakistan's leadership sent President Khan a letter asserting that he had demonstrated "gross neglect, callous inattention, and utter indifference."[122]

Ultimately, these grievances accumulated and spurred an armed insurrection in East Pakistan in 1971; this ultimately led to Bangladesh's secession after West Pakistan's brutal campaign of suppression, which killed hundreds of thousands, failed. While the inadequate cyclone response was not the only source of East Pakistan's rejection of West Pakistan's rule – the government had also made Urdu, a language not widely spoken in the East, the official language – it was one of the important factors that fueled independence efforts.[123]

Hossain has put Pakistan's inadequate response to the cyclone into context. Already postponed from October, the country was scheduled to have its first democratic elections in December 1970, just three weeks after the cyclone. In their campaigning, the opposition Awami League highlighted the "callousness" of the regime in its response to the cyclone and won a landslide in its home province, East Bengal.[124]

Hossain suggests that effective crisis response became part of the nationalist politics of inclusion post-independence, what she calls a "subsistence crisis contract": "That adaptation depended on a transformation of political and social relations so that the power of the ruling elite came to depend substantially on its willingness and capacity to safeguard the population in such times of crisis."[125] She explicitly evokes Alex de Waal's anti-famine social contract, which was part of the ethos that legitimated the Ethiopian regime after the Derg, as discussed in Chapter 4.

[122] Reilly 2009, 180–181.   [123] Ibid., 181.   [124] Hossain 2018, 187.   [125] Ibid., 188.

Since independence, Bangladesh has cultivated a reputation for being especially open to international organizations working in development, earning the moniker of "The Wall Street of Development," given the density of nongovernmental organizations.[126] In April 1991, Bangladesh was again buffeted by an intense cyclone. Mirante notes that the 1991 storm would have been much worse but for foreign aid: "At that time, it took days for outside aid to arrive in the affected areas, but when it did – largely through the US Navy's well-regarded Operation Sea Angel – it saved many lives and prevented a second wave of disease."[127]

Operation Sea Angel, launched within a day of the government's request for aid, involved more than 7,000 US service members and brought aid to isolated islands severely affected by the storm. With the ports at Chittagong largely damaged in its wake and local infrastructure on the islands destroyed, US forces used ships, helicopters, and planes to deliver aid, largely from the sea. The operation also involved multilateral partners such as Japan, the UK, China, Pakistan, and India. Ultimately, the operation was credited with saving some 200,000 lives.[128]

Hossain notes that the subsistence social contract that the Bangladeshi political class enacted and consolidated after independence to secure mass support from the masses also included international donors: "The painful political crisis that ensued eventually led to a tacit agreement between the new political elite, the impoverished masses, and international aid donors to protect against the worst effects of disasters."[129]

By 2007 and Cyclone Sidr, the effects of these improvements in warning and preparedness were apparent. Mitchell et al. compared the outcomes of the Bhola cyclone, the 1999 cyclone, and Cyclone Sidr. They sought to normalize the fatality count by houses destroyed to get at the relative deadliness of the different storms, which, as argued earlier, were similar in strength. They found marked improvements over time in mortality, even as the number of houses destroyed increased.[130]

In India, the 2005 Disaster Management Act is credited with improving the capacity of the state to address disasters, but, as Chhotray notes, disaster relief and the responsibilities of the state to its citizens have long origins dating back to the colonial era. This was often oriented toward famine and such "Famine Codes" mixed a sense of obligation to the state tempered by the desire for citizens to resume normal productive activities

---

[126] Cons 2016, 2.   [127] Mirante 2008a.   [128] GlobalSecurity.Org n.d.
[129] Hossain 2018, 188.   [130] Mitchell et al. 2014, 25–26.

TABLE 6.3 *Comparison of major cyclones in Bangladesh*

Date	Size	Houses destroyed	Evacuated	Deaths	Deaths/houses destroyed
1970 Bhola cyclone	Cat. 3: 4–10 surge	400,000	0	300,000	0.75
1991 cyclone	Cat. 4: 5–8 surge	780,000	2,000,000	138,000	0.18
2007 Sidr cyclone	Cat. 5: >5 m surge	564,000	2,000,000	3,477	0.006

Source: Mitchell et al. 2014

and not become wards of the state. They included the Odisha Relief Code of 1996 that built on earlier famine codes of 1913, 1933, and 1980.[131]

As noted earlier, the Indian government has largely relied on its own resources to address cyclones. In the wake of the 1999 cyclone, India received modest amounts of foreign aid, some $13 million, half of this from the United States.[132] However, the Indian government has long prioritized internal resources to address disaster management, both a function of hostility to "aid imperialism" that might hinder the country's room to maneuver and set its own policies, but also of the country's aspirational self-image as a leading world power. In 2013, India rejected a modest offer from the United States for flood relief in Uttarakhand. This was the result of policy evolution over a decade that saw the country routinely turn down funds for disaster management. A government spokesperson stated at the time: "As a general policy in case of rescue and relief operations, we have followed the practice that we have adequate ability to respond to emergency requirements."[133]

This policy change dates back to 2004 when Prime Minister Singh rejected humanitarian assistance in the wake of the 2004 tsunami. That approach has endured. Thus, in 2018, Prime Minister Narendra Modi rejected a $100 million offer from the United Arab Emirates to help the country respond to a devastating flood in Kerala. The government signaled its willingness to accept foreign funds so long as they were routed through the Indian government directly.[134]

---

[131] Chhotray 2014, 219.   [132] USAID 1999.   [133] Kasturi 2013.
[134] Venkataraman, Raj, and Abi-Habib 2018; Chaulia 2018.

Given the performance of the Indian government since the 2005 Disaster Management Act was passed, it is hard to argue that India does not have the capacity or resources to carry out most response activities on its own, though it has accepted some assistance for technical disaster risk reduction from USAID and other partners for peer responder training and disaster management.[135] India has occasionally accepted modest assistance in the wake of disaster events such as $100,000 from USAID after Cyclone Hudhud buffeted Andhra Pradesh in 2015.[136]

Pal et al. praised the Indian government and the state of Odisha for their superior response to Cyclone Phailin in 2013, compared to the effort in 1999: "This was a remarkable achievement considering the havoc [and] anarchy among [the] administration in the year 1999, when Odisha was challenged by the similar intensity cyclone."[137] They credit government cooperation and community level efforts for having learned the lessons from the 1999 response, with more than 1.2 million ultimately evacuated.[138]

Pal et al. note that in 1999 the Department of Agriculture and Cooperation in the Ministry of Agriculture was the lead agency in charge of disaster response, which then sought to mobilize the Secretaries and Relief Commissioners in the states of Odisha, West Bengal, and Andhra Pradesh. The state level actors were in charge of the evacuation, but efforts were described as mostly "ad hoc."[139] The new structure put in place by the 2005 Disaster Management Act created federal, state, and district authorities with the capacity to prepare for and respond to disasters. The emergency powers of the District Magistrate to take the lead were credited with driving an effective response, notably the capacity to force evacuations to the cyclone shelters.[140] Pal and coauthors credit preparedness and early warning systems to have much improved since the 1999 cyclone. This was echoed by Yadav and Barev in their ground-level vulnerability assessment, which interviewed local people in Odisha state:

It was found that although there were no reliable sources of warning during [the] 1999 super cyclone, which caused about 10,000 fatalities ... during [the] 2013 cyclone Phailin, people received timely proper and reliable early warnings and the death toll was limited to 44. This time, people had enough time to prepare for evacuation at the nearest shelter.[141]

---

[135] Konyndyk 2013; Fleming 2013.   [136] USAID 2016.
[137] Pal, Ghosh, and Ghosh 2017, 350.   [138] Ibid., 351.   [139] Ibid., 352.   [140] Ibid., 357.
[141] Yadav and Barve 2017, 392.

In advance of the 2013 storm, the Indian state evacuated more than one million people to the many purpose-built shelters, most of which had been constructed since the 1999 storm. This response provided the state government with a reputation for improved governance. As Chhotray notes: "The state government, generally accustomed to brickbats for poor governance and development performance, acquired a new halo overnight for its heroic efficiency in one of the poorest and most disaster prone states of India."[142]

In sum, the dynamics of these cases largely confirm my theoretical expectations that weak capacity, exclusive governance, and blocked aid led to large-scale loss of life in Myanmar, while Bangladesh and India have experienced improved state capacity over time and somewhat more inclusive governance. In Bangladesh's case, ample use has been made of international assistance to build capabilities, while India, since the early 1990s, has preferred to handle cyclone preparedness and response itself.

ALTERNATIVE EXPLANATIONS

One possible alternative explanation for the deadliness of the 2008 cyclone in Myanmar is simply lack of experience with cyclones. Unlike other cyclone prone countries in the region, it could be claimed that with respect to Myanmar the government and the population's lack of readiness and failure to respond is a function of the unusual nature of the storm in question. The implication is that repeated exposure, even in countries with poor governance and limited capacity, will encourage a better response.

But does this claim have any support? The EM-DAT international disaster database tracks disaster events. To be included in the database requires that the hazard must already exceed a certain level of impact. Between 1970 and 2018, India had 43 cyclone events, Bangladesh (which is about 5 percent the size of India, geographically) had 37, compared to only 9 in Myanmar.[143] Tasnim et al. make this point: "Compared to neighboring countries like Bangladesh and India, cyclone induced damage during the second half of the twentieth century was relatively low in Myanmar, before Nargis made landfall."[144] Cyclones in April–May normally track northwards in the Bay of Bengal but this storm turned east and ultimately made landfall, "catching residents and authorities

---

[142] Chhotray 2014, 217.   [143] CRED 2019.
[144] Tasnim, Esteban, and Shibayama 2015, 55.

unprepared." The cyclones that had affected Myanmar in recent history had not been nearly as dangerous: "As a result, the residents had little or no awareness about storm surges and they were unaware of the danger they were facing."[145]

Mirante makes a similar point in the literary magazine *Guernica*: "On May 2, 2008 a cyclone hit Burma. They usually don't, or not very hard, but this one, called Nargis ('Daffodil') smashed onshore with an unprecedented fury."[146] Fritz et al. similarly observe: "No previous tropical cyclone track included in the International Best Track Record for Climate Stewardship database has made a direct landfall in Myanmar's Ayeyarwady river delta at an untypically low latitude near 16° N."[147] They credit Myanmar for avoiding large-scale mortality in two previous cyclone episodes in 2004 and 2006 in other parts of the country. Fritz et al. suggest unfamiliarity with cyclones in this part of Myanmar was one reason for its deadliness:

All interviewed eyewitnesses ignored warnings owing to lack of cyclone awareness and evacuation plans, absence of high ground or shelters, and no indigenous knowledge of comparable previous stormsurge flooding in the Ayeyarwady river delta.... In sharp contrast, the residents of the Gwa coastline, frequently struck by cyclones such as Mala, are aware of cyclone hazards and have evacuation plans.[148]

Markus Kostner, who advised the World Bank's engagement in Myanmar after the 2008 cyclone, makes a similar point. He suggests that the country had not previously experienced a cyclone of this magnitude, which may have played a role in the level of preparation and the public response to news of an impending cyclone. Despite some limited early warning from the government, he notes: "[T]he people didn't take it seriously because they couldn't imagine what kind of storm it would be. They never had anything close to Nargis before."[149]

While this perspective has some merit, the lack of familiarity with cyclones is an inadequate explanation. First, there was ample international warning of the storm's direction. The Indian Meteorological Department is charged by the United Nations to provide cyclone warnings to seven countries in the region and provided warnings to the Myanmar government at least thirty-six hours before the cyclone struck. *VOA News* also reported that the cyclone would make landfall in Myanmar rather than India.[150] Second, the government's unwillingness to respond and its

---

[145] Ibid., 59. [146] Mirante 2008a. [147] Fritz et al. 2009, 448.
[148] Fritz et al. 2009, 449.
[149] Markus Kostner, personal communication, December 7, 2020. [150] Herman 2009.

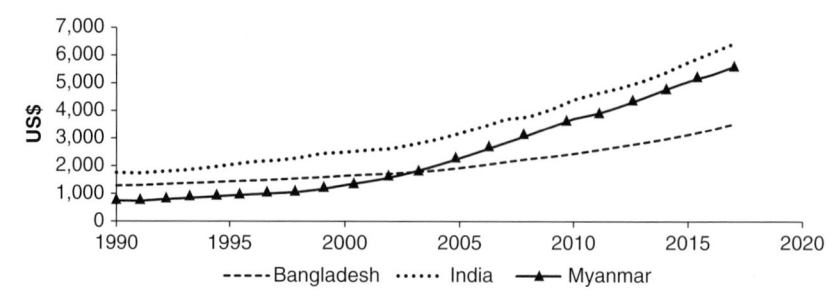

FIGURE 6.6 Real GDP per capita, Myanmar, Bangladesh, and India, 1990–2017 (constant 2011 US$)
Source: World Bank, https://bit.ly/3qL4755

exertion of tight control over foreign assistance created unnecessary suffering. It is unclear what proportion of the deaths occurred in the wake of the storm rather than from immediate exposure, but the government could have done more to prepare the populace before the storm and to respond afterwards. Moreover, as we saw in the case of the 1991 Bangladesh cyclone, familiarity with cyclones by itself is no guarantee of generating a successful response.

One possible explanation for the difference in outcomes between the countries is differential levels of economic growth and income. This argument suggests that India's relative wealth compared to Bangladesh and Myanmar explains its ability to finance disaster response capabilities. Bangladesh's relative success compared to Myanmar – and in comparison with previous cyclones that struck the country – could also be explained by rising income. However, this argument can quickly be found to have some problems. While India was wealthier in terms of GDP per capita than either Bangladesh or Myanmar throughout this period, none of them was especially wealthy.

Myanmar was comparatively wealthier than Bangladesh in terms of GDP per capita in 2008, when it suffered such catastrophic losses from Cyclone Nargis (see Figure 6.6). It had actually been growing at a higher a rate than Bangladesh (Figure 6.7). Thus, Myanmar by rights, was wealthy enough and growing fast enough to have had the resources to prepare for a cyclone.

CONCLUDING THOUGHTS

This chapter has demonstrated that cyclone deaths are not inevitable and that much hinges on governance, both the capacity of the state and its

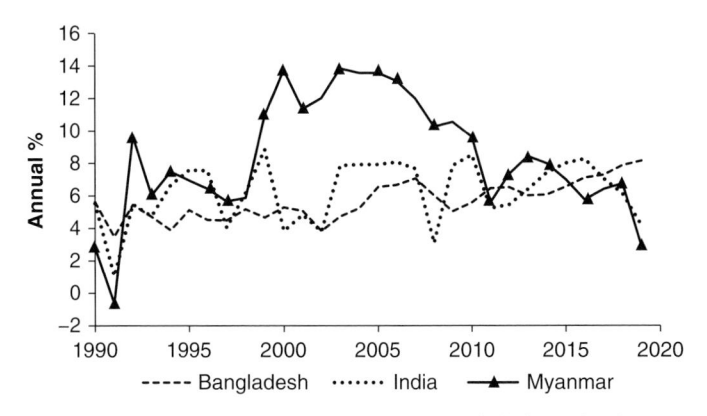

FIGURE 6.7 GDP growth (annual %), Myanmar, Bangladesh, and India, 1990–2019
Source: World Bank, https://bit.ly/3qOqHda

degree of political inclusion, as well as its receptiveness to international assistance where needed. In Myanmar, as in Somalia in Chapter 4, aid was effectively blocked and delayed, meaning the critical hours and days in the wake of the storm were largely lost in terms of foreign help. Given the country's weak capacity, this obstruction likely added tens of thousands to the death toll, though we may never know for sure how many could have been saved if intervention had come earlier. Bangladesh shows how openness to aid can build capacity over time and augment state capacity in crisis moments, making the earlier history of its founding, as well as the 1991 cyclone, painful memories but ones not repeated. In the contemporary era, India has never experienced the magnitude of Bangladesh's trauma, though the 1999 Odisha cyclone was searing enough to spur the federal and Odisha government to act and prepare. India has been less reliant on foreign aid to improve its defenses and response throughout the period studied.

The international dimension of these cases raises questions about the circumstances under which aid is offered. As we saw in Chapter 4, there are moments when aid is not only blocked by receiving countries, but may not be offered because of political differences between donors and the regimes experiencing climate hazards. While altruism and empathy can be powerful drivers of donor responses, there may be broader strategic reasons why the international community cares about negative security outcomes that come from exposure to climate hazards. The next chapter explores those connections.

7

Beyond Internal Conflict

The Practice of Climate Security

We must stop Tuvalu from sinking and the world from sinking with Tuvalu.
— United Nations Secretary-General, António Guterres[1]

The best way to prevent societies from descending into crisis, including but not limited to conflict, is to ensure that they are resilient through investment in inclusive and sustainable development.
— *Pathways for Peace*, United Nations–World Bank Group[2]

In a visit to the Pacific island nation of Tuvalu in 2019, United Nations Secretary-General António Guterres underscored the existential risks of climate change for low-lying island countries. In so doing, the Secretary-General demonstrated that practitioners have a more expansive set of concerns than whether climate change leads to violent conflict – the primary focus of much academic research.[3] Climate-related internal conflict still remains a central focus for practitioners. That said, we still know little about how to prevent climate-related conflicts from starting or how to stop them once they do start, though academics have more understanding now of risk factors.

In the preceding chapters, I reviewed the circumstances under which countries facing climate-related extreme weather events might experience negative security consequences, focusing on internal conflict and humanitarian emergencies. But this discussion begs the question of why external actors might care and what they could prospectively do to prevent such

---

[1] United Nations 2019. [2] United Nations–World Bank Group 2018.
[3] Gleditsch 2021.

outcomes or how they could respond when they do emerge. This chapter surveys how the practice of climate and security has evolved over the last fifteen years, starting in the first section with a review of the variety of climate security challenges policymakers have identified. The second section considers how policymakers have responded to climate security challenges thus far, and the third section looks at what can and should be done to address climate security going forward.[4] I close the chapter with a final section on the relationship between external actors and state development.

THE CHALLENGES

The emergent practice has identified a broad suite of climate security challenges, ranging from operational implications for specific military bases to the existential challenges faced by some countries and regions. The links between climate change and internal conflict – both civil war and communal conflicts – still have a central place in the conversation as evinced by several United Nations Security Council resolutions in relation to ongoing conflicts in Africa. A 2015 report for the G7 identified seven sources of what it described as "compound climate-fragility risks," climate risks that when combined with other sources of state fragility can lead to negative consequences including local resource competition; livelihood insecurity and migration; extreme weather events and disasters; volatile food prices and provision; transboundary water management; sea-level rise and displacement; and unintended effects of climate policies.[5]

This more encompassing set of climate security concerns is partly a function of a greater willingness by practitioners to embrace human security and move beyond a traditional focus on state security. A focus on

---

[4] This chapter builds on Busby 2021. It somewhat reflects my own particular take on the practice of climate and security as someone who started publishing in this space in 2004 with a report for UN Secretary–General Kofi Annan, and extending to advisory work on the first National Intelligence Assessment for the National Intelligence Council in 2008. I was also part of two large research grants from the US Department of Defense under its Minerva Initiative. More recently, I have participated in consultancy projects on the topic for the Consultative Group on International Agricultural Research (CGIAR), USAID, and the United Nations Environment Programme. I have also published widely on climate and security for a variety of think tanks including the Council on Foreign Relations, the Brookings Institution, the Center for Climate and Security, the German Marshall Fund, and the Wilson Center, among others.

[5] Rüttinger et al. 2015, viii–x.

human security constitutes a broadening in two directions, substantively away from conventional security threats (moving beyond armed attacks to encompass environmental change) and in terms of whose security is of concern (moving beyond states to the security of individuals and communities). While the concept has been critiqued for being woolly,[6] a human security lens draws attention to how climate change can lead to negative consequences for people, even if state security is not challenged.[7] The most severe threat, of course, is loss of life, and extreme weather events such as swift-onset storms and even slower-onset droughts can lead to large-scale fatalities.

As discussed in Chapter 2, calls to broaden or redefine security date back to the 1980s.[8] With other concerns like the COVID-19 pandemic also contributing to large-scale death and economic disruption, policymakers have increasingly accepted that health and environmental threats can constitute security concerns. There is, of course, a long-standing debate about the merits of securitizing environmental and other problems, because of the potential for threat inflation, the use of emergency procedures for security problems, and the risks of reinforcing nationalist approaches to collective problem-solving.[9]

This recognition of the security consequences of climate change led the Intergovernmental Panel on Climate Change (IPCC), in its Fifth Assessment Report, to include a chapter on human security, defined as protecting the "vital core" of human lives, which include material and nonmaterial aspects.[10] The IPCC chapter included but was not limited to the links between climate and conflict. While the treatment went further than I would to include threats to cultural survival, the chapter signaled the policy community's broader interests beyond the study of climate–conflict.

Among those security risks are the existential threats to low-lying island countries from sea-level rise, saltwater intrusion, and coastal inundation from storms. These risks constitute both human security concerns and threats to state security. In traditional national security parlance, states worry that armed external attacks might lead to their country ceasing to exist as an independent unit. Some studies suggest a number of Pacific island atolls may become uninhabitable by 2050 if saltwater overtops aquifers and makes it impossible to grow crops or secure

---

[6] Paris 2001, 2004; Busby 2008.   [7] Adger et al. 2021.   [8] Ullman 1983.
[9] Deudney 1990; Hardt 2017; Diez, Lucke, and Wellmann 2016; Rothe 2017.
[10] Adger et al. 2014.

freshwater.[11] Such risks constitute threats to the continued existence of some states, even leaving aside human security impacts. Indeed, some countries have made preliminary preparations for managed retreat by securing land overseas (in Fiji in Kiribati's case).

Even if their existence is not threatened, other countries face extensive risks because of large populations and valuable infrastructure located near coasts. Former US Vice President Al Gore dramatized these risks in his slideshow projections of future climate change.[12] Accurately estimating these risks requires projections of emissions and sea-level rise, adequate representations of elevation, and good population maps and forecasts. A 2019 study corrected some standard biases in digital elevation models to estimate the number of people likely living in expanded flood zones in a variety of emissions scenarios. In the high emissions scenario, it found that some 340 million people would be living below annual flood levels (or below high tide) by the middle of the twenty-first century, up from 250 million today, between 18 and 32 percent in China alone.[13]

The risks to coastal populations extend beyond sea-level rise. One study estimated that 625 million lived in low-elevation coastal zones in 2000 (less than 10 meters above sea level), with that number expanding to between 879 million and nearly 950 million by 2030 under different population growth scenarios.[14] The risks of sea-level rise are magnified by storm surge and hurricanes/cyclonic activity. Large coastal populations in the United States are at grave risk of storms and hurricanes, extending from Texas along the Gulf Coast to Florida and up the eastern seaboard to New York. In 2017, three storms in succession – Harvey, Irma, and Maria – collectively caused more than $250 billion in damages, thousands of deaths, and required the mobilization of tens of thousands of military for humanitarian rescue and response.[15] The island of Puerto Rico, a US possession, had its electricity grid destroyed, with thousands of residents experiencing prolonged power outages over the next year. And, as Chapter 6 noted, the densely populated areas off the Bay of Bengal bordering India, Bangladesh, and Myanmar have experienced intense cyclonic activity, with large-scale loss of life – although deaths caused by these storms have decreased in recent decades in India and Bangladesh. Myanmar experienced catastrophic losses of more than 140,000 lives when Cyclone Nargis battered the Irrawaddy Delta in 2008. The impact of climate change on hurricanes has been a contentious issue among

---

[11] Storlazzi et al. 2018. [12] Gore 2006. [13] Kulp and Strauss 2019.
[14] Neumann et al. 2015. [15] Rice 2018.

scientists, but the linkages have become clearer over time.[16] While some of these risks can be managed with early warning systems, cyclone shelters, and other adaptive responses to climate-proof infrastructure, the enhanced risks of cyclones, along with sea-level rise, may exacerbate the habitability problems of some coastal locations.

These risks of habitability extend beyond islands and coastal areas. Wallace-Wells surveyed a variety of existential challenges to parts of humanity in his evocative essay, and later book, *The Uninhabitable Earth*.[17] As he noted, some studies warn that rising temperatures will make regions of the world in South Asia and the Middle East uninhabitable by stressing crop production, reducing freshwater aquifers, and making regions so hot that it will be difficult for people to spend extended periods of time outdoors. One study, using a midrange scenario for climate change (RCP 4.5),[18] found that under a range of population and economic growth scenarios, between 1.62 billion and 2.49 billion would face mean annual temperatures in excess of 29°C (84°F) by 2070 and would be displaced from what is considered the normal range of conditions habitable for human beings. The Saharan desert, for comparison, has a mean annual temperature of 30°C.[19]

Another study on South Asia examined the intersection of heat and humidity and suggested that the upper bound for human habitability is 35°C with even 31°C dangerous for human beings. A temperature of 34.4°C with 80 percent humidity can feel more like nearly 54°C (129°F) and be quite dangerous, making it difficult for people to cool down their body temperature. A 2015 heat wave in India and Pakistan had a wet-bulb temperature of 50°C and killed 3,500.[20] This study suggested in a high-end emissions scenario (RCP 8.5),[21] nearly 30 percent of the population in the agriculturally rich Indus and Ganges river valleys would face a median temperature of 31°C by 2100, compared to only 2 percent in a RCP 4.5 scenario.[22] This underscores the importance of actions taken today to limit emissions to the lower range scenarios.

---

[16] Mooney 2007; Murakami et al. 2020. [17] Wallace-Wells 2017, 2019.
[18] Representative Concentration Pathway (RCP) 4.5 is a scenario of long-term, global emissions of greenhouse gases, short-lived species, and land-use-land-cover which stabilizes radiative forcing at 4.5 Watts per meter squared. See https://asr.science.energy.gov/publications/program-docs/RCP4.5-Pathway.pdf.
[19] Xu et al. 2020. [20] Leahy 2017.
[21] RCP 8.5 refers to the concentration of carbon that delivers global warming at an average of 8.5 watts per square meter across the planet. See https://climatenexus.org/climate-change-news/rcp-8-5-business-as-usual-or-a-worst-case-scenario/.
[22] Im, Pal, and Eltahir 2017.

Climate risks that threaten human existence extend beyond sea-level rise and temperature increases. As the United States has witnessed in recent years, the spread of populations to forest-rich environments has put many communities at risk of wildfires, a risk accentuated by rising temperatures and disrupted rainfall patterns. In 2018, the Camp Fire in California claimed some 85 lives and led to damages in excess of $16.5 billion.[23] Other climate risks include riverine flooding, which periodically upends the lives and livelihoods of hundreds of millions around the world and is likely getting worse as a result of more variable rains leading to larger downpours. In 2010, Pakistan, for example, experienced severe flooding that led to the displacement of more than 1.5 million with disruptive effects on as many as twenty million people.[24] In July 2020, seasonal flooding in China led to the temporary displacement of more than forty million people, with concerns that the Three Gorges Dam might fail, potentially leading to catastrophic loss of life downstream.[25] While the climate signal in this particular episode is unclear, flood risks are among the various large-scale threats to loss of life from climate-related hazards.

The impacts on loss of life and livelihoods from climate hazards and disasters are clearly significant human security issues, but are they national security issues? One of the concerns of climate-related extreme weather events is population displacement within or across borders. People may be forced from their homes temporarily or permanently as a result of climate change. As discussed in Chapter 3, whether climate migrants are likely to engage in conflict or become targets of violence has been vigorously debated among academics.[26] And, as I explored in Chapter 5, the large population movements triggered by drought in Syria, and whether these movements contributed to the civil war, have been particularly contentious.[27]

Even if the contribution of climate migrants to conflict is unclear, security practitioners are concerned about the dislocative effects of large-scale migration, whether it be pastoralists searching for better grazing land and water potentially coming into contact with settled agriculturalists or internal or cross-border movements of people seeking to escape from extreme weather events. Prior to the COVID-19 pandemic, the rise of populist nationalism had created dangerous political dynamics in the

---

[23] Rice 2019.   [24] Schaffer and Dixon 2010.   [25] Tan 2020.
[26] Reuveny 2007; Raleigh, Jordan, and Salehyan 2008; Koubi et al. 2016b.
[27] Gleick 2014; Kelley et al. 2015; Selby et al. 2017a.

United States and a number of European countries. For example, climate change has been implicated by analysts in the migration decisions of farmers from Central America,[28] and the militarized response of the Trump Administration to those migrants had profound implications for regional security in the region. Although the reasons for migration may not be environmental, displaced populations can also be at risk from climate hazards. After 700,000 Rohingya fled Myanmar in 2017 in response to ethnic cleansing, they found themselves at risk of storms and cyclones in their new location in Cox's Bazar in Bangladesh, raising human security concerns.[29] Moreover, their presence in Bangladesh creates an ongoing source of interstate tension.

Even where populations do not migrate, practitioners worry about the effects on lives and livelihoods, particularly for agriculturally dependent communities, which may face severe food security deficits from growing seasons disrupted by climate extremes and variability. Scientists project major declines in land productivity and crop production under a range of climate scenarios, as crop-growing becomes more difficult when growing conditions deviate from what crops are suited to.[30] In countries with inadequate famine prevention and response capability, these impacts can be severe. As discussed in Chapter 4, in the wake of a drought that began in late 2010, Somalia experienced an estimated 260,000 excess deaths from famine.[31] Somalia's situation, of course, was exacerbated by the long-running civil war between a weak government and the Al-Shabaab insurgency. This example suggests food security is not simply a matter of human security, but is bound up with wider issues of state security. Indeed, one of the core findings of recent climate security research is the risk of internal conflict in agriculturally dependent societies.[32] So, although climate change may initially affect human security, the impacts may escalate to state security threats under certain conditions.

Security risks also extend to food importers who may face higher global prices if climate disruptions increase prices on key foodstuff like wheat. The effects of drought and disrupted agricultural exports on global food prices were implicated in the onset of the Arab Spring protest movements.[33] Although states may insulate their populations from these effects through social policies,[34] academic research has found evidence for

---

[28] Leutert 2018.   [29] Grunebaum 2019.   [30] Iglesias, Quiroga, and Diz 2011.
[31] Checchi and Robinson 2013.   [32] von Uexkull et al. 2016.
[33] Lagi, Bertrand, and Bar-Yam 2011.   [34] Hendrix and Haggard 2015.

increased risks of riots and protests if global prices are passed on to domestic consumers.[35]

Another growing area of concern is the link between climate and public health. Aside from the impacts of extreme weather events on loss of life, the health impacts of climate change have been understudied, at least among security researchers. However, the COVID-19 pandemic reminds us that infectious diseases, in particular, are more than public health challenges. Their impacts can rise to the level of security concerns if they are especially severe and dislocative in terms of loss of life, economic impacts, and societal disruption.

Public health researchers have long been concerned about the connections between climate change and rising temperatures and their contribution to an expanded range of disease vectors like mosquitoes and ticks that spread diseases such as dengue, malaria, and Lyme disease.[36] It is not clear that an increased burden of malaria, for example, would necessarily constitute a security challenge, though there is a literature connecting the heavy burden of malarial disease in parts of the world with slower economic growth.[37] Slow economic growth, in turn, is connected to increased conflict risk.[38]

A number of recent severe infectious disease outbreaks including SARS, Ebola, and COVID-19 reflect zoonotic transfer from animals to humans. Environmental degradation is associated with these new disease outbreaks. The increased interaction of human beings with wild animals and more human incursions into wild lands are seen as drivers of the enhanced risk of zoonotic transfer.[39] It is unclear whether climate change, which will stress ecosystems further, will have a bearing on new infectious diseases, but this will likely become an important area of concern for practitioners and scholars going forward.[40]

Beyond the social and political effects of climate impacts themselves, an emergent concern is the potential for maladaptation and unintended negative effects of responses to climate change that may lead to negative security consequences. For example, Russia reacted to a 2010 drought by banning grain exports, which contributed to the spike in global food prices that preceded the Arab Spring. The Wilson Center published

---

[35] Smith 2014.   [36] Confalonieri et al. 2007; Watts et al. 2021.
[37] Gallup and Sachs 2001.   [38] Collier 2007.
[39] World Health Organization and the Secretariat of and the Convention on Biological Diversity 2020.
[40] Wyns 2020.

a survey of this "backdraft" potential of unintended consequences, which included discussion of land grabs, geoengineering, potential conflict over lithium and other minerals important for clean energy systems, land use issues associated with biofuels and REDD+ forest conservation initiatives, and short-term adaptive responses that may exacerbate environmental damage to forests, water, and land.[41] These potential impacts underscore the distributional impacts of the decisions made to address climate change – both mitigation and adaptation – which could be as, if not more, consequential than the physical impacts of climate change itself.

While this analysis identifies the range of security concerns that policy-makers care about, we still have not answered the question why some states would or should care about the effects of climate change outside their own borders. Here, the concern is that the indirect impacts on others will ultimately affect a country's international assets, allies, or interests, or that distant impacts will ultimately generate negative spillovers that affect the home country.

The proximate effects of climate change on neighboring countries can lead to migration pressures and political instability. For example, the United States is likely to be directly affected by climate-related migration pressures and instability emanating from Mexico, Central America, and the Caribbean. As mentioned earlier, the effects of climate change on agriculture in the dry corridor of Central America may be contributing to migration pressures from the region.[42]

Similarly, Europe is subject to migration pressures and spillover problems from North Africa and the Middle East given the proximity of these regions. The humanitarian crisis that convulsed Syria in the midst of the Syrian civil war sent more than six million people out of the country, overwhelmingly to the neighboring countries of Turkey, Lebanon, and Jordan, though nearly 600,000 ultimately resettled in Germany.[43] The footage of Syrians attempting to flee overland, by train, and by sea across Europe as many countries erected barriers is an evocative reminder of how countries may react to large-scale movements of people. In the European context, such movements can potentially trigger xenophobic political crises and empower proto-fascist political parties, elevating the importance of migration. To the extent that we think climate change had some role in the Syrian crisis, or might play a role in future crises, the spillover

---

[41] Dabelko et al. 2013.   [42] Leutert 2018; Seay-Fleming 2018; Semple 2019.
[43] Todd 2019.

effects of climate change go well beyond humanitarian concerns, as important as they are.

In some cases, although the effects of climate change may generate security concerns internationally, the effects on distant countries may primarily be economic in nature. For example, in 2011, severe seasonal monsoon flooding in Thailand disrupted global supply chains for electronics and semiconductors, with global industrial output declining by 2.5 percent.[44] The floods persisted from July 2011 through to early 2012, leading to more than 800 deaths and amounting to nearly $46 billion in damages.[45] This episode raises the question whether and under what conditions damage to global supply chains could ever rise to the level of security concerns and cease to be merely economic impacts.

Great powers like the United States will have a more expansive set of interests and concerns than countries with a more limited international presence. For countries with such global interests, the effects of climate change on distant locations may be of relevance for a variety of reasons. The kinds of impacts of concern include climate-related effects on overseas bases and embassies; sources of raw materials and/or pieces of global supply chains; important transshipment routes/ports; allies; and countries where impacts could blowback in the form of migration, regional instability, or international terrorism. For example, the risks of sea-level rise in the Pacific pose existential risks for atoll countries and other low-lying islands. A number of them also host critical US national security assets such as the Marshall Islands, Palau, and the US territory of Guam.[46] Locations that are important for any or several of these reasons may have more strategic importance for far-flung powers than others and elicit more international interest.[47]

At the height of the concerns about terrorism in the early 2000s, global disorder anywhere was perceived to be the United States' problem. As the international dimensions of the terrorism problem have receded, the United States may be recalibrating its sense of global interests, particularly with increasing preoccupation with its own internal economic and social challenges. Given questions about the efficacy of drone strikes and other militarized responses to the terrorism problem, it is not obvious whether strategic disengagement from some places would lead to better or worse outcomes, particularly since previous external interventions may have

---

[44] Polycarpou 2014. [45] Pierce 2012. [46] Bhide 2019.
[47] Levy 1995; Busby 2007, 2008, 2016a.

created contexts that are not conducive to local de-escalation of violence without some form of ongoing engagement by outside actors.

At the same time as the United States may be re-examining what role it can and should play with respect to fragile states, other countries, like China, with its state-led Belt and Road Initiative, may be expanding the geography of countries of concern. Like the United States before it, China may find itself drawn into more situations in fragile states, not least to defend Chinese workers and investments. While instrumental calculations of self-interest may explain why some countries care about the impacts of climate change on others, humanitarian concerns about climate impacts on the lives and well-being of people around the world may also inspire international altruism.

THE PRACTICE OF CLIMATE AND SECURITY

Policy discussions about the links between climate and security date back to the mid-2000s but have matured as various governments and international organizations have sought to mainstream climate security concerns. A community of practice has emerged among think tanks, led by organizations such as the US-based Center for Climate and Security and CNA's Military Advisory Board, the German-based adelphi, and the Stockholm International Peace Research Institute (SIPRI).[48]

Among governments, the United States has paid significant attention to climate risks, particularly with respect to its own military bases but also its overseas interests. The Obama administration issued numerous reports highlighting the risks for bases, missions, and training, with thematic reports on the Arctic, food and water security, as well as branch-specific risks for the Navy and the Combatant Commands.[49] Climate and security received presidential attention through executive actions, though these were subsequently rescinded by the Trump administration.[50] However, while attention waned in the Trump era, the specific vulnerabilities of US military bases continued to be investigated.[51] With the election of Joseph Biden in 2020, climate security is already being elevated in his

---

[48] Other efforts include the Planetary Security Initiative (PSI), a Dutch government funded effort that hosted four annual conferences from 2015 to 2019. Clingendael, the Netherlands Institute of International Relations, which led the organization of PSI on behalf of the Dutch government, has also been active on climate and security policy work. A wider international community of practice has been developed through the Climate Security Expert Network and the International Military Council on Climate and Security.
[49] Busby 2016a. [50] Scata 2017; Calma 2019. [51] Department of Defense 2018, 2019.

administration, underscored by his appointment of former Secretary of State John Kerry as his special presidential envoy for climate with a seat at the National Security Council.[52]

Climate security concerns are not limited to the United States. The UK, Germany, several Scandinavian governments, and countries in Oceania have been especially worried. Among advanced industrialized countries, these concerns coalesced in the 2015 report for the G7 mentioned in the section above.[53] At the United Nations Security Council, several member states have sought to create space for discussions on climate and security. The first such discussion, prompted by the UK, occurred in 2007. At the time, the connections between climate and security were not well understood. Some member states, such as Russia and China, resisted calls for the Security Council to broach the topic. Since then, it has more frequently held focused discussions on climate and security, including both formal "open debates" as well as several "Arria-formula" informal sessions.[54] Several ongoing United Nations peacebuilding and peacekeeping missions have, in their mandate renewals, been tasked by the Security Council to report on climate-related security risks, including the Sahel, Lake Chad Basin, Somalia, Mali, and Sudan.[55]

Support from Sweden, Germany, and other states ultimately coalesced in the creation of the Climate Security Mechanism (CSM) at the United Nations in 2018, a tripartite arrangement of the United Nations Development Programme, the United Nations Environmental Programme (UNEP), and the UN's Department of Political and Peacebuilding Affairs. In its first year plus, the CSM sought to mainstream climate security concerns in the wider United Nations,[56] building on more than a decade

---

[52] Conger, Femia, and Werrell 2020. [53] Rüttinger et al. 2015.
[54] Arria-formula meetings are informal gatherings convened by a United Nations Security Council member that can hear testimony from nonmembers such as individuals or NGOs; Climate Security Expert Network n.d. A second open debate occurred in 2011, prompted by Germany when it held the rotating presidency of the Security Council. A Presidential statement from the Security Council expressed concern that climate change could have security implications that the Council should be kept apprised of. Arria-formula events have been held as follows: in 2015 (the first led by Spain and Malaysia, and a second one led by New Zealand); in 2017 (the first led by Ukraine and Germany, and second one, in December, led by a group of countries that included Italy, Sweden, Morocco, the UK, the Netherlands, Peru, Japan, France, the Maldives and Germany); and in July 2018 (led by Sweden). The Dominican Republic, during its month-long presidency of the Council hosted a more formal "open debate" on the topic in January 2019. Another open debate was hosted by Germany in July 2020 following an informal discussion in April of that year.
[55] Eklöw and Krampe 2019. [56] United Nations Climate Security Mechanism 2020.

of UNEP's work on environmental peacebuilding.[57] Other international and regional organizations are also grappling with what role they should play.[58]

TOWARD MORE EFFECTIVE POLICY

The discussion of threats and policy initiatives begs the question of what to do. The US climate security community has developed a group of stakeholders who understand the issue and convene regularly. During the Trump administration, the challenges of even talking about climate change reinforced a tendency to frame the issue in terms of impacts on the US military, its bases and operations.[59] Overmilitarization of the problem may sideline other instruments of national power like development, diplomacy, and humanitarian assistance, which may be more important levers for addressing overseas impacts, and runs the risk of contributing to international cleavages between countries. That emphasis may dissipate during the Biden administration, but the temptation to emphasize the military dimensions of this problem for political purposes remains.

Internationally, we still know little about what works for several climate security risks. For transboundary rivers, we have an appreciation of how river basin management institutions can reduce conflict risks by allocating water, dealing with shocks, and resolving disputes.[60] As in the case of the Indus River Treaty between India and Pakistan, such institutions can endure despite tensions on other matters. The policy agenda here is to build such institutions in basins where they are lacking, and to deepen institutions where they exist. In the contemporary dispute over the construction of the Grand Ethiopian Renaissance Dam, analysts recommend building a more inclusive institutional architecture: Legacy agreements prioritized water access for Egypt and Sudan, leaving other riparians such as Ethiopia with little say over this issue.[61]

For other climate security concerns, particularly with respect to famines and cyclone risks, there are also well-established policies for addressing human security. These include early warning systems, hazard specific interventions such as cyclone shelters, preplacement of emergency supplies, social support mechanisms such as food-for-work schemes and cash

---

[57] Jensen 2019.
[58] Dellmuth et al. 2017; Sherman and Krampe 2020; Krampe and Mobjörk 2018.
[59] The Climate and Security Advisory Group 2019.
[60] Tir and Stinnett 2012; De Stefano et al. 2017. [61] Krampe et al. 2020.

transfers, and disaster risk reduction strategies like better building codes.[62]

In terms of internal conflicts that could potentially be affected by climate change, we know less about what works.[63] As Dellmuth et al. note, "While IGOs [intergovernmental organizations] are increasingly important, little is known about the conditions under which they address climate security challenges, and when they do so effectively."[64] While foresight and horizon-scanning studies of vulnerability are useful, the question is how they should inform subsequent action.[65] A study for Clingendael summarized the problem for practitioners: "Yes, but so what? What is the ministry for development, or foreign affairs, or defence, or environment supposed to do about it?"[66]

Environmental peacebuilding, which overlaps with the climate security field, may be one place where lessons can be learned. Key insights focus on reducing competition between groups over resources, better natural resource management, and enhanced dispute resolution.[67] Other ideas that have been mentioned are for climate-related adaptation/development projects to anticipate potential distributional consequences that might lead to conflict. On security, the guidance has emphasized that climate impacts can upend conflict resolution if peace plans are premised on resource availability or livelihood possibilities that may no longer be tenable. Mainstreaming a climate security perspective in existing development and security programming can help practitioners more accurately diagnose drivers of conflict such as farmer–herder violence.[68] Studies of environmental peacebuilding focus on trust-building and elevating the cooperative possibilities of managing and sharing natural resources across borders or within countries.[69] Support for mediation and traditional conflict resolution mechanisms and promoting inclusive development and sustainable livelihoods are among the recommended strategies to rebuild trust.[70] In seeking to foster cooperation over resources, some environmental peacebuilding approaches emphasize narrow technical cooperation with the hope these activities will foster cooperation on a wider range of issues, though this apolitical approach has its detractors.[71]

---

[62] Cash et al. 2013; de Waal 2018; United Nations Office for Disaster Risk Reduction 2019.
[63] Busby 2018b.   [64] Dellmuth et al. 2017.   [65] Moran et al. 2018.
[66] Smith et al. 2019.   [67] UNEP 2012.   [68] Shaik et al. 2019; Day and Caus 2020.
[69] Swain and Öjendal 2018, 8.
[70] Mobjörk and van Baalen 2016; Krampe 2019; Vivekananda et al. 2019; Mosello and Rüttinger 2020.
[71] Krampe 2017; Ide 2020.

The climate security academic community has started to identify known risk factors for conflict, which include recent conflict, weak state capacity, high agricultural dependence, and high political exclusion.[72] The geography of at-risk countries generally encompasses the "shatter belt" of fragile states that extends from the Sahel across North Africa to the Middle East and North Africa (MENA) region through Turkey to the Philippines.[73] However, as contemporary experiences in Afghanistan, Iraq, and Somalia demonstrate, the international community's ability to build capacity and more inclusive political institutions in fragile states is fraught with difficulties. Some countries may be ripe for change only when an emergent domestic elite is interested in inclusive development, as occurred in Ethiopia in the early 1990s.

There is a better record of targeted international support for state capacity for specific purposes, such as Ethiopia, which built an impressive food security system, and Bangladesh, which invested in cyclone early warning systems and shelters.[74] These developments required local actors with vision and the political space to take advantage of international assistance, a challenging situation in countries with more fragile governance such as Somalia and the countries surrounding Lake Chad.

CLIMATE SECURITY AND STATE DEVELOPMENT

The emergent field of climate and security has gathered significant interest in a little more than a decade. Think tanks and scholars have succeeded in making policy audiences aware of the potential security impacts of climate change, as the first section on challenges demonstrated. Despite some fledging efforts to address climate security concerns, what policymakers are supposed to do with awareness of the risks is less clear, particularly for fragile states.

A focus on fragility has emerged in both the academic and the policy discourse over the last decade or more, an evolution away from the totalizing language of "failed states" in previous years.[75] Indeed, the Fragile States Index that seeks to measure state fragility was previously called the Failed States Index and changed its name only in 2014.[76] Beyond debates about measurement issues and nomenclature,

---

[72] Theisen, Holtermann, and Buhaug 2012; von Uexkull et al. 2016; Busby and von Uexkull 2018.
[73] Buhaug, Gleditsch, and Theisen 2008. [74] Cash et al. 2013; de Waal 2018.
[75] Grono 2010; Newman 2009. [76] Hendry 2014.

international audiences have expressed concern about how the problems of weak/fragile/failing states pose wider regional and potentially global challenges through contagion effects of instability and population movements.

Addressing climate security challenges in these places requires a theory of state development and revisiting some of the themes discussed in Chapter 2. The literature on fragile states has recognized that the classic Weberian state – that has reach over its entire territory, is autonomous from society, and has bureaucratic capacity – may be partly, or completely, lacking in fragile contexts.[77] Scholars such as Alex de Waal have begun to theorize state development in such places as a "patrimonial marketplace," temporary elite compacts of convenience brought about by patronage.[78]

Mancur Olson likened the autocratic state that can extract taxes from its citizens to a "stationary bandit." In stateless, anarchical societies where there is no centralized power, "roving bandits" engage in what he describes as unconstrained "competitive theft" that leaves people insecure and unable – and disincentivized – to invest in long-term development. Unlike roving bandits, Olson argues persuasively, an autocrat, presuming s/he is not immediately threatened with being overthrown in the short-term, will have some concern for the development of the country in the long run, taxing citizens up to a certain point but not so much that they are disincentivized to work and invest. Presaging later work on selectorate theory, Olson argues that democracies, given wider representation of societal interests, tend to allow citizens to retain a larger share of the surplus and will be attentive to and provide more public goods than autocracies.[79]

Fragile states, at some level, do resemble roving banditry or autocracies with insecure tenure, so that there is limited long-term security of investments. Elites either compete to extract resources from people at the local level, or compete for control of the state so that they can use its powers to extract as much as possible while they can. The interesting question that emerges out of this inquiry is how to transition from roving banditry or autocracies with insecure tenure to developmentally oriented autocracies or, better still, democracies. Evocative of Olson's arguments, Charles Tilly described how war and the need for protection helped galvanize projects of state-building in Europe, with wars that ultimately established

---

[77] Giraudy 2012.   [78] de Waal 2009, 102.
[79] Olson 1993. For similar reflections on political order, see Bates 2015.

constraints on rulers and their subjects.[80] In postcolonial settings, Tilly notes, states came into being without those constraints on militaries. Inspired by Tilly's discussion of the symbiotic relationships between war and capital accumulation scholars have assessed, for example, the fraught process of armed militia groups transitioning from "warlord to landlord" in the Democratic Republic of Congo, with groups' private protection over slivers of territory facilitated by their connections to international markets.[81] Avoiding perennial competition between groups for control of either the state or the territory in favor of long-term development continues to elude some countries.

As described in Chapter 4, Ethiopia appeared to resolve this tension in the early 1990s with the Ethiopian People's Revolutionary Democratic Front (EPRDF) coming to power and governing as an inclusive authoritarian government. Ethiopia's searing civil war of the 1980s seemed to produce a cohort of leaders that exhibited an interest in the country's development. Similar developmentally oriented but undemocratic regimes in Uganda and Rwanda came to power after cataclysmic events: Yoweri Museveni in Uganda in 1986 after Idi Amin's wretched rule and Milton Obote's short-lived tenure, and Paul Kagame after the 1994 Rwandan genocide. All three countries – Ethiopia, Uganda, and Rwanda – became favorites of the international community for foreign assistance.

Although Ethiopia continued to grow rapidly, EPRDF control became more tenuous after the 2012 death of its dynamic prime minister, Menes Zelawi. While it took some years, the Oromo people chafed at the minority Tigrayans' control of the levers of government, ultimately leading to the appointment of an Oromo prime minister, Abiy Ahmed, in 2018. While Ahmed's tenure was initially greeted with much fanfare internationally – Ahmed won the Nobel Peace Prize for fostering peace with neighboring Eritrea – the country descended into sectarian violence in 2020 when he sent the military to repress what he perceived to be a Tigrayan insurrection. At the time of writing, the country is at risk of descending into civil war again, with large-scale refugee movements across borders into neighboring countries including Sudan, which itself experienced regime change in 2019 after popular protest brought down its longtime leader, Omar al-Bashir.[82]

The long-lived authoritarian governments of Museveni and Kagame in Uganda and Rwanda, respectively, have also come in for criticism. In 2020, Museveni faced an electoral challenge from musician Bobi Wine

---

[80] Tilly 1985. [81] Raeymaekers 2013. [82] Reuters 2020.

before the 2021 elections, and his government used the pretext of COVID-19 restrictions to arrest Wine, repressing him and his followers much as it had done on a number of occasions in previous years as Wine emerged as a popular opposition figure.[83] In Rwanda's case, the government's "benevolent dictatorship" has been accused of extraterritorial abductions and killings.[84] These are three relatively successful cases of development in East and Central Africa which contrast with the failure of state-building in Somalia described in Chapter 4, but they represent what scholars describe as projects of illiberal state-building.[85]

These episodes underscore the challenges for external actors in being able to help consolidate state capacity and inclusive governance, even in nominally well-performing states. These concerns are not lost on international observers, as reflected by the quote that opened this chapter from the joint United Nations–World Bank report *Pathways for Peace* (2018). That report underscored the argument of this book, concluding: "Exclusion from access to power, opportunity, services, and security creates fertile ground for mobilizing group grievances to violence, especially in areas with weak state capacity or legitimacy or in the context of human rights abuses."[86] The notion of pathways echoes the causal diagrams in Chapter 3. If climate disruptions conjoin with low state capacity and political exclusion to lead to negative security outcomes, what steps can be taken to interrupt negative pathways given those structural vulnerabilities?

Most of the recommended actions focus on prevention with early warning systems and targeted investments in areas thought to be vulnerable to sources of potential grievance, such as border and peripheral areas.[87] Given concerns about political exclusion, the recommendations focus on participatory processes of stakeholder involvement. Such approaches, however, require local interlocutors who are willing to address rather than reinforce societal cleavages. It is difficult to imagine imposed solutions being all that successful. Under such circumstances, as Andrew Radin argues in his book on weak states, international actors are most likely to be successful when they press for changes that do not threaten the core goals of domestic elites.[88]

In terms of crisis moments and shocks, the United Nations–World Bank report calls for managing shocks, early action to address and

---

[83] Dahir 2020; Smith 2020b. [84] Muhumuza 2019.
[85] Jones, de Oliveira, and Verhoeven 2013.
[86] United Nations–World Bank Group 2018, iv. [87] Ibid., xi–xii. [88] Radin 2020, 9.

mediate grievances, and engaging with actors beyond the state to help with peacebuilding and conflict resolution.[89] This last recommendation underscores a point made in the literature on hybrid governance. While the emphasis throughout this book has been state-centric, traditional and subnational authorities and governance systems often exist alongside state institutions in fragile states. Such authorities are often vestiges from precolonial periods that survived despite disruptions from the imposition of new institutions by imperial powers. From this perspective, there is more political order than at first appears by looking solely at state authorities.[90] Such institutional pluralism can be useful for building wider political stability and adjudicating disputes, but it can also be predatory on the lives and livelihoods of citizens.

In his work on Somalia, Ken Menkhaus noted the possibilities of governance without government – informal mechanisms of authority driven by coalitions of business groups, civic associations, and traditional authorities.[91] In the Lebanese context, scholars point to the role of Hezbollah as a rival to state authority in parts of the country, as it provides services for publics, creating an alternative locus of loyalty for citizens and refugee populations alike.[92] While service provision may have positive connotations, other forms of nonstate governance do not. Will Reno observes a so-called "shadow-state," of elite actors who use state assets to create "bureaucratic agencies based on personal ties" outside formal state governance institutions.[93] The "shadow state" may be empowered by external actors, either investors or buyers of resources who confer legitimacy on or support/prop up leaders who plunder the state for their own ends.[94]

These observations underscore the point that local actors, whether they be prospective state-builders or those merely seeking to consolidate local fiefdoms, often rely on external actors for recognition and resources.[95] External actors thus have to be mindful of how their actions potentially empower some groups at the expense of others and magnify disputes. As the cases in the book have shown, there is also a tension between keeping people alive and not rewarding undesirable regimes, groups, or leaders. The international community, and the United States in particular, has

---

[89] United Nations–World Bank Group 2018, xii.
[90] Boege, Brown, and Clements 2009; Meagher 2012; Boege 2019; Naseemullah 2014; Luckham and Kirk 2013; Giraudy 2012; Wunsch 2000.
[91] Menkhaus 2007.   [92] Flanigan and Abdel-Samad 2009.   [93] Reno 2000.
[94] Funke and Solomon 2002.   [95] Sellwood 2011.

periodically slowed or elected not to provide humanitarian and development aid to the Derg government in Ethiopia, Somalia at the height of Al-Shabaab's influence, and the Assad regime in the midst of the drought in Syria. Would different choices potentially have saved lives over the long run?

Alex de Waal identifies two strategies for peacebuilding in weak states that are characterized by patrimonial systems of local allegiances, what he calls "buy-in" and "equilibrium." He sees stability as wrought by elite compacts between local actors so that they can collectively profit more through peaceful resolution of differences rather than violence. In buy-in scenarios, the most powerful actor, typically the national government, provides all the important constituencies with a share of resources to purchase their loyalty. In equilibrium scenarios, local groups control resources and are powerful enough to deter each other. He suggests that where the state continues to be a rich target for distributing sovereign rents, the equilibrium will not be a stable solution without an external actor powerful enough to enforce the peace.[96] In both scenarios, international actors have the potential to upset the balance between different groups by providing resources, whether it be foreign aid, investment, or remittances. Peacebuilding projects might be able to sustain the peace for as long as they remain, but no longer, and should thus be designed to survive external actors' exit: "It follows that a successful international peace engagement will be one that supports the most inclusive and robust buy-in – one that is sufficiently well grounded in the relative value of the parties to survive the withdrawal of its international sponsors."[97]

Concerns about state fragility and the role of external actors thus loom large for United Nations agencies, the World Bank, and bilateral development and security agencies. In early 2020, just as the COVID-19 outbreak took hold, the World Bank was recruiting 100 field advisers for its Fragility, Conflict, and Violence program.[98] In 2019, the US Congress passed the Global Fragility Act which included long-term funding – $200 million a year over five years for a Prevention and Stabilization Fund and $30 million a year for a Complex Crisis Fund. The act was intended to bring together diplomacy, development, and defense in a coherent, multisectoral strategy to end and ward off extremism and violent conflict. The State Department, the United States Agency for

---

[96] On a similar theme of self-enforcing power-sharing agreements in weak states, see Roessler and Ohls 2018.
[97] De Waal 2009, 102.   [98] World Bank 2020.

International Development (USAID), and the Department of Defense were tasked with writing an implementation strategy and identifying focus countries. Although the COVID-19 outbreak slowed down implementation of the act, a strategy document was finally released in December 2020, emphasizing local ownership of solutions, inclusive governance, and integration of US development, diplomacy, and defense instruments.[99] The strategy document echoed many of the themes in this book on state capacity and political inclusion: "The United States will also incorporate peacebuilding approaches to address the drivers of conflict, violence, and instability, such as, *inter alia*, exclusionary politics, entrenched corruption, impunity, or capacity deficits."[100] While it remains to be seen how this act will ultimately be implemented and whether climate security issues will be integrated, the document underscores that concerns about state development and fragility are very much live considerations for practitioners.

CONCLUDING THOUGHTS

For international actors like the United Nations, the World Bank, and the US government, the insights on fragile states need to be brought into the practice of climate security. What would a viable path forward be for a country like Somalia or, say, the conflict in the Lake Chad Basin? What is the sequence of reform? Must the conflict end so the environmental challenges can be dealt with, or is dealing with environmental problems a route to resolving the conflict? Here, we do not have a robust evidence base to draw on from past practice. These are important research questions that would be of great practical value but for which there are not obvious answers.

Already, the emergent operational literature on climate and conflict from USAID and others has underscored the risks of project-level distributional conflicts and incorporated a do-no-harm ethos and a need to anticipate how different groups will fare.[101] However, the task going forward is to develop a more robust evidence base of what works, both in the resolution of existing conflicts and the prevention of new ones and in other negative security outcomes such as humanitarian emergencies.[102]

---

[99] Welsh 2020a, 2020b.   [100] US Department of State 2020, 7.
[101] USAID 2014; Reiling and Brady 2015.
[102] Johnson, Rodríguez, and Quijano Hoyos 2021.

Finally, this discussion begs the question of what to do about the existential climate risks facing some countries and populations. Accepting that some countries need to relocate raises thorny questions: Would relocated countries reestablish sovereignty inside the domain of another country? Would they lose access to resource-rich exclusive economic zones if they abandon their territorial holdings?[103] Before we necessarily accept that some places will become unlivable without massive investment, more in-depth risk assessments of whether and at what cost places could remain viable are warranted. Embracing managed retreat may undermine both local and international efforts to support local adaptation, which is preferable.[104] In the meantime, scholars and practitioners of climate security should create more opportunities to exchange information and ideas, lest academic findings become divorced from what practitioners care about and policymakers support actions researchers believe have little chance of success.[105]

[103] Douglass 2017; Busby 2018a. [104] Barnett 2017. [105] Busby 2018b.

8

The Next Decade of Climate Security Research

> Academic research on climate security also favors forensic analysis – case studies or information on past events, rather than future scenarios that social science methods cannot reliably test.
> — *Caitlin Werrell and Francesco Femia, Center for Strategic Risks*[1]

Academic work in the climate security space is primarily geared toward retrospective explanation or what Werrell and Femia call "forensic analysis." What explains certain patterns or why certain events occurred? What precipitated the Syrian civil war? For a field that purports to be policy relevant, that is helpful to decision-makers, a focus on the past is only useful insofar as it generates lessons that can be learned and applied to help prevent and deal with future negative security outcomes. As this chapter demonstrates, the climate security field and the social sciences writ large have deep skepticism about the methodological rigor associated with talking about the future.

In the last chapter, I examined the suite of concerns that policymakers care about with respect to climate and security, what is being done already, and how the policy agenda going forward could and should productively address this challenge. In this penultimate chapter, I explore the future of academic inquiry concerning climate and security and how the field of international relations ought to change accordingly. I explore several areas where the climate security field will need to develop new approaches and insights, mostly related to the challenges of writing about the future. In the first section, I discuss how academics can be useful

---

[1] Werrell and Femia 2019.

to policy and the need to move beyond the simple phrase of "thread multiplier." In the second section, I look at the significance of the end of "stationarity" and what that means for scholarship going forward. In the third section, I raise the issue of what baselines we use for identifying normal climatic conditions and how far back we can go to identify deep structural drivers. In the fourth section, I examine a related issue. Scholars need to explore more fully the links between human security and state security. While ongoing violence often contributes to human security risks, climate impacts that in the short run lead to large scale loss of life may, in the long run, degrade the economic growth potential of countries, contributing to conflict potential. In the short run, human security impacts may escalate to state security threats, particularly in anocracies that lack either consolidated democratic systems or fully authoritarian ones. In the fifth section, I argue that runaway climate change would make all of these security challenges worse; hence, mitigation – that is, efforts to reduce greenhouse gas emissions – needs to be considered a security concern in its own right. Finally, I suggest that the broader field of political science needs to elevate climate change to become a systemic structural factor like anarchy and think about what this means for the discipline, and for the world.

HOW DO WE MAKE OURSELVES USEFUL?

While policy relevance is not the sole or even the main motivation for many scholars, a number of us have entered the field based on some normative aspirations to make the world better.[2] In so doing, we want to be useful to practitioners, and practitioners often have concrete concerns about particular places. One challenge for climate security scholars, particularly those outside area studies fields, is to be able to say what our findings mean for specific geographies and locations.

In Chapter 2, I critiqued the existing literature on climate security for insufficient attention to causal mechanisms. I argued that the field needed to explore the indirect pathways by which climate and security outcomes are connected, and the contextual factors like political institutions that shape why climate shocks lead to negative security outcomes in some places and not others. I also drew attention to the need to move beyond internal armed conflict as the primary outcome measure of interest. This is

---

[2] Some scholars have deliberately shied away from policy-relevant work. See Kalyvas and Straus 2020.

consistent with critiques other scholars have made in recent years, and the field has started to respond with greater attention to indirect pathways such as migration and the effects on agriculture and economic growth and how these might lead to violence.[3] A number of scholars have explored how institutions and political exclusion shape the context for security outcomes.[4] Scholars have also expanded the set of security outcomes of concern from civil wars to communal conflict to riots, strikes, and demonstrations.[5]

In the last chapter, I suggested we still know little about what works in peacebuilding and what role outside actors can play in influencing the trajectory of countries with highly exclusionary political institutions. While state capacity is arguably somewhat easier to address through targeted interventions, the challenges of state-building and strengthening political inclusion are difficult ones for foreign actors to influence. Understanding the scope for external interventions to enhance state capacity and political inclusion is both a challenge for policy as well as a subject for further academic inquiry.

So too is the need for more work that extends the geographic coverage of climate security research. Much of the climate security scholarship to date has focused on sub-Saharan Africa, which in part reflects a concern about the continent thought most vulnerable to climate change. But, with a number of studies carried out in English-speaking African countries like Kenya, the emphasis also may reflect the "streetlight effect," the relative ease with which scholars can carry out research.[6] We still know little about climate security dynamics in Portuguese-speaking countries in Africa or other countries with more linguistic diversity. While climate security scholarship on Asia has increased in recent years, the region has been understudied in this space despite its vulnerability and the sheer numbers of people who live there.[7] Moreover, although the Syrian conflict has received considerable attention, the Middle East as a whole is chronically understudied, perhaps in part because climate science research on the region is not extensive.[8] In the Intergovernmental Panel on Climate Change (IPCC) Fifth Assessment regional reports, there is no stand-alone report on the Middle East, though there are reports for most regions

---

[3] von Uexkull and Buhaug 2021; Koubi 2019; Theisen 2017; Meierding 2013.
[4] von Uexkull et al. 2016; Ide, Kristensen, and Bartusevičius 2021; Theisen, Holtermann, and Buhaug 2012; Ide et al. 2020.
[5] Smith 2014; Hendrix and Haggard 2015.   [6] Hendrix 2017c; Adams et al. 2018.
[7] Busby and Krishnan 2015.   [8] Bucchignani et al. 2018.

How Do We Make Ourselves Useful?

including Asia, Europe, small islands, and one for polar regions.[9] Even Oceania, the region of low-lying island polities in the Pacific that faces existential risks from climate change, has limited scholarship on climate–security links, perhaps a function of distance and the difficulty of studying a fragmented region of small countries whose blue water reach is as large as the continental United States.[10] These are known lacunae in the field of climate and security.

The language of "threat multiplier" is a way in which the policy community deals with causal complexity but avoids the charge of environmental determinism. Sherri Goodman, through the 2007 CNA Military Advisory Board (MAB) report on climate security, introduced this formulation and, at the time, it served the purpose of recognizing that there is a link between climate change and security without overstating its importance. By stating that climate is a "threat multiplier," the speaker can assert that climate plays a role in making negative security outcomes more likely, but that it is not the sole driver and that its influence happens in concert with other factors.

Unfortunately, the language of threat multiplier does not say what specific combination of factors we should be worried about. If we cannot say more than bad things go together, what can we say about what to do to diminish the negative security consequences of climate change?[11] Here is where scholars can be useful to practitioners by surfacing the most relevant factors thought to combine with climate hazards and contribute to negative security outcomes. An enduring finding from the conflict studies literature is that societies with a history of conflict are the most likely to experience its reoccurrence.[12] As I argued in Chapter 2, emergent research on climate and security has shown that the risk of conflict is greatest in agriculturally dependent societies that exhibit high political exclusion, in other words places where some groups have no representation in government.[13]

As a thought experiment, Nina von Uexkull and I wrote a back-of-the-envelope risk analysis of countries that had a recent history of conflict, where a large percentage of their workforce were employed in agriculture, and that exhibited high levels of political exclusion. We coupled that analysis with data on countries that recently had, or were projected to have, water deficits in the near future. We found that twenty countries in the world – many of them in the Sahel – are perennially at risk of conflict

---

[9] IPCC 2014a. [10] Ray et al. 2020. [11] Busby 2020.
[12] Gates, Nygård, and Trappeniers 2016; Mach et al. 2019.
[13] von Uexkull et al. 2016; Ide et al. 2020; Theisen, Holtermann, and Buhaug 2012.

and/or humanitarian emergencies based on their conflict history and agricultural dependence. Of these twenty countries, eleven faced severe short-term water deficits in late 2018 and early 2019.[14]

While this is a crude measure of risk factors, we think it offers an instructive path forward. By identifying countries most at risk of conflict and/or humanitarian emergencies and assessing their vulnerabilities, context-specific interventions that diminish risks – including food or income support as well as longer-term strategies to diversify income and enhance political inclusion – could be identified. This framework could also serve as an early warning system to identify places chronically at risk due to structural factors and emergent risk factors based on water deficits or other physical hazards.

The combination of drivers we used to analyze risk is one plausible pathway to negative outcomes. Other ideas emerge from climate security research, including the expert elicitation work led by Katharine Mach, which identified low socioeconomic development, diminished state capacity, and intergroup inequality as important drivers of conflict.[15] Indeed, both state capacity and intergroup inequality parallel my argument. A productive approach going forward would be to surface those common risk factors alongside hazard-specific risk factors such as whether a country has an early warning system, adequate numbers of cyclone shelters, pre-positioned relief supplies, adequate building codes, and so on.

Hazard-specific factors offer obvious entry points for policy. If a country does not have an early warning system but is vulnerable to certain hazards, it should develop one. And, if the current system is inadequate, it can be improved with greater precision and foresight. As I argued in the last chapter, we still do not have consolidated wisdom about the points of intervention to disrupt and prevent climate-related conflicts or end them. Obviously, academics who can offer advice on how best to prevent or end farmer–herder conflicts, for example, will likely be in demand from policymakers struggling with these conflicts from the Sahel to the Horn of Africa.

Scholars can go further in proving their usefulness to policy audiences. Risk or vulnerability maps that identify hot spots of concern are helpful points of departure to orient decision-makers about where to focus their energies. But, once places of concern are identified, policy audiences need

---

[14] Busby and von Uexkull 2018. I developed the idea for this approach in Busby 2018b.
[15] Mach et al. 2019.

more granular detail on the geography, ecology, political economy, and security dynamics of particular places. Maps of locations thought vulnerable to climate security effects do not speak for themselves. They require country-specific knowledge and narratives of how the various risk factors are plausibly connected to each other. Academics can offer policy audiences both a theory for why a problem has occurred (or is likely to) and also what instruments might conceivably prevent or ameliorate the problem. Even as scholars seek to prove their relevance to policy, their ability to think through how climate change might depart from past experience will loom increasingly large.

THE FUTURE IS NOT LIKE THE PAST: THE END OF STATIONARITY

This book has explored the conditions under which extreme weather events – with an emphasis on cyclones and droughts – have contributed historically to negative security outcomes. Those extreme weather events are consistent with what we expect to see from anthropogenic climate change but may not themselves be attributable to climate change. Even though the science of attribution studies is getting better, we may never know with certainty if, for example, droughts in the Horn of Africa in the 2010s were made more likely or more extreme as a result of climate change. Here, I want to explore a different issue.

As others have written, much of the emphasis in the field has focused on proxies for climate change that mostly correspond to short-term shocks or deviations in rainfall or temperature. However, this emphasis on the short run is different from long-term changes that we associate with climate change.[16] These changes will unfold over decades and shape the baseline conditions for societies' agriculture, economic development, and even their habitability. We have only just begun to experience these effects.

Moreover, the challenge for security studies scholars is that future climate change may deviate in terms of temperature and rainfall from the normal bounds of what people have been accustomed to for thousands of years. The geography of places affected by weather extremes may also be quite different from previous patterns. Parts of the world may become uninhabitable or barely habitable without major adaptation efforts to insulate and protect people from high temperatures, flooding, and other climatic extremes.

---

[16] von Uexkull and Buhaug 2021; Theisen 2017.

As Jay Gulledge and I have argued in previous work, we are facing the problem of what scientists call the end to "stationarity."[17] The Holocene, the current geologic age which began some 12,000 years ago, has been characterized by "stationarity" or relatively stable global climatic patterns in response to which human societies adapted and evolved, moving from nomadic lifestyles to sedentary agriculture and, ultimately, more urban living that was made possible by relatively predictable environmental conditions. As we wrote, "Stationarity assumes that the range of climate conditions for a given area occurs within a static envelope of variability defined by past extremes. However, climate change means that future climate averages and extremes will differ from those in the past."[18] Under such circumstances, examining the near-term historical record – which is most like the present in terms of social systems – may tell us little about the magnitude or frequency of the kinds of physical effects of climate change we will likely experience. We are likely to see both a change in the means but also greater variance in temperature, rainfall, and other climate parameters, leading to fatter tail effects and more extreme weather.[19]

This makes the kinds of historical analysis carried out by social scientists quite difficult, since the past may tell us less and less about the future. We can mine the deep historical record to see how human societies hundreds or thousands of years ago dealt with such cataclysmic changes in climatic conditions, as some scholars have done. But can we really learn lessons from our ancestors who had such different technological options available to them and who experienced such different social systems?[20]

While there are no easy answers to these questions, some approaches may be more defensible than others. In his trenchant critique of the first generation of environmental security literature, Nils Petter Gleditsch criticized scholars who used the future as evidence of their claims as potentially "slipping into prophesy" and warned that saying: "'There will be water wars in the future' is no more a testable statement than the proverbial 'The End of the World is at Hand,' unless terms such as 'the future' and 'at hand' are clearly specified."[21] However, to be useful, climate security scholars will need to get more comfortable both writing about the future and developing defensible methods. We risk irrelevance

---

[17] Milly 2008; Milly et al. 2015. [18] Busby et al. 2012, 7–8. [19] IPCC 2001.
[20] Diamond 2004; Hsiang, Burke, and Miguel 2013; Zhang et al. 2006.
[21] Gleditsch 1998, 394.

as a field if we say "we don't do windows" – that is, engage in work that talks about a phenomenon whose effects we have yet to (fully) observe.

And, even then, efforts to identify the future with more precision may convey a sense of accuracy that is hard to defend. One 2009 study on climate change and civil wars in Africa projected the number of likely future wars and battle deaths based on extrapolations of past conflict mortality. However, leaving aside whether this move was defensible, the authors had to make assumptions about the future of economic growth and democratization through 2030. While they admirably included two scenarios, one based on extending past trends and another, more optimistic, high economic growth scenario, there is something unsettling and unsatisfying about seemingly precise projections of a 54 percent increase in armed conflict incidence by 2030 and 393,000 battle deaths attributable to rising temperatures.[22]

My concerns may have more to do with the way media seize on such claims for headlines than the research itself, but scholars may have to be more careful in making explicit the assumptions behind such claims (and exercise more oversight of communications professionals writing press releases about their research). They need to convey that such estimates merely give some notional approximation of magnitudes. Importantly, even if we believe scholars have done a good job modeling past relationships, estimates or projections of future change are not destiny. Human agency can alter those patterns and, as recent studies of conflict have shown, attempts to use past patterns of instability and violence to make out of sample predictions to different time periods often perform poorly. The causal associations that led to civil wars in the mid-1990s might be very different from those that contributed to conflict around the time of the Arab Spring. That suggests that predictive science on social systems may be especially fraught. There are methods to try to understand when the drivers of conflict might be changing in different time periods, but looking ahead, we can only surmise what might be important in shaping conflict and instability in the future.[23]

Other studies have engaged in similar analysis examining possible conflict trajectories under different scenarios of economic growth, leaving aside the climate–conflict connections.[24] Analysts have also deployed scenario analysis to explore the potential outcomes of high consequence events but for climate scenarios that are highly unlikely or for which we cannot begin to ascribe probabilities. In a highly publicized analysis for

[22] Burke et al. 2009.   [23] Bowlsby et al. 2020.   [24] Hegre et al. 2016.

the US Department of Defense, futurists Peter Schwartz and Doug Randall explored the potential consequences of an abrupt shutdown in the Gulf Stream that, in their scenario, leads to large-scale cooling over parts of Europe and North America (a shutdown of the Gulf Stream was also the inspiration for a Hollywood disaster movie of the same year, *The Day After Tomorrow*).[25] This particular scenario is highly uncertain, not thought likely, and even if aspects of it were to occur, the process would likely unfold over a much more drawn-out timescale.[26] It is also unclear whether the exercise of thinking through the impacts of global cooling on North America is all that fruitful, given that the almost certain reality in coming decades is the opposite – accelerating warming.

This illustration raises a question about what makes for good work in scenario building.[27] It is certainly reasonable to explore different possible futures with respect to greenhouse gas emissions, ranging from optimistic scenarios where there is high climate mitigation to more pessimistic scenarios where greenhouse gas emissions rise mostly unabated, extending current trajectories. Analysts have carried out that kind of work to explore the effects of sea-level rise on military bases, energy infrastructure, and populations. Even these efforts to overlay different scenarios of sea-level rise and flooding on infrastructure and current populations are potentially problematic given the challenges of developing high-resolution digital elevation models that are globally accurate.[28]

Projections of future climate change have also come in for considerable criticism as the worst-case scenario trajectory – the so-called Representative Concentration Pathway (RCP) 8.5 that was used in the IPCC's Fifth Assessment Report – is frequently misreported as the current business-as-usual case, even though many analysts regard this particular scenario to be above even the current high trajectory of emissions growth.[29] By confusing the worst-case scenario for the status quo or baseline scenario, we may inadvertently reinforce a message that the problem is beyond repair and hopeless, when that is not true.

Needless to say, if we are challenged in being able to make and use scenarios to inform estimates of the physical basis of climate change impacts, then we are even more challenged in trying to extrapolate or

---

[25] Schwartz and Randall 2004.   [26] Woodward 2019.
[27] For a sophisticated exploration of these issues in the climate security and disasters space, see Briggs and Matejova 2019.
[28] Military Expert Panel Members 2018; US Department of Energy 2016, 2015; Kulp and Strauss 2019.
[29] Hausfather and Peters 2020.

guess what impacts those physical changes will have on livelihoods and, in turn, on social and political systems, including the security outcomes discussed in this book. Here, the way forward is not especially clear. For example, what effect do we think a long-term drying or heating trend will have on conflict and other security outcomes? In the last chapter, I referenced studies on human habitability that suggest real challenges in some areas to continue to grow food and for people to work outdoors. The implications of those changes for political and social systems are immense. Stress-testing and war-gaming existing social systems for possible effects would be one approach, but we will likely be surprised how human beings respond to these developments as they unfold.[30]

ON IDENTIFYING BASELINES AND DEEP STRUCTURAL DRIVERS

This reflection on the end of stationarity raises another issue. In terms of retrospective evaluations, much of the work on climate and security has, as I have suggested, focused on triggers for conflict – that is, relatively short-term deviations in rainfall or temperature – as opposed to long-term secular trends. Assessing the causal role of secular trends on later conflict onset or incidence can be particularly challenging. If a drought occurred twenty or thirty years ago, can it possibly be a contributing factor to some contemporary conflict or security outcome? As von Uexkull and Buhaug argue in a review essay on ten years of climate security research: "Attributing observed 'high-frequency' social behavior, such as conflict outbreak, to inert phenomena, such as changes in the physical environment (warming, drying, sea-level rise, etc.), is inherently challenging, and empirical responses to this research priority to date are limited."[31]

Hendrix underscored these concerns in his discussion of "shifting baselines" in our understanding of what constitutes the normal climatic conditions in particular geographic areas. He noted "we know very little about how baseline (i.e., normal) climatic conditions affect conflict, both theoretically and empirically."[32] If we believe that the effects of climatic conditions are mediated by institutions, there are potentially large indirect effects of long-term climatic conditions on security outcomes through economic development and state capacity. Hendrix references studies

---

[30] Committee on Assessing the Impact of Climate Change on Social and Political Stresses et al. 2013.
[31] von Uexkull and Buhaug 2021.   [32] Hendrix 2020.

that found the disease burden at the outset of the colonial era had major effects on what kinds of political institutions imperial powers established in Africa, with large legacy effects on later patterns of political and economic development.[33]

In work on the effects of malaria on economic growth in sub-Saharan Africa, Gallup and Sachs found that per capita income growth in countries with a high incidence of malaria was five times lower than those with low rates of malaria.[34] If we think that low economic growth is one of the best predictors of conflict,[35] then perhaps climatic conditions can diminish the long-term economic prospects of countries, making them more prone to violent conflict and political instability. While statistical methods may be able to account for the effects of long temporal lags, disentangling the long-run structural drivers that contribute to instability and violence in specific places in case studies remains a difficult task.

While some of these connections have been disputed, they raise important questions about whether and how long-term climatic conditions shape the country and regional context in important ways. There is a risk of reviving problematic early twentieth-century scholarship like that of Ellsworth Huntington that ascribed the differential economic success of different countries and regions to the effects of climate differences on people's work ethics.[36] Even Jared Diamond's more sophisticated take on why Europe moved further ahead compared with other regions can be accused of some form of environmental determinism based on disease burdens and capacity to grow food.[37]

Hendrix notes that the existing literature focuses on short-term shocks from normal conditions, but those normal conditions themselves change, making it difficult to define even what constitutes a shock. He relates the stories of two Issofous, prominent Nigeriens with the same surname born in different years, the president born in 1951 and a footballer born in 1978. For the elder Issofou, the Sahel was relatively wet for the first eighteen years of his life, but for the younger Issofou, the Sahel was comparatively dry during the footballer's formative years. Whose reference conditions constitute "normal"? As the climate changes in the future, what conditions should we use as the normal reference baseline?[38] How far back can changes in climatic conditions continue to influence contemporary conditions in ways that we think as causally important?

---

[33] Acemoglu, Johnson, and Robinson 2001. [34] Gallup and Sachs 2001.
[35] Koubi 2017; Collier and Hoeffler 2002. [36] McLeman and Gemenne 2018, 8.
[37] Diamond 1999; Sluyter 2003. [38] Hendrix 2020.

This issue was underscored in discussion of the civil conflict in Darfur, Sudan that began in 2003 where Janjaweed militia backed by the Sudanese government carried out attacks on local farmers. Ban Ki-moon, then United Nations Secretary General, drew on media accounts[39] and described the conflict as being rooted in climate change and a drought that occurred in the 1980s.[40] The scholar Alex de Waal, an expert on the Horn of Africa, contested the connection. Among the reasons he cited was that although the conflict appeared to follow the movement of camel nomads in the 1980s as they sought better pasture lands, "the initial southward migration occurred more than thirty years ago, but large-scale violence only erupted much more recently."[41] In a rejoinder, Thomas Homer-Dixon noted that "the variable of temporality (or what I refer to as 'proximity') is often conflated with the variable of causal power or weight."[42]

Although we often look for temporal contiguity as a sign of a causal connection and causal strength, deep long-lived structural drivers can influence contemporary events. In the context of Darfur, a team of scholars looked to see if there had been a significant rainfall decline before the conflict onset in 2003. They found that rainfall from 1971 to 2002 had been relatively flat preceding the conflict. They concluded that Darfur did experience a "structural break" with rainfall levels declining, but that had taken place more than thirty years earlier and thus would be hard to connect to contemporary conflict events. They noted that other states in the region such as Mali, Niger, Burkina Faso, and Chad had also experienced a similar decline in rainfall thirty years ago, and they raised the question why these places, at the time of writing in 2008, had not experienced internal conflicts of their own.[43] We know, however, that Mali subsequently experienced a coup in 2012, and farmer–herder violence is an ongoing issue throughout the Sahel. This is likely not the last word on the Darfur case, and the episode raises questions both about what baselines to use as representations of normal climatic conditions but also how far back we can or should we go to identify causal influences.

---

[39] Faris 2007.   [40] Ki-moon 2007.   [41] de Waal 2007c.   [42] Homer-Dixon 2007.
[43] Kevane and Gray 2008. Another study, using a different measure of vegetation health, the Normalized Difference Vegetation Index (NVDI), also found that Western and Northern Darfur did not experience a decline in vegetation growth in the preceding twenty-five years before the conflict. In fact, vegetation growth was better than average, though that could mean the region became a draw for other areas experiencing scarcity; Brown 2010.

THE RELATIONSHIP BETWEEN HUMAN SECURITY AND STATE SECURITY

When will human security outcomes escalate to violence rather than simply lead to suffering? As we saw in the Somalia case discussed in Chapter 4, a climate hazard that affects a country already experiencing civil conflict can have even worse human security outcomes, given the difficulties of delivering aid in the midst of war zones. A theme that has not yet gained sufficient research attention is when human security will lead to broader threats to state security. Low socioeconomic development was rated as the most important driver of conflict risk by the expert elicitation led by Katharine Mach.[44] As I noted in Chapter 2, the connections between disasters and economic growth have been explored by economists without definitive results. While hazard events often lead to swift economic recovery because of the need for rebuilding, severe and repeated exposure to climate-related hazards may not only pose a grave risk to human security but may also be a drag on the long-term economic growth potential of countries. This, in turn, can make conflict more likely.

There are more short-term pathways to violence in the days, months, and first years after hazard exposure, but understanding when those threats will escalate to violence and when they will merely lead to impacts on lives and livelihoods requires further study. Here, we may see variation by regime type. In their study of government reaction to food price shocks, Hendrix and Haggard found democracies and anocracies were more likely to experience protests than authoritarian governments. Publics are more likely to be afraid of repression from fully authoritarian regimes and thus be less likely to protest. In democracies, the people are comfortable expressing grievances because of freedoms of expression and assembly. In anocracies, governments possess insufficient coercive power to repress protesters.[45]

Thus, we should see protests in both democracies and anocracies in the wake of climate shocks, following not only food price hikes, but power outages, collapses in access to water and food entitlements, and other disruptions to well-being and livelihoods. We might not observe protest activity if hazards are so severe that people are in survival mode trying to find food, water, and shelter. In mature democracies, we would expect public expressions of protest to be unremarkable. They happen all the time and are rarely a threat to regime survival. Governments generally

---

[44] Mach et al. 2019; see also Collier and Hoeffler 2002.   [45] Hendrix and Haggard 2015.

would not react with repression, nor should the protests escalate to violence.

In anocracies, governments may fear the protests will escalate to challenge the regime, and protesters may be less bound by civic norms to refrain from violence. Moreover, anocracies, particularly those with exclusive institutions, may perform worse than mature democracies in both preparing for and responding to climate shocks. We may see more escalation from human security outcomes to conflict in such settings. This provisional line of argument, or a more sophisticated variation, would be worth exploring in further research. Such research would be useful, not least to surface additional reasons why climate hazard impacts sometimes escalate from loss of life and livelihoods to conflict and what steps can be taken to arrest that trajectory.

MITIGATION AS A SECURITY CONCERN

One topic that has not been discussed in much depth by either climate security scholars or practitioners is how climate mitigation, that is, reducing emissions from greenhouse gases, is itself a security concern. For example, the US policy community on climate and security fairly scrupulously avoided talking about mitigation for many years because it was considered a political subject. There was potentially bipartisan concern about the ways in which unusual weather could affect the US military. But it did not seem strategically wise to allow the climate security issue to become as politicized as the climate issue writ large.

In the early 2010s, the strategic choice to avoid talking about emissions mitigation might have been the right call. At the time, there was not a policy community of retired military, civilians in or out of government, or active-duty military who cared about this issue, either in the United States or elsewhere. Now, in the 2020s, a global community of practice exists. Thus, as a matter of policy, mitigation ought to be elevated as part of the discussion on security threats. Work by the academic community may be important in helping policymakers understand the potential linkages more fully.

If humanity does not address the challenge of reducing greenhouse gas emissions, all of the security consequences of climate change that the policy community and academics worry about – food and water shortages, electricity outages, humanitarian emergencies, base inundation, migration, conflict – will be orders of magnitude worse than they are now.

So, where does that leave us? At the 1992 Earth Summit, the international community drafted the Framework Convention on Climate Change which recognized the goal of stabilizing concentrations of greenhouse gases "at a level that would prevent dangerous anthropogenic (human induced) interference with the climate system."[46] Scientists have subsequently identified 2°C (3.6°F) as an approximation of how much global temperatures can increase above preindustrial levels and yet avoid "dangerous" outcomes. In other words, if global temperatures rise, on average, more than 2°C, scientists worry that unstable climatic dynamics and negative feedback loops may make the problem increasingly unmanageable.

For the last few hundred thousand years, global temperatures stayed within what the economist William Nordhaus described as a "normal range of long-term climatic variation." In a 1975 thought experiment, he sketched out that range to be +/−5°C and noted that the world, at the time, was at the upper end of the range. As a crude guess, he suggested that to stay within that range, global temperature averages could not increase by more than 2°C or 3°C above (then) current average temperatures.[47] In their 2001 summary of the science, the IPCC confirmed that most of the really worrisome outcomes for ecosystems and humanity were likely to occur when global temperature averages exceeded 2°C above preindustrial levels, a slightly different baseline but in keeping with Nordhaus' notional calculation.[48]

The United Nations officially recognized the 2°C threshold in 2010, but the debate did not end there. The 2°C threshold is somewhat arbitrary since the notion of what constitutes "dangerous climate change" is subjective. Scientific understandings of the nature and magnitude of climate change is evolving, and scientists have more confidence in some of the anticipated effects than others. As more scientific evidence has come to light, and as the negative effects of modest warming have already been observed, a number of scientific studies have sought to demonstrate that important negative consequences could be avoided if temperature increases could be kept to a lower threshold of 1.5°C (2.7°F).[49] Thus, while the 2015 Paris Climate Agreement officially reaffirmed the goal of keeping global average temperatures below 2°C, it also embraced an aspirational goal of 1.5°C.[50]

---

[46] UNFCCC n.d.   [47] Nordhaus 2019.   [48] Kaplan 2020; Titley 2017.   [49] IPCC 2018.
[50] Busby 2016b.

Global average temperatures are already more than 1°C (1.8°F) above preindustrial levels. And although the COVID-19 pandemic has temporarily reduced emissions around the world, pre-COVID estimates suggest global average temperatures may exceed the 1.5°C threshold by as early as 2034. It may be challenging to stay below the 2°C increase.[51] But, even if the 2°C threshold is exceeded, keeping temperatures from rising to 2.5°C (4.5°F) or 3°C (5.4°F) above preindustrial levels would still be better than a 4°C (7.2°F) or greater temperature increase, which is within the realm of possibility in high-emissions growth scenarios.[52] These are averages so some parts of the world like the Arctic have, and will continue to experience. more significant deviations in temperature. The key observation here is that every half degree that the world can shave off average temperatures will have an important impact in diminishing the consequences for natural systems, whether it be sea ice loss, drought, sea-level rise, coral reefs, heatwave events, and so on.[53]

To keep global average temperatures from rising by 2°C will require swift action to transition the global economy to net zero carbon emissions by the middle of the twenty-first century or shortly thereafter. That will require a steep reduction in emissions. Harvey estimated that, starting in 2018, emissions would have to fall by 10 percent a year to avoid dangerous climate change.[54] To put this in perspective, emissions during the COVID-19 pandemic were only expected to fall by 8 percent in 2020, amid a global economic shutdown and economic contraction.[55] That same level of reduction or more will be required on an annual basis for the next thirty to forty years without being accompanied by an economic contraction and massive job losses. The level of effort and policy coordination required to reduce emissions globally will be enormous.

What does the mitigation challenge mean for the academic study of climate and security? If the stakes of the mitigation challenge are as high as I claim them to be, then we should expect policymakers to increasingly elevate the importance of a clean energy transition in their foreign policies and grand strategies. Indeed, Joseph Biden was elected president of the United States in 2020, having identified climate change as one of the four critical challenges facing the country in his election campaign, with climate security concerns integral to his understanding of the threat. If more countries finally do prioritize climate change as central to their survival, then the stakes of climate diplomacy will take on greater significance.

[51] Busby 2019b.   [52] IPCC 2014c.   [53] Plumer and Popovich 2018.   [54] Harvey 2018.
[55] Ludden and Brady 2020.

Climate issues will then become bound up with wider issues of great power relations between the United States and China, the world's two largest emitters. According to the Global Carbon Project, the two countries are responsible for more than 40 percent of global emissions of carbon dioxide, the main greenhouse gas. In 2018, China's carbon dioxide emissions amounted to 28 percent of the global total and the United States accounted for 15 percent. The United States was still the world's largest emitter as recently as 2005 and remains the leading emitter of historical emissions dating back to 1870.[56]

In the early 2010s, relations between the United States and China deteriorated with increasing geostrategic competition and concerns about China's hegemonic ambitions and the United States' relative decline in power, particularly in economic terms. Much has been written about the potential for great power war when there are power transitions between a declining hegemon and a rising challenger.[57] Climate change is an area for cooperation between the United States and China because both are vulnerable to its effects, but climate risks get caught up in wider geostrategic rivalry that can impede cooperation and lead to cycles of blame for inaction. In 2009, at the Copenhagen climate negotiations, the United States sought to shift responsibility for the lack of diplomatic progress to China. China has sought to avoid that perception in subsequent negotiations, including the Paris climate negotiations in 2015 where it played a constructive role. When the United States withdrew from the Paris Agreement under the Trump administration, China reaped reputational benefits internationally by simply staying in the agreement, even as its own record on emissions was quite mixed – as leading supplier and deployer of renewables but also the principal builder and financier of coal plants at home and abroad. Perhaps anticipating a likely change in the United States to a Biden administration, in 2020 China announced a pledge to reduce net emissions to zero by 2060.[58]

While this pulling and hauling over mitigation may seem removed from traditional security policy, the elevation of climate change to high politics and part of great power rivalry makes the topic integral to discussions and study of international security hereafter. The policy (and academic) challenge going forward is how to make great power rivalry work for climate protection rather than against it.[59] As Chapter 7 noted, scholars can also

---

[56] Global Carbon Project 2019.
[57] Tammen et al. 2000; Kugler 2006; Gilpin 1981; Allison 2017.   [58] Finamore 2020.
[59] Busby 2018a.

surface the ways in which responses to climate change may be as, if not more, important contributors to conflict than its physical effects and how to minimize those risks. Beyond the mitigation challenge posed by climate change and the significance of great power rivalry for reducing greenhouse gases is how climate change may fundamentally reshape international relations as we know it.

CLIMATE CHANGE IS A STRUCTURAL PARAMETER

Climate change is an emergent structural parameter of international relations, as important, and perhaps ultimately more important, than anarchy in shaping the behavior of states going forward. Not all states will be equally capable of adjusting to its impacts. Like anarchy, climate change will function as a feature, a fact, a set of constraints and opportunities that all states have to react to, and it will create pressures upon them.

We need to theorize how this structural parameter has and will affect world politics. Kenneth Waltz developed a structural theory of realism after ideas about the sovereign formal equality of states had become sedimented for several centuries.[60] While many of us largely accept that anarchy shapes the behavior and confronts states as a fact of life, sovereignty is, of course, a historically contingent phenomenon.[61] Like climate change, anarchy may have more dynamic properties depending upon the degree of hierarchy in the international system and the distribution of friendship among the great powers, both of which can change over time.[62]

What then can we meaningfully say about the structural significance of climate change without lapsing into fortune telling?

(1) All states will be affected by the physical consequences of climate change, some more than others.
(2) Some states will cease to exist because of climate change, most likely some small, low-lying island nations.
(3) Nearly all states will experience humanitarian emergencies from climate change, from extreme storms, fires, drought, and high temperatures. Some effects, such as the increased power of hurricanes, are less well understood by scientists and more contested.[63]

---

[60] Waltz 1979. [61] de Carvalho, Leira, and Hobson 2011. [62] Wendt 1995.
[63] Mooney 2007.

(4) Coastal areas with large urban populations will be especially susceptible to sea-level rise, storm surge, and saltwater intrusion.
(5) Melting sea ice will change borders in the Arctic and open up transit routes.
(6) Water scarcity will increase in aquifers within and between countries, some more than others. For transboundary rivers, those with strong river basin institutions will be best equipped to deal with the inevitable tensions and conflict over sharing of resources. For internal rivers and groundwater resources, countries with strong domestic institutions will fare better than others, all else being equal.
(7) Unreliable and scarce water supplies will disrupt agricultural production, unless there are major technological advances. This is most likely to lead to violence in agriculture-dependent societies with high political exclusion, though violence is not inevitable.
(8) Climate-related humanitarian emergencies will create short-term forced migration, mostly within but sometimes between countries. Long-term habitability will lead to pressures for migration that will be harder to distinguish from economic migration.
(9) Barring major decoupling of economies between and within countries, climate change will pose cascading risks across multiple systems, with disruptions or outages in one part of the world (such as food production and supply chains) radiating to other parts (in terms of higher food prices and delays in manufacturing). Within countries, climate-driven power outages from storms or fire could lead to major disruptions in health care, data and information systems, transport, and anything else that relies on electricity.[64] Sternberg referred to this phenomenon as the "globalization of hazards."[65]
(10) The impacts on human populations are dynamic and potentially will worsen over time as greenhouse gases accumulate in the atmosphere, though they may eventually stabilize as the world decarbonizes and/or if new technologies emerge to take carbon out of the air.

---

[64] Busby 2019a; Briggs and Matejova 2019; Homer-Dixon 2009.
[65] Sternberg 2011, 2012; Femia and Werrell 2016.

Based on these physical and social consequences, for which we have some expertise in identifying likely hot spots of high risk, we can (partially) anticipate what the implications will be for the international community. Impacts will create pressures for adaptation but also latent political demand to address the causes of climate change, namely, greenhouse gas pollution from burning fossils fuels, deforestation, and emissions from short-lived gases like methane. This profound interdependence will create pressures for cooperation, independent of other pressures that pull people apart.

Affected countries will seek aid to deal with and adapt to the consequences as well as compensation for their losses. Although state death, as Tanisha Fazal has argued, has been rare since 1945, it may become an increasingly important phenomenon for low-lying island countries and cause a reckoning for how to handle those displaced.[66] Border politics will become highly salient in places where there are shared vulnerable resources and high population concentrations.

The transition to clean energy, should it occur, will create losers – namely, resource exporters and fossil fuel intensive sectors (and sources of employment) within countries. These countries and sectors will have asymmetric incentives to mobilize to protect their core industries but the broader basis of societal effects may eventually produce winning coalitions that support climate mitigation.[67]

As I noted in Chapter 7, potential responses to climate change – such as land grabs, biofuels policies, geoengineering, forest conservation, efforts to keep fossil fuels in the ground, and taxation policies – may be as, or more, contentious than the impacts of climate change itself. Whether states cooperate to address their shared vulnerability will depend on whether they can overcome, manage, placate, or buy-off opposing forces with minimal social dislocation. Because not all states are equal in their contribution to climate change, states that are seen as responsible for the problem will increasingly be blamed for inaction by other states that are feeling the effects. This will create demands for blame-shifting between major emitters, in the absence of effective collective action.

Climate change will ultimately intersect with another important structural transformation in the balance of power, the rise of China. This

---

[66] Fazal 2007.
[67] These distributional battles will loom large as advocates for climate action contend with legacy industries and sectors resistant to change; see Colgan, Green, and Hale 2020; Aklin and Mildenberger 2020.

country's increasing capacity to destroy, given both its action at home and its ability to finance emissions abroad, will make it a more central target of international demands for action. Inaction or insufficient action will risk condemnation and increase the likelihood that emergent Chinese (regional) hegemony is perceived as illegitimate. Robust action on climate will ease concerns that the transition to multipolarity, or even Chinese hegemony. could be acceptable.[68]

Most of the thinking on the geopolitics of climate change has been carried out thus far by scholars of clean energy transitions.[69] But, going forward, international relations and security scholars should begin to theorize and study these issues in their own right. Companies and countries that make the zero-carbon energy technologies of the future stand to reap enormous profits. To the extent that economic power is fungible into military power, we can expect the beneficiaries of the clean energy transition to enhance both their hard power capabilities but also their soft power reputation, in the way that the space race provided some symbolic advantages to the Soviet Union and the United States.[70]

CONCLUDING THOUGHTS

In this chapter, I sketched out the future of climate security research. To be policy-relevant, I argued that academics need to move beyond the language of threat multiplier to think through what specific threats practitioners should be worried about, and that policymakers need narratives of specific places that tell analytically informed stories about how climate change connects to security outcomes of concern. I suggested that task is difficult, because the field still knows little about how to talk about the long-term future consequences of climate change as the world starts to move beyond known extremes. Moreover, it is made more complicated by the difficulty of establishing baseline normal climatic conditions and recognizing that long-lived climate processes may have enduring effects. I argued that scholars should explore why climate hazards lead to human security impacts that then escalate to state security threats. I made the case that the climate security field also has to accept that climate mitigation itself is a security concern, since the security impacts will become

---

[68] Busby 2019c.
[69] Pascual 2008; Scholten and Bosman 2016; Jasparro and Taylor 2008; Scholten 2018; Bordoff 2020.
[70] Musgrave and Nexon 2018.

unmanageable in a world of unconstrained climate change. Finally, I suggested that the field needs to elevate climate change to a structural parameter of significance on par with anarchy as a constraint that shapes state behavior, and that climate change will increasingly define great power relations and geopolitics throughout the remainder of the twenty-first century.

9

Conclusion

This book has explored the conditions under which the effects of climate change lead to security outcomes in some situations and not others. The argument in its reduced form focuses on just a few parameters – state capacity, political inclusion, and the role of foreign assistance – to explain a variety of outcomes, including humanitarian emergencies and civil conflict. I argued that the combination of low state capacity, exclusive political institutions, and no or one-sided provision of foreign assistance are most likely cases for negative security outcomes. For humanitarian emergencies, the causal chain is relatively short with the climate shock leading to large-scale human suffering, as a result of both lack of preparedness and inadequate response. With respect to civil conflict, Chapter 3 sketched the longer causal pathways to violence through the effects on agriculture, food prices, economic development, and/or migration.

THE BET: WHAT WE CAN LEARN FROM PAIRED CASES

The bet I made in writing the book is that pairing cases that experience similar environmental exposure but very different outcomes would be a productive way to understand the circumstances that lead to violence and negative security outcomes in some contexts but not others. Single cases that illustrate the importance of climate drivers run two risks: first, of fixing a just-so story around climate being the main or featured driver of negative security outcomes,[1] and second, the inability to identify the scope conditions that led to violence or human security impacts in that case.[2]

By selecting paired cases from three different regions – sub-Saharan Africa, the Middle East, and South/Southeast Asia – and taking advantage of within-case variation, I hoped to generate the argument's generalizability and ambition. By drawing on regional experts, I hoped to do justice to these cases, while recognizing the limits given the nature of the way I approached the project. There are scores of scholars with deep country knowledge of all these countries who have spent years interviewing experts and regular folks alike. This book owes those researchers a debt of gratitude for their insights. My hope is that it makes a contribution to their understanding as well, by bringing insights and comparisons from other fields and vantage points that they might not have considered.

There are other cases – of farmer–herder violence in the Sahel, for example, and the effects of climate change on agriculture and migration in the dry corridor of Central America – where I think there is an interesting story to be told about groups outside and somewhat ignored by the reach of fragile states. Those communities' fortunes are falling and failing in part because of climate change and other forms of environmental degradation. The particular plight of low-lying island countries in the Pacific, for which climate change is an existential threat, is another area of interest that I sadly could not discuss in depth here, but merits more treatment. I leave it to other scholars to pick up the mantle and explore these cases, building upon the arguments developed here.

The task of identifying suitable cases for comparison is challenging, and cases are hard to identify for a number of reasons. First, it is rare for two countries to experience the same extent or degree of physical exposure to climate hazards at precisely the same moment in time. While Ethiopia did experience something of a drought during the deep crisis that affected Somalia, the country was not tested in 2011 in the same way it would be in 2015. For that reason, the later Ethiopian drought seemed like a better comparison. A second reason case selection is challenging is that countries, even neighbors that share some environmental and social similarities, may be pretty different from each other in other respects.

---

[1] Elster 2000; 1989; Bates et al. 2000. Just-so stories is a reference to Rudyard Kipling's elaborate fables, constructed to tell the story of how leopards got their spots or camels got their humps; they are detailed and sound plausible but ultimately have no basis in reality.
[2] Levy 1995.

Syria was more dependent on agriculture than Lebanon, and Lebanon enjoyed a diaspora population that could come to its aid more readily in a time of crisis. Lebanon had the memory of a civil war to dampen desire to experience another, while Syria had not yet experienced the kind of bloodshed that may ultimately inform the sensibilities of Syrians of the future when the current conflict ultimately comes to an end. A third challenge for case selection is that the units of analysis are not truly independent if the cases are regional neighbors. Their fates are bound up with each other. Somalia had been invaded by Ethiopia in the early 2000s which set the stage for the rise of Al-Shabaab. Lebanon had been occupied by Syria for nearly three decades until Syria's withdrawal in 2005. While choosing nonneighbors is an option, the risk is identifying countries so culturally or environmentally different that comparisons might feel strained.

EXTERNAL BARRIERS TO PROVIDING FINANCE

One observation that emerged from the cases is that the absence or delay in an international response can come from external actors as much as from domestic blockages – such as Al-Shabaab's refusal to allow foreign aid in the wake of the 2011 drought in Somalia or the Myanmar regime's strict limits on foreign aid in the aftermath of Cyclone Nargis. In the case of Syria, the appeals for assistance in the wake of the droughts in 2007 and 2008 were largely unfulfilled because the Syrian regime was already something of a pariah state internationally, even before the brutal civil war. It is an interesting counterfactual to ask whether the mobilization of $20–30 million in 2008 to forestall such deep suffering in the Syrian countryside would have made a later civil war less likely. In Syria's case, it is likely impossible to know, as there were other drivers of dissatisfaction and sparks that may have set the country alight in any case. In the Somalia example, the threat and fear inspired by the Patriot Act in the United States kept many NGOs from providing aid to that country in the wake of the drought, and the United States was worried about diversion of foreign assistance to Al-Shabaab. Arguably, the combination of external and internal restrictions on the timely delivery of aid contributed to the large-scale loss of life during the Somali drought.

IS THE EMPHASIS TOO STATE-CENTRIC?

In writing this book, I decided to workshop draft chapters at different institutions with a rich history of writing on climate and security. One observation that emerged from those workshops is the relative state-centric focus of the book, which is fair. While my previous work focused on the role of social movements and nonstate actors, I have become convinced that the story of state development is critically important to the lives and well-being of people at scale in different country contexts. While nonstate and subnational actors can sometimes compensate for the lack of central government capacity or the lack of concern for some groups, their ability to provide sustained services and keep the peace over large geographic areas for extended periods of time ultimately seems limited. For better or worse, we live in a world of sovereign states.

There may be nonstate actor groups with more state-like aspirations and capacity – perhaps Hezbollah's presence in eastern Lebanon is an example. Those nonstate actors are important and, as the literature on hybrid governance discussed in Chapter 7 suggests, perennially weak states may ultimately never be able to consolidate power in a way that meets our conventional understanding of sovereign states with a monopoly of force over their whole territory. They may rely on traditional, informal institutions for peacebuilding and conflict resolution. If keeping the peace and stability in countries is underpinned by elite compacts, the nature of the agreements might not simply be one of national political parties contesting for control of the state. Other actors may have more control but over more limited geographic reach, so national leaders may have to find accommodations with powerful local actors.

For external actors, the options are less clear. There was a period in the 1990s when donors worked around the state and tried to support direct service delivery through NGOs and other nonstate actors. Some states that have jealous regard for sovereignty, like India, have increasingly made it difficult for externally supported NGOs to operate. In other country contexts, persistent state weakness ultimately made supporting parallel structures inadequate, and donors made more concerted efforts to build state capacity. Where state actors were imbued with a reasonably inclusive developmental mindset, as they were in Ethiopia and Bangladesh, the results can be promising, though the risk of authoritarianism and sectarian violence remains. In states that lack farsighted local leadership, as has been the case in South Sudan and Somalia, for example, the prospect of protracted conflicts remains. Whether and how the international community can help domestic actors realize that there is more to be gained by setting aside differences and focusing on improving their people's lives remains a grand challenge in fragile states going forward.

PROGRESS UPENDED

In the drafting of the book, some of the paired cases that did not suffer from negative security outcomes in the wake of climate shocks have since experienced them, or come dangerously close to doing so. In 2020, Ethiopia appeared to be poised for civil war. As noted in Chapter 4, the minority Tigrayans which had run the country since the early 1990s ultimately allowed the restive and far more numerous Oromo to appoint a prime minister of their own. However, relations between the Oromo and Tigrayans deteriorated, demonstrating the challenge of building an inclusive state in Ethiopia.[3]

In Lebanon, elites of different sectarian groups seemed to divvy up the spoils of government contracts for their own self-enrichment rather than the public's benefit. These issues came to a head with a crisis of uncollected garbage, wildfires, and finally, a devastating explosion of stored fertilizer in August 2020 that took out many city blocks in Beirut, killing 178 and causing $15 billion in damages. The government promptly resigned.[4] Both of these episodes reveal that countries that may have built state capacity or improved political inclusion can, over time, experience deterioration in those institutional foundations.

In the cases of both Ethiopia and Lebanon, climate-related conditions may be relevant. The Oromo mobilization that brought to power the current leadership began in the wake of the 2015 drought, though other drivers such as the Ethiopian government's intended expansion of the capital into ancestral Oromo lands was a major factor.[5] In Lebanon, the government mishandled hundreds of fires over the course of 2019, prompting major protests, as did similar failures on dealing with the country's garbage. The 2020 explosion, along with the COVID-19 crisis, not only destroyed the government's credibility but further weakened an already feeble economy.[6] We should not overstate the role of climate drivers, but these episodes remind us that reversals and deterioration in country conditions and institutions

---

[3] Woldemariam 2020.   [4] Hubbard 2020.   [5] Busby 2018b.
[6] Nakhoul and Irish 2020.

should keep analysts and policymakers from becoming too comfortable that countries have escaped their unhappy pasts.

A FINAL NOTE ON SCHOLARSHIP

Scholars of environmental politics have long lamented that their issues, including but not limited to climate change, have not featured in the top journals of political science.[7] It seems that the climate and security field has been able to break through in a way that more conventional scholars of environmental politics largely have not. Leading journals of security studies and broader international relations subfield journals have featured a variety of articles on climate and security.[8] Economics and other disciplines have made ample contributions as well, with vigorous debates featuring in more physical science journals like the *Proceedings of the National Academy of Sciences, Science, Nature,* and *Climatic Change*.[9] The field has benefited from innovative methods and new data sources with a productive expansion of themes, topics, and geographies. Despite this wealth of new knowledge, there are gaps and lacunae, even as the field has profited from advanced methods. This book is intended to be a back-to-basics set of qualitative case studies, reminding us of the importance of case selection and causal mechanisms with a careful review of the evidence and attention to alternative explanations. As we look ahead to the next generation of work in this space, I hope that scholars provide and readers demand a healthy mix of methodological innovation and narratives as well as cases of particular places.

---

[7] Green and Hale 2017.
[8] See the special issues of the *Journal of Peace Research* Vol. 49 (1) 2012 and Vol. 58 (1) 2021 as well as *Political Geography* Vol. 43 2014. See also Theisen, Holtermann, and Buhaug 2012; Busby et al. 2013; Busby 2008; Busby et al. 2018; Ide 2020.
[9] Buhaug 2010; Hsiang, Burke, and Miguel 2013; Hsiang and Meng 2014; Burke et al. 2009; Buhaug et al. 2014; Mach et al. 2019.

Bibliography

Aalen, Lovise. 2018. Why is Ethiopia in upheaval? This brief history explains a lot. *Washington Post*. Available at www.washingtonpost.com/news/monkey-cage/wp/2018/02/17/why-is-ethiopia-in-upheaval-this-brief-history-explains-a-lot/. Accessed March 26, 2019.

Ababsa, Myriam. 2015. The end of a world: Drought and agrarian transformation in Northeast Syria (2007–2010). In *Syria from Reform to Revolt: Volume 1: Political Economy and International Relations*, edited by Raymond Hinnebusch and Tina Zintl, 199–222. Syracuse, NY: Syracuse University Press.

Abel, Guy J., Michael Brottrager, Jesus Crespo Cuaresma, and Raya Muttarak. 2019. Climate, conflict and forced migration. *Global Environmental Change* 54: 239–249. Available at https://doi.org/10.1016/j.gloenvcha.2018.12.003. Accessed February 22, 2019.

Abu-Ismail, Khalid, Ali Abdel-Gadir, and Heba El-Laithy. 2011. Poverty and Inequality in Syria (1997–2007). UNDP. Available at www.undp.org/content/dam/rbas/doc/poverty/BG_15_Poverty%20and%20Inequality%20in%20Syria_FeB.pdf. Accessed December 5, 2019.

Acemoglu, Daron, and James Robinson. 2013. *Why Nations Fail: The Origins of Power, Prosperity, and Poverty*. Reprint edition. New York: Crown Business.

Acemoglu, Daron, Simon Johnson, and James A. Robinson. 2001. The colonial origins of comparative development: An empirical investigation. *American Economic Review* 91 (5): 1369–1401. Available at www.aeaweb.org/articles?id=10.1257/aer.91.5.1369. Accessed November 13, 2020.

ACLED. 2019. The Armed Conflict Location & Event Data Project. ACLED. Available at www.acleddata.com/. Accessed June 25, 2021.

Adams, Brad. 2009. The lessons of Cyclone Nargis. *Bangkok Post*. Available at www.hrw.org/news/2009/05/03/lessons-cyclone-nargis. Accessed September 18, 2019.

Adams, Courtland, Tobias Ide, Jon Barnett, and Adrien Detges. 2018. Sampling bias in climate–conflict research. *Nature Climate Change* 8 (3): 200–203.

Available at https://doi.org/10.1038/s41558-018-0068-2. Accessed November 18, 2020.
adelphi. 2015. *Syrian Civil War: The Role of Climate Change. ECC Library.* Available at https://library.ecc-platform.org/conflicts/syrian-civil-war-role-climate-change. Accessed December 7, 2019.
Adger, W. Neil, Nick Brooks, Graham Bentham, Maureen Agnew, and Siri Eriksen. 2004. *New Indicators of Vulnerability and Adaptive Capacity.* Available at http://citeseerx.ist.psu.edu/viewdoc/summary?doi=10.1.1.112.2300. Accessed June 24, 2021.
Adger, W. N., J. M. Pulhin, J. Barnett, et al. 2014. Human security. In *Climate Change 2014: Impacts, Adaptation, and Vulnerability. Part A: Global and Sectoral Aspects. Contribution of Working Group II to the Fifth Assessment Report of the Intergovernmental Panel on Climate Change*, edited by C. B. Field, V. R. Barros, D. J. Dokken, et al., 755–791. New York: Cambridge University Press.
Adger, W. Neil, Ricardo Safra de Campos, Tasneem Siddiqui, et al. 2021. Human security of urban migrant populations affected by length of residence and environmental hazards. *Journal of Peace Research* 58 (1). Available at https://doi.org/10.1177/0022343320973717. Accessed June 25, 2021.
Africa Climate Change Resilience Alliance. 2015. Ethiopia good practice guide: Local disaster risk reduction planning. Africa Climate Change Resilience Alliance. Available at https://reliefweb.int/sites/reliefweb.int/files/resources/cs-risk-reduction-planning-ethiopia-150714-en.pdf. Accessed June 25, 2021.
Aklin, Michaël, and Matto Mildenberger. 2020. Prisoners of the wrong dilemma: Why distributive conflict, not collective action, characterizes the politics of climate change. *Global Environmental Politics* 20 (4): 4–27. Available at https://doi.org/10.1162/glep_a_00578. Accessed November 13, 2020.
Al-Khalidi, Suleiman, and Eric Knecht. 2019. Young, angry Lebanese ditch their differences to target "unjust" system. *Reuters.* Available at www.reuters.com/article/lebanon-protests-youth-idUSL3N2771XN. Accessed December 25, 2020.
Ali, Muna. 2018. The formation of the Somali national disaster management policy. *Adeso Africa.* Available at https://web.archive.org/web/20200215013049/http://adesoafrica.org/what-we-do/blogs/the-formation-of-the-somali-national-disaster-management-policy-/. Accessed July 5, 2021.
Allison, Graham. 2017. *Destined for War: Can America and China Escape Thucydides's Trap?* Brunswick, Australia: Scribe Publications.
Amnesty International. 2020. Lebanon: One year after the October protest movement, impunity reigns. Available at www.amnesty.org/en/latest/news/2020/10/lebanon-one-year-after-the-october-protest-movement-impunity-reigns/. Accessed December 25, 2020.
Art, Robert J. 2003. *A Grand Strategy for America.* Ithaca, NY: Cornell University Press.
Ash, Konstantin. 2019. Post-conflict processes and religion: Lebanon. *Oxford Research Encyclopedia of Politics.* Available at https://doi.org/10.1093/acrefore/9780190228637.013.808. Accessed June 25, 2021.

Ash, Konstantin, and Kevin Mazur. 2020. Identifying and correcting signal shift in DMSP-OLS Data. *Remote Sensing* 12 (14): 2219. Available at https://doi.org/10.3390/rs12142219. Accessed August 7, 2020.

Ash, Konstantin, and Nick Obradovich. 2020. Climatic stress, internal migration, and Syrian civil war onset. *Journal of Conflict Resolution* 64 (1): 3–31. Available at https://doi.org/10.1177/0022002719864140. Accessed December 8, 2019.

Ashwill, Maximillian Shen, David Robinson Lee, Dorte Verner, and Robert Leonard Wilby. 2013. Middle East – Increasing resilience to climate change in the agricultural sector of the Middle East: The cases of Jordan and Lebanon. The World Bank. Available at http://documents.worldbank.org/curated/en/115381468249300050/Middle-East-Increasing-resilience-to-climate-change-in-the-agricultural-sector-of-the-Middle-East-the-cases-of-Jordan-and-Lebanon. Accessed December 4, 2019.

Associated Press. 2008. US ships abort Myanmar mercy mission. *Oklahoman.com*. Available at https://kvia.com/news/2008/06/04/us-ships-abort-myanmar-mercy-mission-2/. Accessed June 25, 2021.

Atarah, Linus. 2012. U.S. Patriot Act kept Somalia starving. *Inter Press Service*. Available at www.ipsnews.net/2012/04/us-patriot-act-kept-somalia-starving/. Accessed March 27, 2019.

Baechler, Günther. 1998. Why environmental transformation causes violence: A synthesis. Woodrow Wilson Center. Available at www.wilsoncenter.org/publication/why-environmental-transformation-causes-violence-synthesis. Accessed May 27, 2021.

Baechler, Günther. 1999a. Environmental degradation in the south as a cause of armed conflict. In *Environmental Change and Security: A European Perspective*, edited by Alexander Carius and Kurt M. Lietzmann, 107–130. Berlin: Springer.

Baechler, Günther. 1999b. *Violence through Environmental Discrimination: Causes, Rwanda Arena, and Conflict Model*. Dordrecht: Springer.

Bahout, Joseph. 2016. The unraveling of Lebanon's Taif Agreement: Limits of sect-based power sharing. Carnegie Endowment for International Peace. Available at https://carnegieendowment.org/2016/05/16/unraveling-of-lebanon-s-taif-agreement-limits-of-sect-based-power-sharing-pub-63571. Accessed December 7, 2019.

Bandyopadhyay, B. K. 2012. *Activities of RSMC, New Delhi*. New Delhi: India Meteorological Office. Available at https://web.archive.org/web/20170629224637/www.wmo.int/pages/prog/www/tcp/documents/3.2_New_Delhi_Bandyopadhyay.pdf. Accessed July 5, 2021.

Banerjee, Neela. 2018. Sea level rise damaging more U.S. Bases, former top military brass warn. *InsideClimate News*. Available at https://insideclimatenews.org/news/26022018/sea-level-rise-military-bases-damaged-national-security-risk-report-admirals-generals. Accessed March 19, 2018.

Banfield, Jessica, and Victoria Stamadianou. 2015. *Towards a Peace Economy in Lebanon*. International Alert. Available at www.international-alert.org/publications/towards-peace-economy-lebanon. Accessed December 5, 2019.

Barber, Charles Victor. 1997. The Case Study of Indonesia. Project on Environmental Scarcities, State Capacity, & Civil Violence. Available at https://homerdixon.com/environmental-scarcities-state-capacity-and-civil-violence/indonesia/. Accessed August 14, 2019.

Barnett, Jon. 2003. Security and climate change. *Global Environmental Change* 13 (1): 7–17.

Barnett, Jonathon. 2017. The dilemmas of normalising losses from climate change: Towards hope for Pacific atoll countries. *Asia Pacific Viewpoint* 58 (1): 3–13. Available at https://doi.org/10.1111/apv.12153. Accessed July 20, 2020.

Barnett, Jon, and W. Neil Adger. 2007. Climate change, human security and violent conflict. *Political Geography* 26 (6): 639–655.

Barnett, Jon, Richard A. Matthew, and Karen L. O'Brien. 2010. Global environmental change and human security: An introduction. In *Global Environmental Change and Human Security*, edited by Richard A. Matthew, Jon Barnett, Bryan McDonald, and Karen L. O'Brien, 3–32. Cambridge, MA: MIT Press.

Barshad, Amos. 2019. In Lebanon, a census is too dangerous to implement. *The Nation*. Available at www.thenation.com/article/lebanon-census. Accessed December 7, 2019.

Bates, Robert H. 2015. *When Things Fell Apart: State Failure in Late-Century Africa*. Cambridge, UK: Cambridge University Press.

Bates, Robert H., Avner Greif, Margaret Levi, Jean-Laurent Rosenthal, and Barry R. Weingast. 2000. The analytic narrative project – Analytic narratives. *American Political Science Review* 94 (3): 696–702. Available at https://doi.org/10.2307/2585843 Accessed July 5, 2021.

BBC. 2011. Horn of Africa tested by severe drought. *BBC News*. Available at www.bbc.com/news/world-africa-14023160. Accessed March 26, 2019.

Benjaminsen, Tor A., Koffi Alinon, Halvard Buhaug, and Jill Tove Buseth. 2012. Does climate change drive land-use conflicts in the Sahel? *Journal of Peace Research* 49 (1): 97–111. Available at https://doi.org/10.1177/0022343311421 7343. Accessed August 10, 2016.

Bennett, Andrew. 2010. Process tracing and causal inference. In *Rethinking Social Inquiry: Diverse Tools, Shared Standards*, edited by Henry E. Brady and David Collier, 207–220. Second edition. Lanham, MD: Rowman & Littlefield Publishers.

Bennett, Andrew. 2016. Do new accounts of causal mechanisms offer practical advice for process tracing? *Qualitative & Multi-Method Research Newsletter* 14 (1–2): 34–39. Available at https://doi.org/10.5281/zenodo.823311. Accessed December 16, 2020.

Bennet, James. 2005. The enigma of Damascus. *The New York Times Magazine*. Available at www.nytimes.com/2005/07/10/magazine/the-enigma-of-damascus.html. Accessed December 5, 2019.

Bergholt, Drago, and Päivi Lujala. 2012. Climate-related natural disasters, economic growth, and armed civil conflict. *Journal of Peace Research* 49 (1): 147–162. Available at https://doi.org/10.1177/0022343311426167. Accessed May 21, 2012.

Berke, Richard L. 1991. U.S. sends troops to aid Bangladesh in cyclone relief. *The New York Times*. Available at www.nytimes.com/1991/05/12/world/us-sends-troops-to-aid-bangladesh-in-cyclone-relief.html. Accessed April 11, 2017.

Bern, C., J. Sniezek, G. M. Mathbor et al. 1993. Risk factors for mortality in the Bangladesh cyclone of 1991. *Bulletin of the World Health Organization* 71 (1): 73–78. Available at www.ncbi.nlm.nih.gov/pmc/articles/PMC2393441/. Accessed July 2, 2019.

Bhide, Jonah. 2019. US defense assets in the Western Pacific. *Climate Security in Oceania*. Available at https://sites.utexas.edu/climatesecurity/2019/11/15/us-defense-assets-in-the-western-pacific/. Accessed December 1, 2020.

Bieber, Florian, and Wondemagegn Tadesse Goshu. 2020. It's not too late to stop the Ethiopian civil war from becoming a broader ethnic conflict. *Foreign Policy*. Available at https://foreignpolicy.com/2020/11/18/its-not-too-late-to-stop-the-ethiopian-civil-war-from-becoming-a-broader-ethnic-conflict/. Accessed December 21, 2020.

Bloch, Matthew, Weiyi Cai, and Blacki Migliozzi. 2019. Live tracking map: Cyclone Fani batters India. *The New York Times*. Available at www.nytimes.com/interactive/2019/05/02/world/asia/india-cyclone-fani-map.html. Accessed September 15, 2019.

BMAS. 2014. Explaining extreme events of 2013 from a climate perspective [special report]. *Bulletin of the American Meteorological Society* 75 (9). Available at www2.ametsoc.org/ams/index.cfm/publications/bulletin-of-the-american-meteorological-society-bams/explaining-extreme-events-of-2013-from-a-climate-perspective/. Accessed June 21, 2021.

Boege, Volker. 2019. To avoid conflict, responses to climate change in Oceania must heed customary actors and institutions. *New Security Beat*. Available at www.newsecuritybeat.org/2019/10/avoid-conflict-responses-climate-change-oceania-heed-customary-actors-institutions/. Accessed December 8, 2020.

Boege, Volker, M. Anne Brown, and Kevin P. Clements. 2009. Hybrid political orders, not fragile states. *Peace Review* 21 (1): 13–21. Available at https://doi.org/10.1080/10402650802689997. Accessed December 8, 2020.

Bordoff, Jason. 2020. Everything you think about the geopolitics of climate change Is wrong. *Foreign Policy*. Available at https://foreignpolicy.com/2020/10/05/climate-geopolitics-petrostates-russia-china/. Accessed November 20, 2020.

Bowlsby, Drew, Erica Chenoweth, Cullen Hendrix, and Jonathan D. Moyer. 2020. The future is a moving target: Predicting political instability. *British Journal of Political Science* 50 (4): 1405–1417. Available at https://doi.org/10.1017/S0007123418000443. Accessed June 21, 2021.

Bradbury, Mark. 2010. *State-building, Counterterrorism, and Licensing Humanitarianism in Somalia*. Tufts University Feinstein International Center. Available at https://fic.tufts.edu/publication-item/state-building-counterterrorism-and-licensing-humanitarianism-in-somalia/. Accessed September 12, 2019.

Bretthauer, Judith M. 2015. Conditions for peace and conflict: Applying a fuzzy-set qualitative comparative analysis to cases of resource scarcity.

Journal of Conflict Resolution 59 (4): 593–616. Available at https://doi.org/10.1177/0022002713516841. Accessed June 21, 2021.

Briggs, Chad M., and Miriam Matejova. 2019. *Disaster Security*. Cambridge, UK: Cambridge University Press.

Brooks, Nick, W. Neil Adger, and P. Mick Kelly. 2005. The determinants of vulnerability and adaptive capacity at the national level and the implications for adaptation. *Global Environmental Change* 15 (2): 151–163.

Brown, Ian A. 2010. Assessing eco-scarcity as a cause of the outbreak of conflict in Darfur: A remote sensing approach. *International Journal of Remote Sensing* 31 (10): 2513–2520. Available at https://doi.org/10.1080/01431161003674592. Accessed October 16, 2020.

Brown, Oli. 2009. Rising temperatures, rising tensions: Climate change and the risk of violent conflict in the Middle East. *IISD*. Available at www.iisd.org/library/rising-temperatures-rising-tensions-climate-change-and-risk-violent-conflict-middle-east. Accessed December 7, 2019.

Bucchignani, Edoardo, Paola Mercogliano, Hans-Jürgen Panitz, and Myriam Montesarchio. 2018. Climate change projections for the Middle East–North Africa domain with COSMO-CLM at different spatial resolutions. *Advances in Climate Change Research* 9 (1): 66–80. Available at www.sciencedirect.com/science/article/pii/S1674927817300552. Accessed November 18, 2020.

Bueno de Mesquita, Alastair Smith Bruce, Randolph M. Siverson, and James D. Morrow. 2003. *The Logic of Political Survival*. Cambridge, MA.: MIT Press.

Buhaug, Halvard. 2010. Climate not to blame for Africa's civil wars. *Proceedings of the National Academy of Sciences* 107 (38): 16477–16482.

Buhaug, Halvard. 2014. Concealing agreements over climate–conflict results. *Proceedings of the National Academy of Sciences* 111 (6): E636. Available at https://doi.org/10.1073/pnas.1323773111 Accessed June 21, 2021.

Buhaug, Halvard, Nils Petter Gleditsch, and Ole Magnus Theisen. 2008. *Implications of Climate Change for Armed Conflict*. The World Bank. Available at www.prio.org/utility/DownloadFile.ashx?id=595&type=publicationfile. Accessed June 21, 2021.

Buhaug, Halvard, Lars-Erik Cederman, and Kristian Skrede Gleditsch. 2014. Square pegs in round holes: Inequalities, grievances, and civil war. *International Studies Quarterly* 58 (2): 418–431. Available at www.jstor.org/stable/24017836 Accessed June 21, 2021.

Buhaug, H., J. T. Nordkvelle, T. Bernauer et al. 2014. One effect to rule them all? A comment on climate and conflict. *Climatic Change* 127 (3–4): 391–397. Available at https://doi.org/10.1007/s10584-014-1266-1. Accessed June 21, 2021.

Burke, Marshall B, Edward Miguel, Shanker Satyanath et al. 2009. Warming increases the risk of civil war in Africa. *Proceedings of the National Academy of Sciences* 106 (49): 20670–20674. Available at https://doi.org/10.1073/pnas.0907998106. Accessed June 21, 2021.

Busby, Joshua. 2007. *Climate Change and National Security: An Agenda for Action*. Council on Foreign Relations. Available at www.cfr.org/report/climate-change-and-national-security. Accessed June 21, 2021.

Busby, Joshua W. 2008. Who cares about the weather? Climate change and U.S. national security. *Security Studies* 17 (3): 468–504.

Busby, Joshua. 2016a. Climate Change and U.S. national security: Sustaining security amidst unsustainability. In *Sustainable Security: Rethinking American National Security Strategy*, edited by Jeremi Suri and Benjamin Valentino, 196–232. New York: Oxford University Press.

Busby, Joshua W. 2016b. After Paris: Good enough climate governance. *Current History* 15 (777): 3–9.

Busby, Joshua. 2017a. *Water and U.S. National Security*. Council on Foreign Relations. Available at www.cfr.org/sites/default/files/pdf/2017/01/Discussion_Paper_Busby_Water_and_US_Security_OR.pdf. Accessed June 21, 2021.

Busby, Joshua W. 2017b. Environmental security. In *Handbook of International Security*, edited by William C. Wohlforth and Alexandra Gheciu, 471–486. Oxford: Oxford University Press.

Busby, Joshua. 2018a. Warming world. *Foreign Affairs*. Available at www.foreignaffairs.com/articles/2018-06-14/warming-world. Accessed June 21, 2021.

Busby, Joshua. 2018b. Taking stock: The field of climate and security. *Current Climate Change Reports* 4 (4): 338–346. Available at https://doi.org/10.1007/s40641-018-0116-z. Accessed June 21, 2021.

Busby, Joshua. 2019a. A clear and present danger: Climate risks, the energy system, and U.S. national security. In *Impact of Climate Risk on the Energy System*, edited by Amy Myers Jaffe, 54–64. New York: Council on Foreign Relations. Available at www.cfr.org/report/impact-climate-risk-energy-system. Accessed December 19, 2020.

Busby, Joshua. 2019b. As the stakes rise, climate action loses momentum. *Current History* 118 (804): 36–38.

Busby, Joshua. 2019c. Climate change as anarchy: The need for a new structural theory of IR. *Duck Of Minerva*. Available at https://duckofminerva.com/2019/04/climate-change-as-anarchy-the-need-for-a-new-structural-theory-of-ir.html. Accessed November 20, 2020.

Busby, Joshua W. 2019d. The field of climate and security: A scan of the literature. *Social Science Research Council*. Available at www.ssrc.org/publications/view/the-field-of-climate-and-security-a-scan-of-the-literature. Accessed June 15, 2020.

Busby, Joshua. 2020. It's time we think beyond "threat multiplier" to address climate and security. *New Security Beat*. Available at www.newsecuritybeat.org/2020/01/its-time-threat-multiplier-address-climate-security/. Accessed November 20, 2020.

Busby, Joshua W. 2021. Beyond internal conflict: The emergent practice of climate security. *Journal Of Peace Research* 58 (1): 186–194.

Busby, Joshua W., and Nisha Krishnan. 2015. Widening the scope to Asia: Climate change and security. In *The U.S. Asia–Pacific Rebalance, National Security and Climate Change*, edited by Francesco Femia and Caitlin E. Werrell, 23–30. Washington, DC: Center for Climate and Security. Available at https://climateandsecurity.files.wordpress.com/2015/11/ccs_us_asia_pacific-rebalance_national-security-and-climate-change.pdf. Accessed June 21, 2021.

Busby, Joshua, and Nina von Uexkull. 2018. Climate shocks and humanitarian crises. *Foreign Affairs*. Available at www.foreignaffairs.com/articles/world/2018-11-29/climate-shocks-and-humanitarian-crises. Accessed February 22, 2019.

Busby, Joshua W., Jay Gulledge, Todd G. Smith, and Kaiba White. 2012. Of climate change and crystal balls: The future consequences of climate change in Africa. *Air & Space Power Journal Africa and Francophonie* (3): 4–44. Available at www.thefreelibrary.com/Of+climate+change+and+crystal+balls%3A+the+future+consequences+of...-a0363189121. Accessed June 21, 2021.

Busby, Joshua W., Todd G. Smith, Kaiba L. White, and Shawn M. Strange. 2013. Climate Change and insecurity: Mapping vulnerability in Africa. *International Security* 37 (4): 132–172. Available at http://dx.doi.org/10.1162/ISEC_a_00116. Accessed February 11, 2015.

Busby, Joshua, Clionadh Raleigh, and Idean Salehyan. 2014. The political geography of climate vulnerability, conflict, and aid in Africa. In *Peace and Conflict 2014*, edited by Paul K. Huth, Jonathan Wilkenfeld, and David A. Backer, 71–91. Boulder, CO: Routledge.

Busby, Joshua W., Todd G. Smith, and Nisha Krishnan. 2014. Climate security vulnerability in Africa mapping 3.0. *Political Geography* 43: 51–67. Available at https://doi.org/10.1016/j.polgeo.2014.10.005. Accessed June 21, 2021.

Busby, Joshua, Todd G. Smith, Nisha Krishnan et al. 2018. In harm's way: Climate security vulnerability in Asia. *World Development* 112: 88–118. Available at https://doi.org/10.1016/j.worlddev.2018.07.007. Accessed June 21, 2021.

Bussell, Jennifer. 2014. *Institutional Capacity for Natural Disasters: Case Studies in Africa*. Strauss Center for International Security and Law, University of Texas at Austin. Available at www.strausscenter.org/wp-content/uploads/Institutional-Capacity-for-Natural-Disasters-Case-Studies-in-Africa-2014.pdf. Accessed July 5, 2021.

Butler, Christopher K., and Scott Gates. 2012. African range wars: Climate, conflict, and property rights. *Journal of Peace Research* 49 (1): 23–34. Available at https://doi.org/10.1177/0022343311426166 Accessed June 21, 2021.

Butter, David. 2015. *Syria's Economy: Picking up the Pieces*. Syria From Within/Chatham House. Available at https://syria.chathamhouse.org/research/syrias-economy-picking-up-the-pieces. Accessed December 5, 2019.

Calma, Justine. 2019. The State Department could gut Obama's last remaining executive action on climate change. *Mother Jones*. Available at www.motherjones.com/environment/2019/01/the-state-department-could-gut-obamas-last-remaining-executive-action-on-climate-change. Accessed July 20, 2020.

Cammett, Melani. 2014. *Compassionate Communalism: Welfare and Sectarianism in Lebanon*. First edition. Ithaca, NY: Cornell University Press.

Cammett, Melani. 2015. Sectarianism and the ambiguities of welfare in Lebanon. *Current Anthropology* 56 (S11): S76–S87. Available at https://doi.org/10.1086/682391. Accessed June 21, 2021.

Cammett, Melani, and Sukriti Issar. 2010. Bricks and mortar clientelism: Sectarianism and the logics of welfare allocation in Lebanon. *World Politics* 62 (3): 381–421. Available at https://doi.org/10.1017/S0043887110000080 Accessed June 21, 2021.

Campbell, Kurt M., Jay Gulledge, J.R. McNeill et al. 2007. *The Age of Consequences*. Center for Strategic & International Studies. Available at www.csis.org/analysis/age-consequences. Accessed June, 21, 2021.

Cao, Xun, and Hugh Ward. 2015. Winning coalition size, state capacity, and time horizons: An application of modified selectorate theory to environmental public goods provision. *International Studies Quarterly* 59 (2): 264–279. Available at https://doi.org/10.1111/isqu.12163. Accessed June 29, 2021.

Carr, Edward R. 2011. Drought does not equal famine. *New Security Beat*. Available at www.newsecuritybeat.org/2011/07/drought-does-not-equal-famine/. Accessed June 21, 2021.

Casey, Michael. 2008. Why the cyclone in Myanmar was so deadly. *Associated Press*. Available at https://web.archive.org/web/20081020031216/http://news.nationalgeographic.com/news/2008/05/080508-AP-the-perfect_2.html. Accessed September 15, 2019.

Cash, Richard A., Shantana R. Halder, Mushtuq Husain et al. 2013. Reducing the health effect of natural hazards in Bangladesh. *The Lancet* 382 (9910): 2094–2103. Available at https://doi.org/10.1016/S0140-6736(13)61948-0. Accessed June 21, 2021.

Cavallo, Eduardo, Sebastian Galiani, Ilan Noy, and Juan Pantano. 2013. Catastrophic natural disasters and economic growth. *Review of Economics and Statistics* 95 (5): 1549–1561. Available at http://dx.doi.org/10.1162/REST_a_00413. Accessed March 10, 2017.

CBS News. 2014. First time in 800,000 years: April's CO2 levels above 400 ppm. Available at www.cbsnews.com/news/first-time-in-800000-years-aprils-co2-levels-above-400-ppm/. Accessed June 6, 2015.

Cederman, Lars-Erik, Brian Min, and Andreas Wimmer. 2009. The Ethnic Power Relations (EPR) dataset. Available at https://dataverse.harvard.edu/dataset.xhtml?persistentId= doi:10.7910/DVN/NDJUJM. Accessed July 5, 2021.

Cederman, Lars-Erik, Kristian Skrede Gleditsch, and Halvard Buhaug. 2013. *Inequality, Grievances, and Civil War*. New York: Cambridge University Press.

Center for Systemic Peace. 2020. INSCR data page. Center for Systemic Peace. Available at www.systemicpeace.org/inscrdata.html. Accessed March 26, 2019.

Chaulia, Sreeram. 2018. India right to spurn foreign disaster relief. *Nikkei Asian Review*. Available at https://asia.nikkei.com/Opinion/India-right-to-spurn-foreign-disaster-relief. Accessed September 18, 2019.

Checchi, Francesco, and W. Courtland Robinson. 2013. *Mortality among Populations of Southern and Central Somalia Affected by Severe Food Insecurity and Famine during 2010–2012*. FAO and FEWS NET. Available at www.fsnau.org/downloads/Somalia_Mortality_Estimates_Final_Report_8May 2013_upload.pdf. Accessed June 21, 2021.

Chhotray, Vasudha. 2014. Disaster relief and the Indian state: Lessons for just citizenship. *Geoforum* 54: 217–225. Available at https://doi.org/10.1016/j.geoforum.2014.01.013. Accessed June 21, 2021.

Cho, Renee. 2016. El Niño and global warming – What's the connection? *State of the Planet*. Available at https://blogs.ei.columbia.edu/2016/02/02/el-nino-and-global-warming-whats-the-connection/. Accessed September 2, 2019.

Ciccone, Antonio. 2011. Economic shocks and civil conflict: A comment. *American Economic Journal: Applied Economics* 3 (4): 215–227. Available at http://doi.org/10.1257/app.3.4.215. Accessed June 21, 2021.

CIESIN, Columbia University. 2018. Gridded Population of the World (GPW), v4. SEDAC. Available at https://sedac.ciesin.columbia.edu/data/collection/gpw-v4/maps/services. Accessed September 15, 2019.

Climate and Development Knowledge Network. 2017. Science summary: The drought in Ethiopia, 2015. *ReliefWeb*. Available at https://reliefweb.int/report/ethiopia/science-summary-drought-ethiopia-2015. Accessed June 18, 2018.

Climate Central. 2014. 2014 Extreme weather: Looking for climate ties. Available at www.climatecentral.org/news/2014-extreme-weather-attribution-18150. Accessed June 4, 2015.

Climate Hazards Group. n.d. CHIRPS. Available at www.chc.ucsb.edu/data/chirps. Accessed June 21, 2021.

Climate Home News. 2013. Climate change linked to 2011 East Africa drought. Available at www.climatechangenews.com/2013/02/21/climate-change-linked-to-2011-east-africa-drought/. Accessed March 27, 2019.

Climate Security Expert Network. n.d. Short history of UNSC engagement on climate-related security risks. Available at https://climate-security-expert-network.org/topic-5. Accessed July 16, 2020.

CNA Corporation. 2007. National security and the threat of climate change. Available at www.cna.org/cna_files/pdf/National%20Security%20and%20the%20Threat%20of%20Climate%20Change.pdf. Accessed June 21, 2021.

CNN. 2019. Syrian civil war fast facts. Available at www.cnn.com/2013/08/27/world/meast/syria-civil-war-fast-facts/index.html. Accessed November 16, 2019.

Colgan, Jeff, Jessica F. Green, and Thomas Hale. 2020. Asset evaluation and the existential politics of climate change. *International Organization* 75 (2): 586–610. Available at http://doi.org/10.1017/S0020818320000296. Accessed June 21, 2021.

Collier, David. 2011. Understanding process tracing. *PS: Political Science & Politics* 44 (4): 823–830. Available at http://doi.org/10.1017/S1049096511001429. Accessed June 21, 2021.

Collier, Paul. 2007. *The Bottom Billion: Why the Poorest Countries Are Failing and What Can Be Done About It*. Oxford, New York: Oxford University Press.

Collier, Paul, and Anke Hoeffler. 2002. On the incidence of civil war in Africa. *Journal of Conflict Resolution* 46 (1): 13–28.

Collier, Paul, and Anke Hoeffler. 2004. Greed and grievance in civil war. *Oxford Economic Papers* 56 (4): 563–595. Available at http://econpapers.repec.org/

article/oupoxecpp/v_3a56_3ay_3a2004_3ai_3a4_3ap_3a563-595.htm. Accessed April 18, 2017.

Committee on Assessing the Impact of Climate Change on Social and Political Stresses, Board on Environmental Change and Society, Division of Behavioral and Social Sciences and Education et al. 2013. *Climate and Social Stress: Implications for Security Analysis*. Washington, DC: National Academies Press.

Confalonieri, Ulisses, Bettina Menne, Rais Akhtar et al. 2007. Human health. In *Climate Change 2007: Impacts, Adaptation and Vulnerability. Contribution of Working Group II to the Fourth Assessment Report of the Intergovernmental Panel on Climate Change*, edited by M. L. Parry, O. F. Canziani, J. P. Palutikof et al., 42. Cambridge, UK: Cambridge University Press.

Conger, John, Francesco Femia, and Caitlin Werrell. 2020. *A Very Strong Signal: 5 Key Takeaways on John Kerry's Climate Envoy Role and Seat on the National Security Council*. The Center for Climate & Security. Available at https://climateandsecurity.org/2020/11/a-very-strong-signal-5-key-takeaways-on-john-kerrys-climate-envoy-role-and-seat-on-the-national-security-council/. Accessed December 8, 2020.

Cons, Jason. 2016. *Exploring State Vulnerability and Climate Change: Bangladesh's Resilience Plan*. Strauss Center for International Security and Law, University of Texas at Austin. Available at www.strausscenter.org/cepsa-research-briefs?download=632:bangladesh-s-resilience-plan. Accessed June 21, 2021.

CRED. 2019. *EM-DAT: The OFDA/CRED International Disaster Database*. Université catholique de Louvain, Brussels.

Dabelko, Geoffrey D., Lauren Herzer Risi, Schuyler Null, Meaghan Parker, and Russell Sticklor. 2013. *Backdraft: The Conflict Potential of Climate Change Adaptation and Mitigation*. Available at www.wilsoncenter.org/publication/backdraft-the-conflict-potential-climate-change-adaptation-and-mitigation. Accessed June 21, 2021.

Dahir, Abdi Latif. 2020. Uganda releases opposition leader after clashes kill at least 28. *The New York Times*. Available at www.nytimes.com/2020/11/20/world/africa/Uganda-Bobi-Wine-protests.html. Accessed December 7, 2020.

Dalby, Simon. 2009. *Security and Environmental Change*. First edition. Cambridge, UK: Polity.

Daoudy, Marwa. 2020. *The Origins of the Syrian Conflict: Climate Change and Human Security*. Cambridge, UK: Cambridge University Press.

Dasgupta, Susmita, Benoit Laplante, Siobhan Murray, and David Wheeler. 2011. Exposure of developing countries to sea-level rise and storm surges. *Climatic Change* 106 (4): 567–579. Available at https://doi.org/10.1007/s10584-010-9959-6. Accessed July 2, 2021.

Dasgupta, Susmita, Mainul Huq, Zahirul Huq Khan et al. 2014. Cyclones in a changing climate: The case of Bangladesh. *Climate and Development* 6 (2): 96–110. Available at https://doi.org/10.1080/17565529.2013.868335. Accessed July 2, 2021.

Day, Adam, and Jessica Caus. 2020. *Conflict Prevention in the Era of Climate Change: Adapting the UN to Climate-Security Risks*. New York: United

Nations University Centre for Policy Research. Available at https://cpr.unu.edu/climate-security.html. Accessed December 8, 2020.

de Carvalho, Benjamin, Halvard Leira, and John M. Hobson. 2011. The big bangs of IR: The myths that your teachers still tell you about 1648 and 1919. *Millennium* 39 (3): 735–758. Available at https://doi.org/10.1177/030582981 1401459. Accessed November 13, 2020.

de Châtel, Francesca. 2014. The role of drought and climate change in the Syrian uprising: Untangling the triggers of the revolution. *Middle Eastern Studies* 50 (4): 521–535. Available at http://dx.doi.org/10.1080/00263206.2013.850076. Accessed June 21, 2021.

De Juan, Alexander, and André Bank. 2015. The Ba'athist blackout? Selective goods provision and political violence in the Syrian civil war. *Journal of Peace Research* 52 (1): 91–104. Available at https://doi.org/10.1177/002234331455 9437. Accessed December 8, 2019.

de Ree, Joppe, and Eleonora Nillesen. 2009. Aiding violence or peace? The impact of foreign aid on the risk of civil conflict in sub-Saharan Africa. *Journal of Development Economics* 88 (2): 301–313. Available at https://doi.org/10.1016/j.jdeveco.2008.03.005. Accessed June 23, 2021.

de Soysa, Indra. 2000. The resource curse: Are civil wars driven by rapacity or paucity? In *Greed and Grievance: Economic Agendas in Civil Wars*, edited by Mats Berdal and David M. Malone, 113–136. Ottawa: Lynne Rienner.

De Stefano, Lucia, James Duncan, Shlomi Dinar et al. 2012. Climate change and the institutional resilience of international river basins. *Journal of Peace Research* 49 (1): 193–209. Available at http://jpr.sagepub.com/content/49/1/193. Accessed June 12, 2015.

De Stefano, L., Jacob D. Petersen-Perlman, Eric A. Sproles, Jim Eynard, and Aaron T. Wolf. 2017. Assessment of transboundary river basins for potential hydro–political tensions. *Global Environmental Change* 45: 35–46. Available at https://doi.org/10.1016/j.gloenvcha.2017.04.008. Accessed June 21, 2021.

de Waal, Alexander. 1991. *Evil Days: Thirty Years of War and Famine in Ethiopia*. New York: Human Rights Watch.

de Waal, Alexander. 1997. *Famine Crimes: Politics & the Disaster Relief Industry in Africa*. London: African Rights & The International African Institute, in association with James Currey and Indiana University Press.

de Waal, Alex. 2007a. Class and power in a stateless Somalia. *Social Science Research Council*. Available at https://items.ssrc.org/crisis-in-the-horn-of-africa/class-and-power-in-a-stateless-somalia/. Accessed December 21, 2020.

de Waal, Alex. 2007b. Is climate change the culprit for Darfur? Available at https://africanarguments.org/2007/06/is-climate-change-the-culprit-for-darfur/. Accessed June 25, 2021.

de Waal, Alex. 2009. Mission without end? Peacekeeping in the African political marketplace. *International Affairs (Royal Institute of International Affairs 1944-)* 85 (1): 99–113. Available at www.jstor.org/stable/27694922. Accessed July 23, 2020.

de Waal, Alex. 2016. Is the era of great famines over? *The New York Times*. Available at www.nytimes.com/2016/05/09/opinion/is-the-era-of-great-famines-over.html. Accessed June 13, 2016.

de Waal, Alex. 2018. *Mass Starvation: The History and Future of Famine*. First edition. Cambridge, UK; Malden, MA: Polity.

de Waal, Alex. 2020. Violence in Ethiopia doesn't stay there. *Foreign Policy*. Available at https://foreignpolicy.com/2020/11/19/violence-in-ethiopia-doesnt-stay-there/. Accessed December 21, 2020.

Del Rosso, Stephen J Jr. 1995. The insecure state: Reflections on "the state" and "security" in a changing world. *Daedalus* 124 (2): 175–207.

Dellmuth, Lisa, Maria-Therese Gustafsson, Niklas Bremberg, and Malin Mobjörk. 2017. *IGOs and Global Climate Security Challenges: Implications for Academic Research and Policymaking*. SIPRI. Available at www.sipri.org/sites/default/files/2017-12/fs_1712_igos_and_climate_security_0.pdf. Accessed June 21, 2021.

Department of Defense. 2014. *2014 Climate Change Adaptation Roadmap*.

Department of Defense. 2018. *Department of Defense Climate-Related Risk to DoD Infrastructure Initial Vulnerability Assessment Survey (SLVAS) Report*. Available at www.acq.osd.mil/eie/Downloads/Congress/Climate-Related%20Risk%20to%20DoD%20Infrastructure%20(SLVAS)%20Report.pdf. Accessed June 21, 2021.

Department of Defense. 2019. *Report on Effects of a Changing Climate to the Department of Defense*. Available at https://media.defense.gov/2019/Jan/29/2002084200/-1/-1/1/CLIMATE-CHANGE-REPORT-2019.PDF. Accessed June 21, 2021.

Deudney, Daniel. 1990. The case against linking environmental degradation and national security. *Millennium* 19 (3): 461–476.

Devlin, Colleen, and Cullen S. Hendrix. 2014. Trends and triggers redux: Climate change, rainfall, and interstate conflict. *Political Geography* 43: 27–39. Available at https://doi.org/10.1016/j.polgeo.2014.07.001. Accessed June 21, 2021.

Diamond, Jared M. 1999. *Guns, Germs, and Steel: The Fates of Human Societies*. First edition. New York: W. W. Norton & Company.

Diamond, Jared. 2004. *Collapse: How Societies Choose to Succeed or Fail*. New York: Viking.

Diez, Thomas, Franziskus von Lucke, and Zehra Wellmann. 2016. *The Securitisation of Climate Change: Actors, Processes and Consequences*. First edition. Abingdon, UK: Routledge.

Diya, Sabhanaz Rashid, and Jennifer Bussell. 2017. Disaster Preparedness in Bangladesh. Strauss Center for International Security and Law, University of Texas at Austin. Available at www.strausscenter.org/cepsa-research-briefs?download=645:disaster-preparedness-in-bangladesh. Accessed June 21, 2021.

Dorosh, Paul, and Shahidur Rashid. 2015. Ethiopia's 2015 drought: No reason for a famine. *IFPRI*. Available at www.ifpri.org/blog/ethiopias-2015-drought-no-reason-famine. Accessed March 27, 2019.

Douglass, Collin. 2017. Sea Level Rise and Deterritorialized States and Migration. The Center for Climate & Security. Available at https://climateandsecurity.org/

2017/11/briefer-sea-level-rise-and-deterritorialized-states/. Accessed December 1, 2020.

Dukhan, Haian. 2014. Tribes and tribalism in the Syrian uprising. *Syria Studies* 6 (2): 1–28. Available at https://ojs.st-andrews.ac.uk/index.php/syria/article/view/897. Accessed December 25, 2020.

Dukhan, Haian. 2018. *State and Tribes in Syria: Informal Alliances and Conflict Patterns.* First edition. London; New York: Routledge.

Egorova, Aleksandra, and Cullen Hendrix. 2014. Can Natural Disasters Precipitate Peace? Research Brief – August 2014. Strauss Center for International Security and Law, University of Texas at Austin. Available at www.strausscenter.org/wp-content/uploads/researchbrief22-ccaps_f-for-web1.pdf. Accessed July 5, 2021.

Eklöw, Karolina, and Florian Krampe. 2019. *Climate-Related Security Risks and Peacebuilding in Somalia.* SIPRI. Available at www.sipri.org/publications/2019/sipri-policy-papers/climate-related-security-risks-and-peacebuilding-somalia. Accessed June 18, 2020.

Eklund, Lina, and Darcy Thompson. 2017. Differences in resource management affects drought vulnerability across the borders between Iraq, Syria, and Turkey. *Ecology and Society* 22 (4): 9. Available at https://doi.org/10.5751/ES-09179-220409 Accessed June 21, 2020.

Elster, Jon. 1989. Social norms and economic theory. *The Journal of Economic Perspectives* 3 (4): 99–117. Available at www.jstor.org/stable/1942912. Accessed November 20, 2020.

Elster, Jon. 2000. Rational choice history: A case of excessive ambition. *American Political Science Review* 94 (3): 685–695.

Emanuel, Kerry. 2005. *Divine Wind: The History and Science of Hurricanes.* Oxford: Oxford University Press.

Encyclopedia Britannica. n.d.a. Ethiopia: Soils. Available at www.britannica.com/place/Ethiopia/Soils#ref1034070. Accessed June 21, 2021.

Encyclopedia Britannica. n.d.b. Somalia: Land. Available at www.britannica.com/place/Somalia/Land#ref419582. Accessed June 21, 2021.

Erian, Wadid, Bassem Katlan, and Ouldbdey Babah. 2010. Drought Vulnerability in the Arab Region: Special Case Study: Syria. UNISDR. Available at www.preventionweb.net/english/hyogo/gar/2011/en/bgdocs/Erian_Katlan_&_Babah_2010.pdf. Accessed June 21, 2021.

ESRI. n.d. Bhola Cyclone in 1970. Available at www.arcgis.com/apps/MapJournal/index.html?appid=f3cdea1974eb48d584e1ca8c76bc6a92. Accessed September 13, 2019.

Esty, Daniel C., Jack A. Goldstone, Ted Robert Gurr et al. 1999. State Failure Task Force Report: Phase II Findings. Environmental Change and Security Project Report. Woodrow Wilson Center. Available at www.wilsoncenter.org/sites/default/files/media/documents/event/Phase2.pdf. Accessed July 5, 20121.

ETH Zurich. 2015. *Ethnicity in Haiti.* Available at https://growup.ethz.ch/atlas/Haiti. Accessed June 21, 2021.

ETH Zurich. 2018. Ethnic power relations. *International Conflict Research.* Available at https://icr.ethz.ch/data/epr/core/. Accessed September 16, 2019.

Ethiopia National Meteorological Agency. n.d. Seasonal forecast. Available at https://web.archive.org/web/20200227145342/www.ethiomet.gov.et/other_forecasts/seasonal_forecast. Accessed July 5, 2021.

European Commission's Directorate-General for European Civil Protection and Humanitarian Aid Operations. 2019. India, Bangladesh – Tropical Cyclone FANI update (DG ECHO, UN OCHA, IMD, BMD) (ECHO Daily Flash of 07 May 2019). *ReliefWeb*. Available at https://reliefweb.int/report/india/india-bangladesh-tropical-cyclone-fani-update-dg-echo-un-ocha-imd-bmd-echo-daily-flash. Accessed September 15, 2019.

Fanack Water. 2019. Water resources in Syria. Available at https://water.fanack.com/syria/water-resources/. Accessed December 4, 2019.

FAO. 2019a. GIEWS country brief on Syrian Arab Republic. Available at www.fao.org/giews/countrybrief/country.jsp?code=SYR. Accessed November 25, 2019.

FAO. 2019b. GIEWS country brief on Lebanon. Available at www.fao.org/giews/countrybrief/country.jsp?lang=en&code=LBN. Accessed November 25, 2019.

Faris, Stephan. 2007. The real roots of Darfur. *The Atlantic Monthly*.

Faris, Stephan. 2009. *Forecast: The Surprising – and Immediate – Consequences of Climate Change*. New York: Henry Holt and Company.

Fazal, Tanisha M. 2007. *State Death: The Politics and Geography of Conquest, Occupation, and Annexation*. Princeton, NJ: Princeton University Press.

Fearon, James, and David D. Laitin. 2003. Ethnicity, insurgency, and civil war. *American Political Science Review* 1: 75–90.

Federal Republic of Somalia. 2017. *Aid Flows in Somalia: Analysis of aid flow data*. Available at https://reliefweb.int/sites/reliefweb.int/files/resources/Aid%20Flows%20Booklet%20-%202017.pdf. Accessed June 21, 2021.

Feitelson, Eran, and Amit Tubi. 2017. A main driver or an intermediate variable? Climate change, water and security in the Middle East. *Global Environmental Change* 44: 39–48. Available at https://doi.org/10.1016/j.gloenvcha.2017.03.001. Accessed June 21, 2021.

Femia, Francesco, and Caitlin Werrell. 2012. Syria: Climate change, drought and social unrest. The Center for Climate & Security. Available at https://climateandsecurity.org/2012/02/29/syria-climate-change-drought-and-social-unrest/. Accessed November 26, 2018.

Femia, Francesco, and Caitlin Werrell. 2016. *The Climate and Security Imperative*. Routledge (E-book).

FEWSNET. 2011. Expanding famine across southern Somalia. Available at http://fews.net/sites/default/files/documents/reports/FSNAU_FEWSNET_2007 11press%20release_final.pdf. Accessed June 21, 2021.

FEWSNET. 2015. Food security emergency in central/eastern Ethiopia follows worst drought in more than 50 years. Available at http://fews.net/sites/default/files/documents/reports/FEWS%20NET_WFP_Ethiopia%20Alert_20151204.pdf. Accessed June 21, 2021.

Finamore, Barbara. 2020. What China's plan for net-zero emissions by 2060 means for the climate. *The Guardian*. Available at www.theguardian.com/commentisfree/2020/oct/05/china-plan-net-zero-emissions-2060-clean-technology. Accessed December 22, 2020.

Financial Tracking Service. 2019a. Lebanon crisis 2006 (Flash appeal). Available at https://fts.unocha.org/appeals/214/summary. Accessed December 8, 2019.

Financial Tracking Service. 2019b. Syria drought appeal 2008. Available at https://fts.unocha.org/appeals/300/summary. Accessed December 9, 2019.

Financial Tracking Service. 2019c. Syria drought response plan (revised) (July 2009–June 2010). Available at https://fts.unocha.org/appeals/313/summary. Accessed December 9, 2019.

Findley, Michael G. 2018. Does foreign aid build peace? *Annual Review of Political Science* 21 (1): 359–384. Available at https://doi.org/10.1146/annurev-polisci-041916-015516. Accessed November 12, 2018.

Fjelde, Hanne, and Nina von Uexkull. 2012. Climate triggers: Rainfall anomalies, vulnerability and communal conflict in Sub-Saharan Africa. *Political Geography* 31 (7): 444–453. Available at https://doi.org/10.1016/j.polgeo.2012.08.004. Accessed June 22, 2021.

Flanigan, Shawn Teresa, and Mounah Abdel-Samad. 2009. Hezbollah's social Jihad: Nonprofits as resistance organizations. *Middle East Policy* 16 (2): 122–138. Available at https://go.gale.com/ps/i.do?p=AONE&sw=w&issn=10611924&v=2.1&it=r&id=GALE%7CA203482071&sid=googleScholar&linkaccess=abs. Accessed December 8, 2020.

Fleming, James. 2013. Pounds of prevention: Focus on India. *USAID Impact Blog*. Available at https://blog.usaid.gov/2013/10/pounds-of-prevention-focus-on-india/. Accessed September 18, 2019.

Floodlist News. 2019. India and Bangladesh – Tropical Cyclone Fani brings storm surge and severe wind damage. Available at http://floodlist.com/asia/india-bangladesh-tropical-cyclone-fani-may-2019. Accessed September 13, 2019.

Flores, Alejandro Quiroz, and Alastair Smith. 2010. Surviving Disasters. Available at www.researchgate.net/publication/268203178_Surviving_Disasters. Accessed July 5, 2021.

Food and Agriculture Organization. 2005. Somalia. Irrigation in Africa in figures – AQUASTAT survey 2005. Available at www.fao.org/nr/water/aquastat/countries_regions/SOM/SOM-CP_eng.pdf. Accessed June 22, 2021.

Food and Agriculture Organization. 2018. *Drought Characteristics and Management in North Africa and the Near East*. Rome: FAO.

Food and Agriculture Organization. 2019a. Aquastat. Available at www.fao.org/nr/water/aquastat/data/query/index.html?lang=en. Accessed June 22. 2021.

Food and Agriculture Organization. 2019b. Earth observation: Agricultural stress index. Available at www.fao.org/giews/earthobservation/asis/index_1.jsp?lang=en. Accessed December 4, 2019.

Food and Agriculture Organization. 2019c. Earth observation: Lebanon. Available at www.fao.org/giews/earthobservation/country/index.jsp?lang=en&code=LBN. Accessed December 4, 2019.

Food and Agriculture Organization. 2019d. Earth observation: Syrian Arab Republic. Available at www.fao.org/giews/earthobservation/country/index.jsp?lang=en&code=SYR. Accessed December 4, 2019.

Foster, Gavin L., Dana L. Royer, and Daniel J. Lunt. 2017. Future climate forcing potentially without precedent in the last 420 million years. *Nature*

Communications 8: 14845. Available at https://doi.org/10.1038/ncomms14845. Accessed June 22, 2021.

Fountain, Henry. 2015. Researchers link Syrian conflict to a drought made worse by climate change. *The New York Times*. Available at www.nytimes.com/2015/03/03/science/earth/study-links-syria-conflict-to-drought-caused-by-climate-change.html. Accessed December 2, 2016.

Fountain, Henry. 2017. Sydney's swelter has a climate change link, scientists say. *The New York Times*. Available at www.nytimes.com/2017/03/02/science/australia-heat-climate-change.html. Accessed April 11, 2017.

Frank, Neil L., and S. A. Husain. 1971. The deadliest tropical cyclone in history? *Bulletin of the American Meteorological Society* 52 (6): 438–445. Available at https://doi.org/10.1175/1520-0477(1971)052<0438:TDTCIH>2.0.CO;2. Accessed June 22, 2021.

Freeman, Laura. 2017. Environmental change, migration, and conflict in Africa: A critical examination of the interconnections. *The Journal of Environment & Development* 26 (4): 351–374. Available at https://doi.org/10.1177/1070496517727325. Accessed June 18, 2018.

Friedman, Thomas L. 2013. Without water, revolution. *The New York Times*. Available at www.nytimes.com/2013/05/19/opinion/sunday/friedman-without-water-revolution.html. Accessed December 1, 2016.

Friedman, Tom. 2014. Climate wars – Syria with Tom Friedman. Available at https://theyearsproject.com/story/climate-wars/. Accessed June 22, 2021.

Fritz, Hermann M., Chris Blount, Swe Thwin, Moe Kyaw Chan, and Nyein Chan. 2011. Observations and modeling of Cyclone Nargis storm surge in Myanmar. ASCE conference paper. Available at https://doi.org/10.1061/41185(417)1. Accessed June 22, 2021.

Fröhlich, Christiane J. 2016. Climate migrants as protestors? Dispelling misconceptions about global environmental change in pre-revolutionary Syria. *Contemporary Levant* 1 (1): 38–50. Available at http://dx.doi.org/10.1080/20581831.2016.1149355. Accessed June 22, 2021.

Fukuyama, Francis. 2013. What is governance? *Governance* 26 (3): 347–368. Available at https://doi.org/10.1111/gove.12035. Accessed June 22, 2021.

Funk, Chris, Sharon E. Nicholson, Martin Landsfeld et al. 2015. The Centennial Trends Greater Horn of Africa precipitation dataset. *Scientific Data* 2: 150050. Available at https://doi.org/10.1038/sdata.2015.50. Accessed July 8, 2021.

Funke, Nikki, and Hussein Solomon. 2002. *The Shadow State in Africa: A Discussion*. Addis Ababa: Development Policy Management Forum.

Gallagher, Mary E., and Jonathan K. Hanson. 2015. Power tool or dull blade? Selectorate theory for autocracies. *Annual Review of Political Science* 18 (1): 367–385. Available at https://doi.org/10.1146/annurev-polisci-071213-041224. Accessed November 10, 2018.

Gallup, John L., and Jeffrey D. Sachs. 2001. The economic burden of malaria. *The American Journal of Tropical Medicine and Hygiene* 64 (1–2 Suppl.): 85–96.

Gates, Scott, Håvard Mokleiv Nygård, and Esther Trappeniers. 2016. *Conflict Recurrence*. Conflict Trends, PRIO. Available at www.prio.org/Publications/Publication/?x=9056. Accessed November 20, 2021.

George, Alexander L., and Andrew Bennett. 2005. *Case Studies and Theory Development in the Social Sciences*. Cambridge, MA: MIT Press.

Gill, Peter. 2012. *Famine and Foreigners: Ethiopia Since Live Aid*. First edition. Oxford: Oxford University Press.

Gilpin, Robert. 1981. *War and Change in World Politics*. Cambridge; New York: Cambridge University Press.

Giraudy, Agustina. 2012. Conceptualizing state strength: Moving beyond strong and weak states. *Revista de ciencia política (Santiago)* 32 (3): 599–611. Available at http://dx.doi.org/10.4067/S0718-090X2012000300005. Accessed June 22, 2021.

Gleditsch, Nils Petter. 1998. Armed conflict and the environment: A critique of the literature. *Journal of Peace Research* 35 (3): 381–400.

Gleditsch, Nils Petter. 2012. Whither the weather? Climate change and conflict. *Journal of Peace Research* 49 (1): 3–9. Available at https://doi.org/10.1177/0022343311431288. Accessed June 22, 2021.

Gleick, Peter H. 2014. Water, drought, climate change, and conflict in Syria. *Weather, Climate, and Society* 6 (3): 331–340. Available at www.jstor.org/stable/24907379. Accessed June 22, 2021.

Gleick, Peter H. 2017. Climate, water, and conflict: Commentary on Selby et al. 2017. *Political Geography* 60: 248–250.

Gleditsch, Nils Petter. 2021. This time is different! Or is it? *Neomalthusians and Environmental Optimists in the Age of Climate Change* 58 (1), Available at https://doi.org/10.1177/0022343320969785. Accessed June 22, 2021.

Gleditsch, Nils Petter, and Ragnhild Nordås. 2014. Conflicting messages? The IPCC on conflict and human security. *Political Geography* 43: 82–90. Available at https://doi.org/10.1016/j.polgeo.2014.08.007 Accessed June 22, 2021.

Gleditsch, Nils Petter, Ragnhild Nordås, and Idean Salehyan. 2007. *Climate Change and Conflict: The Migration Link*. Coping with Crisis Working Paper Series, International Peace Academy. Available at www.ipacademy.org/asset/file/169/CWC_Working_Paper_Climate_Change.pdf. Accessed June 22, 2021.

Global Carbon Project. 2019. *Global Carbon Budget Report 2019*. Available at www.globalcarbonproject.org/global/images/carbonbudget/Infographic_Emissions2019.pdf. Accessed June 22, 2021.

Global Facility for Disaster Risk Reduction. n.d. Somalia. Available at www.gfdrr.org/en/somalia. Accessed August 29, 2019.

Global Risk Data Platform. 2013. United Nations Environment Programme. Available at https://preview.grid.unep.ch/index.php?preview=map&lang=eng. Accessed June 22, 2021.

GlobalSecurity.Org. n.d. Operation Sea Angel / Productive Effort. Available at www.globalsecurity.org/military/ops/sea_angel.htm. Accessed September 18, 2019.

Goldsmith, Leon. 2018. Syria's Alawis: Structure, perception and agency in the Syrian security dilemma. In *The Syrian Uprising*, edited by Raymond A. Hinnebusch and Omar Imady, 141–158. Abingdon, UK: Routledge.

Goobjoog News. 2015. Nominated ministers and their clans. Available at http://goobjoog.com/english/nominated-ministers-and-their-clans/. Accessed March 26, 2019.

Gore, Albert. 2006. *An Inconvenient Truth: The Planetary Emergency of Global Warming and What We Can Do about It*. Emmaus, PA: Rodale Press.

Gottlieb, Gregory. 2018. 10 years after, Cyclone Nargis still holds lessons for Myanmar. *The Conversation*. Available at http://theconversation.com/10-years-after-cyclone-nargis-still-holds-lessons-for-myanmar-95039. Accessed September 15, 2019.

Government of Bangladesh. 1991. Density of Population – 1991. Available at https://reliefweb.int/map/bangladesh/bangladesh-density-population-1991. Accessed July 5, 2021.

Government of Bangladesh. 1998. *Bangladesh: End of IDNDR Assessment Report (1998)*. Available at www.preventionweb.net/english/hyogo/progress/reports/v.php?id=32451&pid:223. Accessed September 18, 2019.

Government of Bangladesh. 2009. *Bangladesh: National progress Report on the Implementation of the Hyogo Framework for Action (2007–2009)*. Available at www.preventionweb.net/english/hyogo/progress/reports/v.php?id=7485&pid:223. Accessed September 18, 2019.

Government of Bangladesh. 2015. *Bangladesh: National Progress Report on the Implementation of the Hyogo Framework for Action (2013–2015)*. Available at www.preventionweb.net/english/hyogo/progress/reports/v.php?id=40155&pid:223. Accessed September 18, 2019.

Government of Ethiopia. 2013. *Ethiopia: National Policy and Strategy on Disaster Risk Management – Policy, Plans & Statements*. Available at www.preventionweb.net/english/professional/policies/v.php?id=42435. Accessed September 11, 2019.

Government of Ethiopia. 2015. *National Progress Report on the Implementation of the Hyogo Framework for Action (2013–2015)*. Available at www.preventionweb.net/files/41877_ETH_NationalHFAprogress_2013-15.pdf. Accessed June 22, 2021.

Government of India. 2014. *India: National Progress Report on the Implementation of the Hyogo Framework for Action (2013–2015)*. Available at www.preventionweb.net/english/hyogo/progress/reports/v.php?id=40210&pid:223. Accessed September 18, 2019.

Government of India. 2016. Primary census abstract, 2001 – Orissa. *data.gov.in*. Available at https://data.gov.in/resources/primary-census-abstract-2001-orissa. Accessed September 13, 2019.

Government of India. 2019. List of districts of Orissa. Available at www.census2011.co.in/census/state/districtlist/orissa.html. Accessed September 13, 2019.

Government of Myanmar. 2010. *Myanmar: National Progress Report on the Implementation of the Hyogo Framework for Action (2009–2011) – Interim*. Available at www.preventionweb.net/english/hyogo/progress/reports/v.php?id=16315&pid:223. Accessed June 22, 2021.

Government of the Netherlands. 2019. *Climate Change Profile: Lebanon*. Available at https://reliefweb.int/report/lebanon/climate-change-profile-lebanon. Accessed December 4, 2019.

Green, Jessica F., and Thomas N. Hale. 2017. Reversing the marginalization of global environmental politics in international relations: An opportunity for the discipline. *PS: Political Science & Politics* 50 (2): 473–479. Available at https://doi.org/10.1017/S1049096516003024. Accessed June 22, 2020.

Grono, Nick. 2010. Fragile states and conflict. *International Crisis Group*. Available at www.crisisgroup.org/global/fragile-states-and-conflict. Accessed December 8, 2020.

Grunebaum, Dave. 2019. Cyclone raises fears about vulnerability of Rohingya refugee camps. *Voice of America*. Available at www.voanews.com/south-central-asia/cyclone-raises-fears-about-vulnerability-rohingya-refugee-camps. Accessed July 22, 2020.

Guha-Sapir, Debarati, and Florian Vogt. 2009. Cyclone Nargis in Myanmar: Lessons for public health preparedness for cyclones. *American Journal of Disaster Medicine* 4 (5): 273–278.

Gundel, Joakim. 2009. *Clans in Somalia*. Austrian Red Cross. Available at www.refworld.org/pdfid/4b29f5e82.pdf%20. Accessed June 22, 2021.

Habib, Arjumand, Md. Shahidullah, and Dilder Ahmed. 2012. The Bangladesh cyclone preparedness program. A vital component of the nation's multi-hazard early warning system. In *Institutional Partnerships in Multi-Hazard Early Warning Systems*, edited by M. Golnarahhi, 29–62. Berlin: Springer-Verlag.

Haldevang, Max de. 2017. The enigma of Assad: How a painfully shy eye doctor turned into a murderous tyrant. *Quartz*. Available at https://qz.com/959806/the-enigmatic-story-of-how-syrias-bashar-al-assad-turned-from-a-painfully-shy-eye-doctor-into-a-murderous-tyrant/. Accessed December 5, 2019.

Halimuddin, Sandi. 2013. U.S. counterterrorism laws block international humanitarian aid. *World Policy*. Available at https://worldpolicy.org/2013/12/19/u-s-counterterrorism-laws-block-international-humanitarian-aid/. Accessed March 27, 2019.

Hanson, Jonathan K. 2018. State capacity and the resilience of electoral authoritarianism: Conceptualizing and measuring the institutional underpinnings of autocratic power. *International Political Science Review* 39 (1): 17–32. Available at https://doi.org/10.1177/0192512117702523. Accessed November 11, 2018.

Hanson, Jonathan K., and Rachel Sigman. 2013. *Leviathan's Latent Dimensions: Measuring State Capacity for Comparative Political Research*. Unpublished manuscript.

Haque, Ubydul, Masahiro Hashizume, Korine N Kolivras et al. 2012. Reduced death rates from cyclones in Bangladesh: What more needs to be done? *Bulletin of the World Health Organization* 90 (2): 150–156. Available at https://doi.org/10.2471/BLT.11.088302. Accessed June 22, 2021.

Hardt, Judith Nora. 2017. *Environmental Security in the Anthropocene: Assessing Theory and Practice*. First edition. London; New York: Routledge.

Hartzell, Caroline A., Matthew Hoddie, Robert J. Art, and Robert Jervis. 2016. Crafting peace through power sharing. In *International Politics: Enduring Concepts and Contemporary Issues*, edited by Robert J. Art, and Robert Jervis Thirteenth edition, 442–453. New York: Pearson.

Harvey, Hal. 2018. Carbon math. *Twitter*. Available at https://twitter.com/hal_harvey/status/1068175177874202624. Accessed January 7, 2020.

Hauge, Wenche, and Tanja Ellingsen. 2001. Causal pathways to conflict. In *Environmental Conflict*, edited by Paul F. Diehl and Nils Petter Gleditsch, 36–57. Boulder, CO: Westview Press.

Hausfather, Zeke, and Glen P. Peters. 2020. Emissions – the "business as usual" story is misleading. *Nature* 577 (7792): 618–620. Available at https://media.nature.com/original/magazine-assets/d41586-020-00177-3/d41586-020-00177-3.pdf. Accessed October 13, 2020.

Hegre, Håvard, Halvard Buhaug, Katherine V. Calvin et al. 2016. Forecasting civil conflict along the shared socioeconomic pathways. *Environmental Research Letters* 11 (5): 054002. Available at https://doi.org/10.1088/17489326/11/5/054002. Accessed June 22, 2021.

Hendrix, Cullen S. 2010. Measuring state capacity: Theoretical and empirical implications for the study of civil conflict. *Journal of Peace Research* 47 (3): 273–285. Available at www.jstor.org/stable/20752162. Accessed March 22, 2019.

Hendrix, Cullen S. 2013. Climate Change, Global Food Markets, and Urban Unrest. Strauss Center for International Security and Law, University of Texas at Austin. Available at www.strausscenter.org/wp-content/uploads/CCAPS-Research-Brief-No.-7_final.pdf. Accessed July 5, 2021.

Hendrix, Cullen. 2016. Putting environmental stress (back) on the mass atrocities agenda. Center for Climate & Security. Available at https://climateandsecurity.org/2016/10/putting-environmental-stress-back-on-the-mass-atrocities-agenda/. Accessed June 22, 2021.

Hendrix, Cullen. 2017a. Climate change and the Syrian civil war. *Political Violence at a Glance*. Available at http://politicalviolenceataglance.org/2017/09/19/climate-change-and-the-syrian-civil-war/. Accessed November 26, 2018.

Hendrix, Cullen S. 2017b. A comment on "climate change and the Syrian civil war revisited." *Political Geography* 60: 251–252.

Hendrix, Cullen S. 2017c. The streetlight effect in climate change research on Africa. *Global Environmental Change* 43: 137–147. Available at https://doi.org/10.1016/j.gloenvcha.2017.01.009. Accessed June 22, 2021.

Hendrix, Cullen. 2020. Shifting baselines and their implications for climate change and conflict research. Unpublished paper, University of Denver.

Hendrix, Cullen S., and Sarah M. Glaser. 2007. Trends and triggers: Climate change and civil conflict in Sub-Saharan Africa. *Political Geography* 26 (6): 695–715. Available at https://doi.org/10.1016/j.polgeo.2007.06.006. Accessed June 22, 2021.

Hendrix, Cullen S, and Stephan Haggard. 2015. Global food prices, regime type, and urban unrest in the developing world. *Journal of Peace Research* 52 (2): 143–157. Available at https://doi.org/10.1177/0022343314561599. Accessed June 18, 2018.

Hendrix, Cullen S., and Idean Salehyan. 2012. Climate change, rainfall, and social conflict in Africa. *Journal of Peace Research* 49 (1): 35–50. Available at https://doi.org/10.1177/0022343311426165. Accessed June 22, 2021.

Hendrix, Cullen, Scott Gates, and Halvard Buhaug. 2016. Environment and conflict. In *What Do We Know about Civil Wars?*, edited by T. David Mason

and Sara M. Mitchell, 231–246. Reprint edition. Lanham, MD: Rowman & Littlefield.
Hendry, Krista. 2014. From failed to fragile: Renaming the index. *Fund for Peace*. Available at https://web.archive.org/web/20190311010721/http://library.fundforpeace.org/fsi14-namechange. Accessed December 8, 2020.
Herman, Steve. 2009. Indian forecasters gave Burma advance warning of Cyclone Nargis. *Voice of America*. Available at www.voanews.com/archive/indian-forecasters-gave-burma-advance-warning-cyclone-nargis. Accessed September 18, 2019.
Hinnebusch, Raymond. 2012. Syria: From "authoritarian upgrading" to revolution? *International Affairs* 88 (1): 95–113. Available at https://doi.org/10.1111/j.1468-2346.2012.01059.x. Accessed June 22, 2021.
Hinnebusch, Raymond. 2019. What went wrong: Understanding the trajectory of Syria's conflict. In *Syria: From National Independence to Proxy War*, edited by Linda Matar and Ali Kadri, 29–52. Cham: Springer.
Hinnebusch, Raymond, and Omar Imady. 2018. *The Syrian Uprising*. First edition. London; New York: Routledge.
Hoerling, Martin, Jon Eischeid, Judith Perlwitz et al. 2012. On the increased frequency of Mediterranean drought. *Journal of Climate* 25 (6): 2146–2161. Available at www.jstor.org/stable/26191 30. Accessed December 6, 2019.
Homer-Dixon, Thomas F. 1991. On the threshold: Environmental changes as causes of acute conflict. *International Security* 16 (2): 76–116.
Homer-Dixon, Thomas F. 1994. Environmental scarcities and violent conflict: Evidence from cases. *International Security* 19 (1): 5–40.
Homer-Dixon, Thomas F. 1999. *Environment, Scarcity, and Violence*. Princeton, NJ: Princeton University Press.
Homer-Dixon, Thomas. 2007. Cause and Effect. Available at https://homerdixon.com/cause-effect/. Accessed June 22, 2021.
Homer-Dixon, Thomas. 2009. The rise of complex terrorism. *Foreign Policy*. Available at https://foreignpolicy.com/2009/11/16/the-rise-of-complex-terrorism/. Accessed March 2, 2019.
Homer-Dixon, Thomas F., and Jessica Blitt, eds. 1998. *Ecoviolence: Links among Environment, Population, and Security*. Lanham, MD: Rowman & Littlefield.
Homer-Dixon, Thomas F., and Marc Levy. 1995a. Correspondence. *International Security* 20 (3): 189–198.
Homer-Dixon, Thomas F., and Marc A. Levy. 1995b. Environment and security. *International Security* 20 (3): 189–198. Available at https://doi.org/10.2307/2539143. Accessed June 22, 2021.
Horn, Heather. 2012. To know a tyrant: Inside Bashar al-Assad's transformation from "reformer" to Killer. *The Atlantic*. Available at www.theatlantic.com/international/archive/2012/09/to-know-a-tyrant-inside-bashar-al-assads-transformation-from-reformer-to-killer/262486. Accessed November 27, 2019.
Hossain, Naomi. 2018. The 1970 Bhola cyclone, nationalist politics, and the subsistence crisis contract in Bangladesh. *Disasters* 42 (1): 187–203. Available at https://doi.org/10.1111/disa.12235. Accessed August 8, 2018.
Hsiang, Solomon M., and K. C. Meng. 2014. Reconciling disagreement over climate–conflict results in Africa. *Proceedings of the National Academy of*

Sciences 111 (6): 2100–2103. Available at https://doi.org/10.1073/pnas.1316006111. Accessed June 22, 2021.

Hsiang, Solomon M., Kyle C. Meng, and Mark A. Cane. 2011. Civil conflicts are associated with the global climate. *Nature* 476 (7361): 438–441. Available at http://dx.doi.org/10.1038/nature10311. Accessed June 22. 2021.

Hsiang, Solomon M., Marshall Burke, and Edward Miguel. 2013. Quantifying the influence of climate on human conflict. *Science* 341 (6151): 1237557. Available at https://doi.org /10.1126/science.1235367. Accessed June 22, 2021.

Hubbard, Ben. 2020. Lebanon's government resigns amid widespread anger over blast. *The New York Times*. Available at www.nytimes.com/2020/08/10/world/middleeast/lebanon-government-resigns-beirut.html. Accessed September 4, 2020.

Hubbard, Ben, and Hwaida Saad. 2020. Lebanon's currency plunges, and protesters surge into streets. *The New York Times*. Available at www.nytimes.com/2020/06/11/world/middleeast/lebanon-protests.html. Accessed December 16, 2020.

Human Rights Watch. 2020. Lebanon: New coalition to defend free speech. *Human Rights Watch*. Available at www.hrw.org/news/2020/07/13/lebanon-new-coalition-defend-free-speech. Accessed December 25, 2020.

Ide, Tobias. 2018. Climate war in the Middle East? Drought, the Syrian civil war and the state of climate–conflict research. *Current Climate Change Reports* 4: 347–354. Available at https://doi.org/10.1007/s40641-018-0115-0. Accessed June 22, 2021.

Ide, Tobias. 2020. The dark side of environmental peacebuilding. *World Development* 127: 104777. Available at https://doi.org/10.1016/j.worlddev.2019.104777. Accessed June 22, 2021.

Ide, Tobias, Michael Brzoska, Jonathan F. Donges, and Carl-Friedrich Schleussner. 2020. Multi-method evidence for when and how climate-related disasters contribute to armed conflict risk. *Global Environmental Change* 62: 102063. Available at https://doi.org/10.1016/j.gloenvcha.2020.102063. Accessed June 22, 2021.

Ide, Tobias, Anders Kristensen, and Henrikas Bartusevičius. 2021. First comes the river, then comes the conflict? A qualitative comparative analysis of flood-related political unrest. *Journal of Peace Research* 58 (1). Available at https://doi.org/10.1177/0022343320966783. Accessed June 22, 2021.

Iglesias, Ana, Sonia Quiroga, and Agustin Diz. 2011. Looking into the future of agriculture in a changing climate. *European Review of Agricultural Economics* 38 (3): 427–447.

Ignatius, David. 1983. How to rebuild Lebanon. *Foreign Affairs*. Available at www.foreignaffairs.com/articles/lebanon/1983-06-01/how-rebuild-lebanon. Accessed December 5, 2019.

Ikeuchi, Hiroaki, Yukiko Hirabayashi, Dai Yamazaki et al. 2017. Compound simulation of fluvial floods and storm surges in a global coupled river-coast flood model: Model development and its application to 2007 Cyclone Sidr in Bangladesh. *Journal of Advances in Modeling Earth Systems* 9 (4): 1847–1862. Available at https://doi.org/10.1002/2017MS000943. Accessed June 22, 2021.

Im, Eun-Soon, Jeremy S. Pal, and Elfatih A. B. Eltahir. 2017. Deadly heat waves projected in the densely populated agricultural regions of South Asia. *Science Advances* 3 (8): e1603322. Available at https://doi.org/10.1126/sciadv.1603322. Accessed June 22, 2021.

India Meteorological Department. 2000. *Report on Cyclonic Disturbances over North Indian Ocean during 1999.* Available at http://rsmcnewdelhi.imd.gov.in/uploads/report/27/27_510109_35_66a331_1999.pdf. Accessed June 22, 2021.

India Meteorological Department. 2010. *Report on Cyclonic Disturbances over North Indian Ocean during 2009.* Available at http://rsmcnewdelhi.imd.gov.in/uploads/report/27/27_4e34f3_rsmc-2009.pdf. Accessed June 22, 2021.

India Meteorological Department. 2013. *Very Severe Cyclonic Storm, PHAILIN over the Bay of Bengal (08–14 October 2013): A Report.* Available at http://rsmcnewdelhi.imd.gov.in/uploads/report/26/26_38a1d4_phailin.pdf. Accessed June 22, 2021.

India Meteorological Department. 2018a. *Best Tracks Data (1982–2018).* Available at www.rsmcnewdelhi.imd.gov.in/index.php?option=com_content&view=article&id=48&Itemid=194&lang=en. Accessed June 22, 2021.

India Meteorological Department. 2018b. *Cyclone eAtlas – Tracks of Cyclones and Depressions over the North Indian Ocean 1891–2020.* Available at http://14.139.191.203/Login.aspx. Accessed July 5, 2021.

Indian Red Cross Society. 2019. *Odisha FANI cyclone Assessment Report – India. ReliefWeb.* Available at https://reliefweb.int/report/india/odisha-fani-cyclone-assessment-report. Accessed September 15, 2019.

Indiaonlinepages.com. 2019. *Population of Orissa 2019.* Available at www.indiaonlinepages.com/population/orissa-population.html. Accessed September 13, 2019.

International Crisis Group. 2018. *Al-Shabaab Five Years After Westgate: Still a Menace in East Africa.* Available at www.crisisgroup.org/africa/horn-africa/kenya/265-al-shabaab-five-years-after-westgate-still-menace-east-africa. Accessed June 22, 2021.

IPCC. 2001. *Climate Change 2001: The Scientific Basis.* Available at https://archive.ipcc.ch/ipccreports/tar/wg1/fig2-32.htm. Accessed November 20, 2020.

IPCC. 2011. *Managing the Risks of Extreme Events and Disasters to Advance Climate Change Adaptation (SREX).* Available at www.ipcc.ch/report/managing-the-risks-of-extreme-events-and-disasters-to-advance-climate-change-adaptation/. Accessed June 22, 2021.

IPCC. 2014a. *AR5 Climate Change 2014: Impacts, Adaptation, and Vulnerability.* Available at www.ipcc.ch/report/ar5/wg2/. Accessed November 18, 2020.

IPCC. 2014b. *IPCC 5th Assessment Report: Summary for Policymakers.* Available at www.ipcc.ch/site/assets/uploads/2018/02/ipcc_wg3_ar5_summary-for-policymakers.pdf. Accessed June 22, 2021.

IPCC. 2014c. *Topic 2: Future Changes, Risks and Impacts. IPCC 5th Assessment Synthesis Report.* Available at http://ar5-syr.ipcc.ch/topic_futurechanges.php. Accessed November 19, 2020.

IPCC. 2018. *Special Report: Global Warming of 1.5 °C.* Available at www.ipcc.ch/sr15/. Accessed June 11, 2019.

IPCC Working Group I. 2014. *Climate Change 2013: The Physical Science Basis. Contribution of Working Group I to the Fifth Assessment Report of the Intergovernmental Panel on Climate Change*. Cambridge, UK: Cambridge University Press.

ISciences. n.d. Water Security Indicator Model. *ISCIENCES*. Available at www.isciences.com/water-security-indicator-model/. Accessed June 22, 2021.

Ismail, Yasin. 2018. Somalia's clan politics. *World Policy*. Available at http://worldpolicy.org/2018/03/13/somalia-clan-politics/. Accessed March 27, 2019.

Ismay, John. 2020. What is ammonium nitrate, blamed in the Beirut explosion? *The New York Times*. Available at www.nytimes.com/2020/08/05/world/middleeast/beirut-explosion-ammonium-nitrate.html. Accessed December 16, 2020.

Iwasaki, Shimpei. 2016. Linking disaster management to livelihood security against tropical cyclones: A case study on Odisha state in India. *International Journal of Disaster Risk Reduction* 19: 57–63. Available at https://doi.org/10.1016/j.ijdrr.2016.08.019. Accessed June 22, 2021.

Jackson, Ashley, and Abdi Aynte. 2013. *Talking to the Other side: Humanitarian Negotiations with Al-Shabaab in Somalia*. London: Overseas Development Institute.

Jama, A. 2018. The 4.5 formula. *Medium*. Available at https://medium.com/@ajRabi_67780/the-4-5-formula-6f0820646cc5. Accessed March 26, 2019.

Jasparro, Christopher, and Jonathan Taylor. 2008. Climate change and regional vulnerability to transnational security threats in Southeast Asia. *Geopolitics* 13: 232–256.

Jensen, David. 2019. Environmental Cooperation for Peacebuilding. UNEP.

Johnson, McKenzie F., Luz A. Rodríguez, and Manuela Quijano Hoyos. 2021. Intrastate environmental peacebuilding: A review of the literature. *World Development* 137: 105150. Available at https://doi.org/10.1016/j.worlddev.2020.105150. Accessed June 22, 2021.

Jones, Adam. 2010. *Genocide: A Comprehensive Introduction*. Second edition. London; New York: Routledge.

Jones, Will, Ricardo Soares de Oliveira, and Harry Verhoeven. 2013. *Africa's Illiberal State-Builders*. RSC Working Paper Series 89. Available at www.rsc.ox.ac.uk/publications/africas-illiberal-state-builders. Accessed December 8, 2020.

Kahl, Colin H. 1998. Population growth, environmental degradation, and state-sponsored violence: The case of Kenya, 1991–93. *International Security* 23 (2): 80–119.

Kahl, Colin H. 2002. Demographic change, natural resources and violence: The current debate. *Journal of International Affairs* 56 (1): 257–282.

Kahl, Colin H. 2006. *States, Scarcity, and Civil Strife in the Developing World*. Princeton, NJ: Princeton University Press.

Kalsi, S. R. 2006. Orissa super cyclone – A Synopsis. *Mausam* 57 (1): 1–20. Available at http://metnet.imd.gov.in/mausamdocs/15711_F.pdf. Accessed June 22, 2021.

Kalyvas, Stathis, and Scott Straus. 2020. Stathis Kalyvas on 20 years of studying political violence. *Violence: An International Journal* 1 (2). Available at https://doi.org/10.1177/2633002420972955. Accessed June 22, 2021.

Kaplan, Robert. 1994. The coming anarchy. *Atlantic Monthly*.
Kaplan, Sarah. 2020. What does "dangerous" climate change really mean? *Washington Post*. Available at www.washingtonpost.com/climate-solutions/2020/01/22/what-does-dangerous-climate-change-really-mean/. Accessed November 19, 2020.
Kasturi, Charu Sudan. 2013. Foreign aid? No, thanks. *The Telegraph*. Available at www.telegraphindia.com/india/foreign-aid-no-thanks/cid/278963. Accessed September 18, 2019.
Kaufmann, D., A. Kraay, and M. Mastruzzi. 2010. *Worldwide Governance Indicators: Methodology and Analytical Indicators*. Available at http://info.worldbank.org/governance/wgi/#reports. Accessed June 22, 2021.
Keating, Michael, and Matt Waldman, eds. 2019. *War and Peace in Somalia: National Grievances, Local Conflict and Al-Shabaab*. New York, NY: Oxford University Press.
Kelley, Colin P., Shahrzad Mohtadi, Mark A. Cane, Richard Seager, and Yochanan Kushnir. 2015. Climate change in the Fertile Crescent and implications of the recent Syrian drought. *Proceedings of the National Academy of Sciences* 112 (11): 3241–3246. Available at https://doi.org/10.1073/pnas.1421533112. Accessed June 22, 2021.
Kelley, Colin P., Shahrzad Mohtadi, Mark Cane, Richard Seager, and Yochanan Kushnir. 2017. Commentary on the Syria case: Climate as a contributing factor. *Political Geography* 60: 245–247.
Kelman, Ilan. 2006. Island security and disaster diplomacy in the context of climate change. *Les Cahiers de la Sécurité* 63: 61–94. Available at www.disasterdiplomacy.org/kelman2006cce.pdf. Accessed June 22, 2021.
Keola, Souknilanh, Magnus Andersson, and Ola Hall. 2015. Monitoring economic development from space: Using nighttime light and land cover data to measure economic growth. *World Development* 66: 322–334. Available at https://doi.org/10.1016/j.worlddev.2014.08.017. Accessed June 22, 2021.
Kevane, Michael, and Leslie Gray. 2008. Darfur: Rainfall and conflict. *Environmental Research Letters* 3: 034006. Available at https://iopscience.iop.org/article/10.1088/1748-9326/3/3/034006/meta. Accessed 22 June, 2021.
Khaddour, Kheder, and Kevin Mazur. 2013. The struggle for Syria's regions. *Middle East Report* 43: 2–11. Available at https://merip.org/2014/01/the-struggle-for-syrias-regions/. Accessed June 22, 2021.
Khaddour, Kheder, and Kevin Mazur. 2017. *Eastern Expectations: The Changing Dynamics in Syria's Tribal Regions*. Carnegie Middle East Center. Available at https://carnegie-mec.org/2017/02/28/eastern-expectations-changing-dynamics-in-syria-s-tribal-regions-pub-68008. Accessed December 25, 2020.
Khalil, Gazi Md. 1993. The catastrophic cyclone of April 1991: Its Impact on the economy of Bangladesh. *Natural Hazards* 8 (3): 263–281. Available at https://doi.org/10.1007/BF00690911. Accessed June 28, 2019.
Ki-moon, Ban. 2007. A climate culprit in Darfur. *Washington Post*. Available at www.washingtonpost.com/wp-dyn/content/article/2007/06/15/AR2007061501857_pf.html. Accessed June 22, 2021.
Kilcullen, David, and Nate Rosenblatt. 2014. The rise of Syria's urban poor: Why the war for Syria's future will be fought over the country's new urban villages.

PRISM 4: 32–41. Available at www.jstor.org/stable/26469775. Accessed December 25, 2020.

Knutson, Tom. 2018. Summary statement on tropical cyclones and climate change. WMO Task Team on Tropical Cyclones and Climate Change. Available at https://ane4bf-datap1.s3-eu-west-1.amazonaws.com/wmocms/s3fs-public/ckeditor/files/WMO_Task_Team_Chair_Summary_Statement_TC_Climate_Sept_20_2018.pdf?IWzbdQKxXdDzJSlULy6kWtc8JR8FuW04. Accessed June 22, 2021.

Konyndyk, Jeremy. 2013. A tale of two cyclones. *The Huffington Post*. Available at www.huffingtonpost.com/jeremy-konyndyk/a-tale-of-two-cyclones_b_4160741.html. Accessed April 19, 2016.

Koren, Ore. 2018. Food abundance and violent conflict in Africa. *American Journal of Agricultural Economics* 100 (4): 981–1006. Available at https://doi.org/10.1093/ajae/aax106. Accessed June 22, 2021.

Koren, Ore. 2019a. Climate change and conflict. *Political Violence at a Glance*. Available at http://politicalviolenceataglance.org/2019/02/04/climate-change-and-conflict/. Accessed February 22, 2019.

Koren, Ore. 2019b. Food resources and strategic conflict. *Journal of Conflict Resolution* 63 (10): 2236–2261. Available at https://doi.org/10.1177/0022002719833160. Accessed December 19, 2020.

Koren, Ore, and Benjamin E. Bagozzi. 2017. Living off the land: The connection between cropland, food security, and violence against civilians. *Journal of Peace Research* 54 (3): 351–364. Available at https://doi.org/10.1177/0022343316684543. Accessed February 22, 2019.

Koren, Ore, and Anoop K. Sarbahi. 2018. State capacity, insurgency, and civil war: A disaggregated analysis. *International Studies Quarterly* 62 (2): 274–288. Available at https://doi.org/10.1093/isq/sqx076. Accessed June 22, 2021.

Kota, Suechika. 2012. Undemocratic Lebanon?: The power-sharing arrangements after the 2005 independence intifada. *Journal of Ritsumeikan, Social Sciences and Humanities* 4–6: 104–132.

Koubi, Vally. 2017. Climate change, the economy, and conflict. *Current Climate Change Reports* 3 (4): 200–209.

Koubi, Vally. 2019. Climate change and conflict. *Annual Review of Political Science* 22 (1): 343–360. Available at https://doi.org/10.1146/annurev-polisci-050317-070830. Accessed August 19, 2019.

Koubi, Vally, Thomas Bernauer, Anna Kalbhenn, and Gabriele Spilker. 2012. Climate variability, economic growth, and civil conflict. *Journal of Peace Research* 49 (1): 113–127. Available at https://doi.org/10.1177/0022343311427173. Accessed June 22, 2021.

Koubi, Vally, Gabriele Spilker, Lena Schaffer, and Thomas Bernauer. 2016a. Environmental stressors and migration: Evidence from Vietnam. *World Development* 79: 197–210. Available at https://doi.org/10.1016/j.worlddev.2015.11.016. Accessed June 22, 2021.

Koubi, Vally, Gabriele Spilker, Lena Schaffer, and Tobias Böhmelt. 2016b. The role of environmental perceptions in migration decision-making: Evidence from both migrants and non-migrants in five developing countries. *Population and*

Environment 38 (2): 134–163. Available at https://doi.org/10.1007/s11111-01 6-0258-7. Accessed June 22, 2021.

Krampe, Florian. 2017. Toward sustainable peace: A new research agenda for post-conflict natural resource management. *Global Environmental Politics* 17 (4): 1–8. Available at https://doi.org/10.1162/GLEP_a_00431. Accessed July 23, 2020.

Krampe, Florian. 2019. Climate Change, Peacebuilding and Sustaining Peace. IPI Global Observatory. Available at https://theglobalobservatory.org/2019/09/climate-change-peacebuilding-and-sustaining-peace/. Accessed June 22, 2021.

Krampe, Florian, and Malin Mobjörk. 2018. Responding to climate-related security risks: Reviewing regional organizations in Asia and Africa. *Current Climate Change Reports* 4 (4): 330–337. Available at https://doi.org/10.1007/s40641-018-0118-x. Accessed December 8, 2020.

Krampe, Florian, Luke van de Goor, Anniek Barnhoorn, Elizabeth Smith, and Dan Smith. 2020. *Water Security and Governance in the Horn of Africa*. SIPRI. Available at www.sipri.org/publications/2020/sipri-policy-papers/water-security-and-governance-horn-africa. Accessed June 23, 2020.

Kugler, Jacek. 2006. The Asian ascent: Opportunity for peace or precondition for war? *International Studies Perspectives* 7 (1): 36–42. Available at www.jstor.org/stable/44218450. Accessed June 22, 2021.

Kulp, Scott A., and Benjamin H. Strauss. 2019. New elevation data triple estimates of global vulnerability to sea-level rise and coastal flooding. *Nature Communications* 10 (1): 4844. Available at https://doi.org/10.1038/s41467-019-12808-z. Accessed June 22, 2021.

Kumar, Hari, Jeffrey Gettleman, and Sameer Yasir. 2019a. How do you save a million people from a cyclone? Ask a poor state in India. *The New York Times*. Available at www.nytimes.com/2019/05/03/world/asia/cyclone-fani-india-evacuations.html. Accessed June 20, 2019.

Kumar, Hari, Jeffrey Gettleman, and Sameer Yasir. 2019b. "The worst is over": A sigh of relief in India, mostly spared by cyclone. *The New York Times*. Available at www.nytimes.com/2019/05/04/world/asia/india-cyclone.html. Accessed June 20, 2019.

Kumar, T. P. L. N. Srinivasa, M. Murty et al. 2015. Modeling storm surge and its associated inland inundation extent due to very severe cyclonic storm Phailin. *Marine Geodesy* 38 (4): 345–360. Available at https://doi.org/10.1080/01490419.2015.1053640. Accessed June 28, 2019.

Lagi, Marco, Karla Z. Bertrand, and Yaneer Bar-Yam. 2011. *The Food Crises and Political Instability in North Africa and the Middle East*. ArXiv e-print. Available at http://arxiv.org/abs/1108.2455. Accessed September 30, 2013.

Larkin, Emma. 2011. *No Bad News for the King: The True Story of Cyclone Nargis and Its Aftermath in Burma*. New York: Penguin Books.

Lawson, Fred H. 2016. Explaining the spread of ethnosectarian conflict: Syria's civil war and the resurgence of Kurdish militancy in Turkey. *Nationalism and Ethnic Politics* 22 (4): 478–496. Available at https://doi.org/10.1080/13537113.2016.1239460. Accessed August 7, 2020.

Leahy, Stephen. 2017. Parts of Asia may be too hot for people by 2100. *National Geographic News*. Available at www.nationalgeographic.com/news/2017/08/south-asia-heat-waves-temperature-rise-global-warming-climate-change/. Accessed July 21, 2020.

Leenders, Reinoud. 2012. Collective action and mobilization in Dar'a: An anatomy of the onset of Syria's popular uprising. *Mobilization: An International Quarterly* 17 (4): 419–434. Available at https://doi.org/10.17813/maiq.17.4.gj8km668p1861rhj. Accessed December 25, 2020.

Lemma, Melisew Dejene, and Logan Cochrane. 2019. Policy coherence and social protection in Ethiopia: Ensuring no one is left behind. *Societies* 9 (1): 19. Available at https://doi.org/10.3390/soc9010019. Accessed June 22, 2021.

Leutert, Stephanie. 2018. How climate change is affecting rural Honduras and pushing people north. *Washington Post*. Available at www.washingtonpost.com/news/global-opinions/wp/2018/11/06/how-climate-change-is-affecting-rural-honduras-and-pushing-people-north/. Accessed July 22, 2020.

Levy, Marc A. 1995. Is the environment a national security Issue? *International Security* 20 (2): 35–62.

Levy, Marc A., Catherine Thorkelson, Charles Vörösmarty, Ellen Douglas, and Macartan Humphreys. 2005. *Freshwater Availability Anomalies and Outbreak of Internal War: Results from a Global Spatial Time Series Analysis*. Available at www.ciesin.columbia.edu/documents/waterconflict.pdf. Accessed June 22, 2021.

Library of Congress. 1988. Syria – Labor force. Available at www.country-data.com/cgi-bin/query/r-13517.html. Accessed December 4, 2019.

Linke, Andrew M., Frank D. W. Witmer, John O'Loughlin, J. Terrence McCabe, and Jaroslav Tir. 2018a. The consequences of relocating in response to drought: Human mobility and conflict in contemporary Kenya. *Environmental Research Letters* 13 (9): 094014. Available at https://doi.org/10.1088%2F1748-9326%2Faad8cc. Accessed December 4, 2019.

Linke, Andrew M., Frank D. W. Witmer, John O'Loughlin, J. Terrence McCabe, and Jaroslav Tir. 2018b. Drought, local institutional contexts, and support for violence in Kenya. *Journal of Conflict Resolution* 62 (7): 1544–1578. Available at https://doi.org/10.1177/0022002717698018. Accessed September 5, 2018.

Luckham, Robin, and Tom Kirk. 2013. Understanding security in the vernacular in hybrid political contexts: A critical survey. *Conflict, Security & Development* 13 (3): 339–359. Available at https://doi.org/10.1080/14678802.2013.811053. Accessed December 8, 2020.

Ludden, Jennifer, and Jeff Brady. 2020. Greenhouse gas emissions predicted to fall nearly 8% – Largest decrease ever. *NPR.org*. Available at www.upr.org/post/greenhouse-gas-emissions-predicted-fall-nearly-8-largest-decrease-ever. Accessed June 22, 2021.

Lührmann, Anna, Valeriya Mechkova, Sirianne Dahlum et al. 2018. State of the world 2017: Autocratization and exclusion? *Democratization* 25 (8): 1321–1340. Available at https://doi.org/10.1080/13510347.2018.1479693. Accessed November 10, 2018.

Lyon, Bradfield. 2011. *Quantifying Drought – Some Basic Concepts*. Memo for workshop on Mapping and Modeling Climate Security Vulnerability. LBJ

School of Public Affairs. Available at https://strausscenter.org/wp-content/uploads/pdf/climateworkshop/lyon_memo_for_web.pdf. Accessed June 23, 2021.

Mabey, Nick. 2008. *Delivering Climate Security. International Security Responses to a Climate Changed World*. London: Routledge.

MacFarquhar, Neil. 2010. World food program suspends aid to southern Somalia. *The New York Times*. Available at www.nytimes.com/2010/01/06/world/africa/06somalia.html. Accessed March 27, 2019.

Mach, Katharine J., Caroline M. Kraan, W. Neil Adger et al. 2019. Climate as a risk factor for armed conflict. *Nature* 571 (7764): 193–197. Available at https://doi.org/10.1038/s41586-019-1300-6. Accessed June 23, 2021.

Macias, Lee. 2013. *Complex Emergencies*. Austin, TX: Strauss Center for International Security and Law.

Mahmood, Zeshan. 2019. Locator map of the former East Pakistan provincial region (1947–1971). Available at www.quora.com/How-many-districts-are-in-our-Bangladesh?redirected_qid=17873058. Accessed July 5, 2021.

Mahoney, James. 2012. The logic of process tracing tests in the social sciences. *Sociological Methods & Research* 41 (4): 570–597. Available at https://doi.org/10.1177/0049124112437709. Accessed June 23, 2021.

Malthus, T. R. 1798. *An Essay on the Principle of Population, As It Affects the Future Improvement of Society. With Remarks on the Speculations of Mr. Godwin, M. Condorcet and Other Writers*. London: J. Johnson.

Mann, Michael. 1984. The autonomous power of the state: Its origins, mechanisms and results. *European Journal of Sociology/Archives Européennes de Sociologie/Europäisches Archiv für Soziologie* 25 (2): 185–213. Available at www.jstor.org/stable/23999270. Accessed August 12, 2019.

Marin-Ferrer, Montserrat, L. Vernaccini, and K. Poljansek. 2017. INFORM: Index for Risk Management – Concept and Methodology Version 2017. Available at https://drmkc.jrc.ec.europa.eu/inform-index/Portals/0/InfoRM/INFORM%20Concept%20and%20Methodology%20Version%202017%20Pdf%20FINAL.pdf. Accessed July 5, 2021.

Marshall, Monty G., and Ted Robert Gurr. 2016. *Political Regime Characteristics and Transitions, 1800–2016*. Polity IV Project. Available at www.systemicpeace.org/inscr/p4manualv2016.pdf. Accessed June 23, 2021.

Masters, Jeff. 2007. Tropical Cyclone Sidr devastates Bangladesh. *Weather Underground*. Available at www.wunderground.com/blog/JeffMasters/tropical-cyclone-sidr–devastates-bangladesh.html. Accessed September 15, 2019.

Mathews, Jessica Tuchman. 1989. Redefining security. *Foreign Affairs* 68 (2): 162–177.

Maxwell, Daniel, and Merry Fitzpatrick. 2012. The 2011 Somalia famine: Context, causes, and complications. *Global Food Security* 1 (1): 5–12. Available at https://doi.org/10.1016/j.gfs.2012.07.002. Accessed June 23, 2021.

Maxwell, Daniel, and Nisar Majid, eds. 2016. *Famine in Somalia: Competing Imperatives, Collective Failures, 2011–12*. First edition. New York: Oxford University Press.

Maystadt, Jean-François, and Olivier Ecker. 2014. Extreme weather and civil war: Does drought fuel conflict in Somalia through livestock price shocks? *American Journal of Agricultural Economics* 96 (4): 1157–1182. Available at https://doi.org/10.1093/ajae/aau010. Accessed June 23, 2021.

Mazur, Kevin. 2020a. Dayr al-Zur from revolution to ISIS: Local networks, hybrid identities, and outside authorities. In *Syria: Borders, Boundaries, and the State*, edited by Matthieu Cimino, 151–195. Basingstoke, UK: Palgrave Macmillan.

Mazur, Kevin. 2020b. Networks, informal governance, and ethnic violence in a Syrian city. *World Politics* 72 (3): 481–524. Available at https://doi.org/10.1017/S0043887120000052. Accessed June 23, 2021.

Mazur, Kevin. 2021. *Revolution in Syria: Identity, Networks, and Repression*. Cambridge, UK: Cambridge University Press.

McLauchlin, Theodore. 2018. The loyalty trap: Regime ethnic exclusion, commitment problems, and civil war duration in Syria and beyond. *Security Studies* 27 (2): 296–317. Available at https://doi.org/10.1080/09636412.2017.1386938. Accessed August 7, 2020.

McLeman, Robert, and François Gemenne. 2018. *Routledge Handbook of Environmental Displacement and Migration*. London: Routledge.

McVeigh, Tracy. 2011. Charity president says aid groups are misleading the public on Somalia. *The Observer*. Available at www.theguardian.com/global-development/2011/sep/03/charity-aid-groups-misleading-somalia. Accessed September 11, 2019.

Meagher, Kate. 2012. The strength of weak states? Non-state security forces and hybrid governance in Africa. *Development and Change* 43 (5): 1073–1101. Available at https://doi.org/10.1111/j.1467-7660.2012.01794.x. Accessed December 8, 2020.

Meier, Patrick, Doug Bond, and Joe Bond. 2007. Environmental influences on pastoral conflict in the Horn of Africa. *Political Geography* 26 (6): 716–735. Available at https://doi.org/10.1016/j.polgeo.2007.06.001. Accessed June 23, 2021.

Meierding, Emily. 2013. Climate change and conflict: Avoiding small talk about the weather. *International Studies Review* 15 (2): 185–203. Available at www.jstor.org/stable/24032947. Accessed June 23, 2021.

Meldrum, Andrew. 2011. Somalia famine: Aid blocked by Al Shabaab rebels. *Public Radio International*. Available at www.pri.org/stories/2011-07-28/somalia-famine-aid-blocked-al-shabaab-rebels. Accessed March 27, 2019.

Menkhaus, Ken. 2007. Governance without government in Somalia: Spoilers, state building, and the politics of coping. *International Security* 31 (3): 74–106. Available at https://doi.org/10.1162/isec.2007.31.3.74. Accessed September 2, 2019.

Menkhaus, Ken. 2012. No access: Critical bottlenecks in the 2011 Somali famine. *Global Food Security* 1 (1): 29–35. Available at https://doi.org/10.1016/j.gfs.2012.07.004. Accessed June 23, 2021.

Menkhaus, Ken. 2014. State failure, state-building, and prospects for a "functional failed state" in Somalia. *The ANNALS of the American*

Academy of Political and Social Science 656 (1): 154–172. Available at https://doi.org/10.1177/0002716214547002. Accessed August 29, 2019.

Menkhaus, Ken. 2018. *Elite Bargains and Political Deals Project: Somalia Case Study*. Stabilisation Unit. Available at https://assets.publishing.service.gov.uk/government/uploads/system/uploads/attachment_data/file/766049/Somalia_case_study.pdf. Accessed June 23, 2021.

Miguel, Edward, Shankar Satyanath, and Ernest Sergenti. 2004. Economic shocks and civil conflict: An instrumental variables approach. *Journal of Political Economy*, 112 (4): 725–753.

Military Expert Panel Members. 2018. *Military Expert Panel Report: Sea Level Rise and the U.S. Military's Mission*. Second Edition. Washington, DC: Center for Climate and Security.

Mill, John Stuart. 2002. *A System of Logic: Ratiocinative and Inductive*. Hawaii, HI: University Press of the Pacific.

Milly, P. C. D. 2008. Stationarity is dead: Whither water management? *Science* 319 (573–574).

Milly, P. C. D., Julio Betancourt, Malin Falkenmark et al. 2015. On critiques of "Stationarity is dead: Whither water management?" *Water Resources Research* 51 (9): 7785–7789. Available at https://doi.org/10.1002/2015WR017408. Accessed June 23, 2021.

Mirante, Edith. 2008a. The ruby ape: A deadly storm and a lethal regime. Part 1. *Guernica*. Available at www.guernicamag.com/edith_mirante_the_ruby_ape_a_d/. Accessed June 23, 2021.

Mirante, Edith. 2008b. Edith Mirante: The ruby ape: A deadly storm and a lethal regime. Part 2. *Guernica*. Available at www.guernicamag.com/edith_mirante_the_ruby_ape_a_d_1/. Accessed June 23, 2021.

Mitchell, Tom, Debarati Guha-Sapir, Julia Hall et al. 2014. *Setting, Measuring and Monitoring Targets for Disaster Risk Reduction: Recommendations for Post-2015 International Policy Frameworks*. ODI. Available at www.odi.org/publications/8448-setting-measuring-monitoring-targets-disaster-risk-reduction-recommendations-post-2015-international-policy-frameworks. Accessed September 18, 2019.

Mobjörk, Malin, and Sebastian van Baalen. 2016. *Climate Change and Violent Conflict in East Africa: Implications for Policy*. SIPRI. Available at www.sipri.org/publications/2016/other-publications/climate-change-and-violent-conflict-east-africa-implications-policy. Accessed June 23, 2020.

Mooney, Chris. 2007. *Storm World: Hurricanes, Politics, and the Battle over Global Warming*. Orlando, FL: Harcourt.

Moran, Ashley, Josh Busby, Clionadh Raleigh et al. 2018. *Intersection of Global Fragility and Climate Risks*. Washington, DC: USAID.

Moran, Daniel, ed. 2011. *Climate Change and National Security: A Country-Level Analysis*. Washington, DC: Georgetown University Press.

Morisetti, Neil. 2014. Failure to set a robust 2030 climate target will hurt our national security. *EURACTIV.com*. Available at www.euractiv.com/section/sustainable-dev/opinion/failure-to-set-a-robust-2030-climate-target-will-hurt-our-national-security/. Accessed May 23, 2017.

Mosello, Beatrice, and Lukas Rüttinger. 2020. Linking adaptation and peacebuilding: Lessons learned and the way forward. adelphi. Available at www.adelphi.de/en/publication/linking-adaptation-and-peacebuilding. Accessed December 8, 2020.

Muhumuza, Rodney. 2019. 25 years after genocide, Rwanda's Kagame is praised, feared. *AP NEWS*. Available at https://apnews.com/article/a97d40a146284383a717aa2ec42eb39b. Accessed December 7, 2020.

Mules, Ineke. 2020. Ethiopia: A timeline of the Tigray crisis. *Deutsche Welle*. Available at www.dw.com/en/ethiopia-a-timeline-of-the-tigray-crisis/a-55632181. Accessed December 21, 2020.

Murakami, Hiroyuki, Thomas L. Delworth, William F. Cooke et al. 2020. Detected climatic change in global distribution of tropical cyclones. *Proceedings of the National Academy of Sciences* 117 (20): 10706–10714. Available at https://doi.org/10.1073/pnas.1922500117. Accessed December 22, 2020.

Murphy, Kara. 2006. *The Lebanese Crisis and Its Impact on Immigrants and Refugees*. Migration Policy Institute. Available at www.migrationpolicy.org/article/lebanese-crisis-and-its-impact-immigrants-and-refugees. Accessed December 17, 2020.

Musgrave, Paul, and Daniel H. Nexon. 2018. Defending hierarchy from the moon to the Indian Ocean: Symbolic capital and political dominance in early modern China and the Cold War. *International Organization* 72 (3): 591–626. Available at https://doi.org/10.1017/S0020818318000139. Accessed June 23, 2021.

Myers, Norman. 1989. Environment and Security. *Foreign Policy* (74): 23–41.

Nakhoul, John, and John Irish. 2020. Analysis: "No free lunch" for Lebanon any more, donor states warn. *Reuters*. Available at https://in.reuters.com/article/lebanon-crisis-idINKBN27X233. Accessed November 20, 2020.

Naseemullah, Adnan. 2014. Shades of sovereignty: Explaining political order and disorder in Pakistan's northwest. *Studies in Comparative International Development* 49 (4): 501–522. Available at https://doi.org/10.1007/s12116-014-9157-z. Accessed December 8, 2020.

Nasser, Rassie, Zaki Mehchy, and Khalid Abu Ismail. 2013. *Socioeconomic Impacts Of The Syrian Crisis*. Syrian Centre for Policy Research. Available at www.scpr-syria.org/socioeconomic-impacts-of-the-syrian-crisis/. Accessed December 8, 2019.

National Drought Mitigation Center. 2016. NU agencies lead project to help MENA region respond to drought. Available at https://drought.unl.edu/Publications/News.aspx?id=254. Accessed December 7, 2019.

National Hurricane Center, National Oceanic and Atmospheric Administration. n.d. Saffir–Simpson Hurricane Wind Scale. Available at www.nhc.noaa.gov/aboutsshws.php. Accessed September 13, 2019.

Nel, Philip, and Marjolein Righarts. 2008. Natural disasters and the risk of violent civil conflict. *International Studies Quarterly* 52 (1): 159–185.

Neumann, Barbara, Athanasios T. Vafeidis, Juliane Zimmermann, and Robert J. Nicholls. 2015. Future coastal population growth and exposure to sea-level rise and coastal flooding – A global assessment. *PLoS ONE* 10 (3): e0118571. Available at https://doi.org/10.1371/journal.pone.0118571. Accessed June 23, 2021.

Newman, Edward. 2009. Failed states and international order: Constructing a post-Westphalian world. *Contemporary Security Policy* 30 (3): 421–443. Available at https://doi.org/10.1080/13523260903326479. Accessed December 8, 2020.

Nixon, Ron, and Matt Stevens. 2017. Harvey, Irma, Maria: Trump administration's response compared. *The New York Times*. Available at www.nytimes.com/2017/09/27/us/politics/trump-puerto-rico-aid.html. Accessed August 14, 2019.

NOAA Hurricane Research Division. 2015. 45th anniversary of the Bhola Cyclone. Hurricane Research Division. Available at https://noaahrd.wordpress.com/2015/11/13/45th-anniversary-of-the-bhola-cyclone/. Accessed September 13, 2019.

Noor, Abdisalan M., Victor A. Alegana, Peter W. Gething, Andrew J. Tatem, and Robert W. Snow. 2008. Using remotely sensed night-time light as a proxy for poverty in Africa. *Population Health Metrics* 6 (5). Available at https://doi.org/10.1186/1478-7954-6-5. Accessed June 23, 2021.

Nordås, Ragnild, and Nils Petter Gleditsch. 2007. Climate change and conflict. *Political Geography* 26 (6): 627–638.

Nordhaus, William. 2019. Can we control carbon dioxide? (from 1975). *American Economic Review* 109 (6): 2015–2035. Available at https://doi.org/10.1257/aer.109.6.2015. Accessed November 19, 2020.

North, Douglass C., John Joseph Wallis, and Barry R. Weingast. 2009. *Violence and Social Orders: A Conceptual Framework for Interpreting Recorded Human History*. First edition. Cambridge; New York: Cambridge University Press.

North, Douglass C., John Joseph Wallis, Steven B. Webb, and Barry R. Weingast, eds. 2012. *In the Shadow of Violence: Politics, Economics, and the Problems of Development*. Cambridge, UK: Cambridge University Press.

Northon, Karen. 2017. NASA, NOAA data show 2016 warmest year on record globally. *NASA*. Available at www.nasa.gov/press-release/nasa-noaa-data-show-2016-warmest-year-on-record-globally. Accessed April 11, 2017.

Null, Schuyler, and Lauren Herzer Risi. 2016. *Navigating Complexity: Climate, Migration, and Conflict in a Changing World*. USAID. Available at www.wilsoncenter.org/publication/navigating-complexity-climate-migration-and-conflict-changing-world. Accessed December 6, 2019.

OCHA Financial Tracking Service. n.d.a. About FTS/ using FTS data. Available at https://fts.unocha.org/content/about-fts-using-fts-data. Accessed September 18, 2019.

OCHA Financial Tracking Service. n.d.b. Financial Tracking Service. Available at https://fts.unocha.org/data-search. Accessed September 18, 2019.

Olson, Mancur. 1993. Dictatorship, democracy, and development. *The American Political Science Review* 87 (3): 567–576. Available at https://doi.org/10.2307/2938736. Accessed June 23, 2021.

Orrell, John. 2010. Hurricane Katrina response: National Guard's "finest hour." www.army.mil. Available at www.army.mil/article/44368. Accessed April 12, 2017.

Ortiz-Ospina, Esteban, and Max Roser. 2016. Taxation. *Our World in Data*. Available at https://ourworldindata.org/taxation. Accessed December 7, 2019.

OSDMA. 2019. Odisha State Disaster Management Authority. Available at www.osdma.org/ViewDetails.aspx?vchglinkid=GL007&vchplinkid=PL040&vchslinkid=SL013. Accessed September 18, 2019.

Pal, Indrajit, Tuhin Ghosh, and Chandan Ghosh. 2017. Institutional framework and administrative systems for effective disaster risk governance – Perspectives of 2013 Cyclone Phailin in India. *International Journal of Disaster Risk Reduction* 21: 350–359. Available at https://doi.org/10.1016/j.ijdrr.2017.01.002. Accessed June 23, 2021.

Pankhurst, Alula. 1992. *Resettlement and Famine in Ethiopia: The Villager's Experience*. Manchester; New York: Manchester University Press.

Parfitt, Tom. 2010. Vladimir Putin bans grain exports as drought and wildfires ravage crops. *The Guardian*. Available at www.theguardian.com/world/2010/aug/05/vladimir-putin-ban-grain-exports. Accessed December 5, 2019.

Paris, Roland. 2001. Human security: Paradigm shift or hot air? *International Security* 26 (2): 87.

Paris, Roland. 2004. Still an inscrutable concept. *Security Dialogue* 35 (3): 370–372.

Pascual, Carlos. 2008. *The Geopolitics of Energy: From Security to Survival*. Brookings Institution. Available at www.brookings.edu/papers/2008/01_energy_pascual.aspx. Accessed June 23, 2021.

Paul, Katie. 2008. Burma: Forecasting cyclones and weather disasters. *Newsweek*. Available at www.newsweek.com/burma-forecasting-cyclones-and-weather-disasters-89839. Accessed September 15, 2019.

Pearlman, Wendy. 2016. Narratives of fear in Syria. *Perspectives on Politics* 14 (1): 21–37. Available at https://doi.org/10.1017/S1537592715003205. Accessed June 23, 2021.

Peduzzi, B., H. Chatenoux, A. Dao et al. 2010. *The Global Risk Analysis for the 2009 Global Assessment Report on Disaster Risk Reduction*. International Disaster and Risk Conference, Davos, Switzerland. Available at https://archive-ouverte.unige.ch/unige:19305. Accessed June 23, 2021.

Peduzzi, P., H. Dao, C. Herold, and F. Mouton. 2009. Assessing global exposure and vulnerability towards natural hazards: The Disaster Risk Index. *Natural Hazards and Earth Systems Sciences* 9 (4): 1149–1159. Available at https://doi.org/10.5194/nhess-9-1149-2009. Accessed June 23, 2021.

Peluso, Nancy Lee, and Michael Watts, eds. 2001. *Violent Environments*. Ithaca, NY: Cornell University Press.

Penna, Anthony N., and Jennifer S. Rivers. 2013. *Natural Disasters in a Global Environment*. Malden, MA: John Wiley & Sons.

Percival, Val, and Thomas Homer-Dixon. 1998. Environmental scarcity and violent conflict: The case of South Africa. *Journal of Peace Research* 35 (3): 279–298. Available at https://doi.org/10.1177/0022343398035003002. Accessed June 23, 2021.

Peters, Katie, Nuhia Eltinay, and Kerrie Holloway. 2019. *Disaster Risk Reduction, Urban Informality and a "Fragile Peace": The Case of Lebanon*. ODI. Available at www.odi.org/publications/11412-disaster-risk-reduction-urban-informality-and-fragile-peace-case-lebanon. Accessed December 18, 2020.

Phys.org. 2019. What is a storm surge and why is it so dangerous? Available at https://phys.org/news/2019-05-storm-surge-dangerous.html. Accessed September 13, 2019.

Pierce, Justin. 2012. The 2011 Thai floods: Changing the perception of risk in Thailand. *AIR Currents*. Available at www.air-worldwide.com/SiteAssets/Publications/AIR-Currents/2012/attachments/The-2011-Thai-Floods–Changing-the-Perception-of-Risk-in-Thailand. Accessed December 8, 2020.

Plumer, Brad, and Nadja Popovich. 2018. Why half a degree of global warming is a big deal. *The New York Times*. Available at www.nytimes.com/interactive/2018/10/07/climate/ipcc-report-half-degree.html. Accessed November 19, 2020.

Polk, William R. 2013a. Your labor day Syria reader, part 2: William Polk. *The Atlantic*. Available at www.theatlantic.com/international/archive/2013/09/your-labor-day-syria-reader-part-2-william-polk/279255/. Accessed September 26, 2013.

Polk, William R. 2013b. Understanding Syria: From pre-civil war to post-Assad. *The Atlantic*. Available at www.theatlantic.com/international/archive/2013/12/understanding-syria-from-pre-civil-war-to-post-assad/281989/. Accessed December 5, 2019.

Polycarpou, Lakis. 2014. Floods, companies and supply chain risk. *State of the Planet*. Available at https://blogs.ei.columbia.edu/2014/11/17/floods-companies-and-supply-chain-risk/. Accessed December 8, 2020.

Press Trust of India. 2013. Cyclone Phailin triggers India's biggest evacuation operation in 23 years. *NDTV.com*. Available at www.ndtv.com/india-news/cyclone-phailin-triggers-indias-biggest-evacuation-operation-in-23-years-537522. Accessed April 19, 2016.

Price-Smith, Andrew T. 2001. *The Health of Nations: Infectious Disease, Environmental Change, and Their Effects on National Security and Development*. First edition. Cambridge, MA: MIT Press.

Price-Smith, Andrew T. 2009. *Contagion and Chaos Disease, Ecology, and National Security in the Era of Globalization*. Cambridge, MA: MIT Press.

PRS Group. 2012. *International Country Risk Guide Methodology*. Available at www.prsgroup.com/wp-content/uploads/2012/11/icrgmethodology.pdf. Accessed June 23, 2021.

Punton, Melanie, and Katharina Welle. 2015. *Straws-in-the-wind, Hoops and Smoking Guns: What Can Process Tracing Offer to Impact Evaluation?* Centre for Development Impact. Available at www.adcoesao.pt/sites/default/files/avaliacao/4_3straws-in-the-wind_hoops_and_smoking_guns_what_can_process_tracing_offer_to_impact_evaluation_abril_2015.pdf. Accessed June 23, 2021.

Purvis, Nigel, and Josh Busby. 2004. *The Security Implications of Climate Change for the UN System*. Environmental Change and Security Project, Wilson Center. Available at www.brookings.edu/research/the-security-implications-of-climate-change-for-the-un-system/. Accessed June 23, 2021.

Quinn, Audrey, and Jackie Roche. 2014. Syria's climate-fueled conflict, in one stunning comic strip. *Mother Jones*. Available at www.motherjones.com/

politics/2014/05/syria-climate-years-living-dangerously-symbolia/. Accessed December 5, 2019.
Quiroz Flores, Alejandro. 2015. Protecting people from natural disasters: Political institutions and ocean-originated hazards. *Political Science Research and Methods* 6 (1): 111–134. Available at https://doi.org/10.1017/psrm.2015.72. Accessed June 23, 2021.
Radin, Andrew. 2020. *Institution Building in Weak States*. Washington, DC: Georgetown University Press.
Raeymaekers, Timothy. 2013. Post-war conflict and the market for protection: The challenges to Congo's hybrid peace. *International Peacekeeping* 20 (5): 600–617. Available at https://doi.org/10.1080/13533312.2013.854591. Accessed December 8, 2020.
Raleigh, Clionadh. 2010. Political marginalization, climate change, and conflict in African Sahel states. *International Studies Review* 12 (1): 69–86. Available at http://dx.doi.org/10.1111/j.1468-2486.2009.00913.x. Accessed June 22, 2021.
Raleigh, Clionadh, and Dominic Kniveton. 2012. Come rain or shine: An analysis of conflict and climate variability in East Africa. *Journal of Peace Research* 49 (1): 51–64. Available at https://doi.org/10.1177/0022343311427754. Accessed June 23, 2021.
Raleigh, Clionadh, Lisa Jordan, and Idean Salehyan. 2008. *Assessing the Impact of Climate Change on Migration and Conflict*. The Social Dimensions of Climate Change, World Bank Group. Available at www.semanticscholar.org/paper/Assessing-the-Impact-of-Climate-Change-on-Migration-Raleigh-Jordan/bad883876c6d11f71d7fe1e360f20f5fe137909f. Accessed June 23, 2021.
Raphaeli, Nimrod. 2006. The Syrian economy under Bashar al-Assad. *MEMRI*. Available at www.memri.org/reports/syrian-economy-under-bashar-al-assad. Accessed December 5, 2019.
Ray, Caleb, Matos, Christopher, Matthew Preisser, and Molly Ellsworth. 2020. *Vulnerability Mapping in Oceania*. LBJ School of Public Affairs, University of Texas at Austin. Available at https://sites.utexas.edu/climatesecurity/files/2020/05/LBJ_Oceania_Mapping.pdf. Accessed June 23, 2021.
Ray-Bennett, Nibedita S. 2018. Disasters, deaths, and the Sendai goal one: Lessons from Odisha, India. *World Development* 103: 27–39. Available at https://doi.org/10.1016/j.worlddev.2017.10.003. Accessed June 23, 2021.
Real Climate. 2017. El Niño, global warming, and anomalous U.S. winter warmth. *RealClimate*. Available at www.realclimate.org/index.php/archives/2007/01/el-nino-global-warming-and-anomalous-winter-warmth/. Accessed April 11, 2017.
Reiling, Kirby, and Cynthia Brady. 2015. Climate Change and Conflict: An Annex to the USAID Climate-Resilient Development Framework. USAID. Available at www.usaid.gov/sites/default/files/documents/1866/ClimateChangeConflictAnnex_2015%2002%2025%2C%20Final%20with%20date%20for%20Web.pdf. Accessed June 23, 2021.
Reilly, Benjamin. 2009. *Disaster and Human History*. Jefferson, NJ: McFarland.
Reilly, James A. 1982. Israel in Lebanon, 1975–1982. MERIP. Available at https://merip.org/1982/09/israel-in-lebanon-1975-1982/. Accessed June 23, 2021.

Reno, William. 2000. Shadow states and the political economy of civil wars. In *Greed and Grievance: Economic Agendas in Civil Wars*, edited by David Malone and Mats R. Berdal, 43–68. Boulder, CO: Lynne Rienner.

Republic of Lebanon Ministry of Finance. 2012. *Wheat and Bread Subsidies 2007–2011*. Available at www.finance.gov.lb/en-us/Finance/Rep-Pub/DRI-MOF/Thematic%20Reports//Thematic%20Report%20Wheat%20subsidy%20Final%202.pdf. Accessed June 23, 2021.

Reuters. 2020. Ethiopia scorns guerrilla war fears, U.N. team shot at in Tigray. *Reuters*. Available at www.reuters.com/article/ethiopia-conflict-idUSKBN28H0VT. Accessed December 7, 2020.

Reuveny, Rafael. 2007. Climate change-induced migration and violent conflict. *Political Geography* 26 (6): 656–673.

Reuveny, Rafael, and Will H. Moore. 2009. Does environmental degradation influence migration? Emigration to developed countries in the late 1980s and 1990s. *Social Science Quarterly* 90 (3): 461–479.

Riachi, Raymond. 2014. Beyond rehashed policies: Lebanon must tackle its water crisis head-on. The Lebanese Center for Policy Studies. Available at http://lcps-lebanon.org/featuredArticle.php?id=27. Accessed December 18, 2020.

Rice, Doyle. 2018. 2017's three monster hurricanes – Harvey, Irma and Maria – among five costliest ever. *USA TODAY*. Available at www.usatoday.com/story/weather/2018/01/30/2017-s-three-monster-hurricanes-harvey-irma-and-maria-among-five-costliest-ever/1078930001/. Accessed July 21, 2020.

Rice, Doyle. 2019. USA had world's 3 costliest natural disasters in 2018, and Camp Fire was the worst. *USA TODAY*. Available at www.usatoday.com/story/news/2019/01/08/natural-disasters-camp-fire-worlds-costliest-catastrophe-2018/2504865002/. Accessed July 21, 2020.

Ritchie, Hannah, and Max Roser. 2017. Water use and stress. *Our World in Data*. Available at https://ourworldindata.org/water-use-stress. Accessed December 6, 2019.

Roessler, Philip, and David Ohls. 2018. Self-enforcing power sharing in weak states. *International Organization* 72 (2): 423–454. Available at https://doi.org/10.1017/S0020818318000073. Accessed June 23, 2021.

Roser, Max. 2013. Employment in agriculture. *Our World in Data*. Available at https://ourworldindata.org/employment-in-agriculture. Accessed December 4, 2019.

Roser, Max, and Mohamed Nagdy. 2013. Military spending. *Our World in Data*. Available at https://ourworldindata.org/military-spending. Accessed December 26, 2020.

Ross, Michael L. 2015. What have we learned about the resource curse? *Annual Review of Political Science* 18 (1): 239–259. Available at https://doi.org/10.1146/annurev-polisci-052213-040359. Accessed December 19, 2020.

Rothe, Delf. 2017. *Securitizing Global Warming*. First edition. Abingdon, UK: Routledge.

Rüttinger, Lukas, Dan Smith, Gerald Stang, Dennis Tänzler, and Janani Vivekananda. 2015. *A New Climate for Peace*. Available at www.newclimateforpeace.org/. Accessed July 19, 2020.

Sachs, Susan. 2000. Assad patronage puts a small sect on top in Syria. *The New York Times*. Available at www.nytimes.com/2000/06/22/world/assad-patronage-puts-a-small-sect-on-top-in-syria.html. Accessed December 7, 2019.

Salama, Peter, Grainne Moloney, Oleg O. Bilukha et al. 2012. Famine in Somalia: Evidence for a declaration. *Global Food Security* 1 (1): 13–19. Available at https://doi.org/10.1016/j.gfs.2012.08.002. Accessed June 23, 2021.

Saleeby, Suzanne. 2012. Sowing the seeds of dissent: Economic grievances and the Syrian social contract's unraveling. *Jadaliyya*. Available at www.jadaliyya.com/Details/25271. Accessed December 18, 2020.

Salehyan, Idean. 2014. Climate change and conflict: Making sense of disparate findings. *Political Geography* 43: 1–5. Available at https://doi.org/10.1016/j.polgeo.2014.10.004. Accessed June 23, 2021.

Salehyan, Idean, and Kristian Skrede Gleditsch. 2006. Refugees and the spread of civil war. *International Organization* 60 (2): 335–366.

Salehyan, Idean, and Cullen S. Hendrix. 2014. Climate shocks and political violence. *Global Environmental Change* 28: 239–250. Available at https://doi.org/10.1016/j.gloenvcha.2014.07.007. Accessed June 23, 2021.

Salem, Paul. 2016. Seven reasons why Lebanon survives – and three reasons why it might not. *Lawfare*. Available at www.lawfareblog.com/seven-reasons-why-lebanon-survives-and-three-reasons-why-it-might-not. Accessed December 5, 2019.

Samenow, Jason. 2019. Undeniable warming: The planet's hottest five years on record in five images. *Washington Post*. Available at www.washingtonpost.com/weather/2019/02/06/undeniable-warming-planets-hottest-five-years-record-five-images/. Accessed August 23, 2019.

Scata, Joel. 2017. Pres. Trump exposes communities, military to extreme weather. *NRDC*. Available at www.nrdc.org/experts/joel-scata/pres-trump-exposes-communities-military-extreme-weather. Accessed July 20, 2020.

Schaffer, Teresita, and Rebecca Anne Dixon. 2010. Pakistan Floods: Internally Displaced People and the Human Impact. Center for Strategic and International Studies, Washington, DC. Available at http://csis.org/publication/pakistan-floods-internally-displaced-people-and-human-impact. Accessed June 24, 2021.

Scheman, Paul. 2017. Ethiopia is facing a killer drought. But it's going almost unnoticed. *Washington Post*. Available at www.washingtonpost.com/news/worldviews/wp/2017/05/01/ethiopia-is-facing-a-killer-drought-but-its-going-almost-unnoticed/. Accessed March 27, 2019.

Schmidt, Roberto. 2011. US pledges $5 million for drought-hit Somalis. *AFP*. Available at https://web.archive.org/web/20110724064439/https://news.yahoo.com/us-pledges-5-million-drought-hit-somalis-185930616.html. Accessed March 27, 2019.

Schmidt, Søren. 2018. The power of "sultanism": Why Syria's non-violent protests did not lead to a democratic transition. In *The Syrian Uprising*, edited by Raymond Hinnebusch and Omar Imady, 30–43. Abingdon, UK: Routledge.

Scholten, Daniel, ed. 2018. *The Geopolitics of Renewables*. Cham, Switzerland: Springer International.

Scholten, Daniel, and Rick Bosman. 2016. The geopolitics of renewables; exploring the political implications of renewable energy systems. *Technological Forecasting and Social Change* 103: 273–283. Available at https://doi.org/10.1016/j.techfore.2015.10.014. Accessed June 23, 2021.

Schultz, Kenneth A., and Justin S. Mankin. 2019. Is temperature exogenous? The impact of civil conflict on the instrumental climate record in Sub-Saharan Africa. *American Journal of Political Science* 63 (4): 723–739. Available at https://doi.org/10.1111/ajps.12425. Accessed June 24, 2021.

Schwartz, Daniel, Tom Deligiannis, and Thomas Homer-Dixon. 2001. The environment and violent conflict. In *Environmental Conflict*, edited by Paul F. Diehl and Nils Petter Gleditsch, 273–294. Boulder, CO: Westview Press.

Schwartz, Peter, and Doug Randall. 2004. An abrupt climate change scenario and its implications for United States national security. *GBN*. Available at https://irp.fas.org/agency/dod/schwartz.pdf. Accessed June 24, 2021.

Seal, Andrew, and Rob Bailey. 2013. The 2011 famine in Somalia: Lessons learnt from a failed response? *Conflict and Health* 7 (22): 22. Available at https://doi.org/10.1186/1752-1505-7-22. Accessed June 23, 2021.

Seay-Fleming, Carrie. 2018. Beyond violence: Drought and migration in Central America's northern triangle. *New Security Beat*. Available at www.newsecuritybeat.org/2018/04/violence-drought-migration-central-americas-northern-triangle/. Accessed December 1, 2020.

Selby, Jan, Omar S. Dahi, Christiane J. Fröhlich, and Mike Hulme. 2017a. Climate change and the Syrian civil war revisited. *Political Geography* 60: 232–244.

Selby, Jan, Omar S. Dahi, Christiane J. Fröhlich, and Mike Hulme. 2017b. Climate change and the Syrian civil war revisited: A rejoinder. *Political Geography* 60: 253–255.

Sellwood, Elizabeth. 2011. State-*Building and Political Change:* Options *for* Palestine 2011, Center for International Cooperation, New York University. Available at https://cic.es.its.nyu.edu/sites/default/files/sellwood_statebuilding_1.pdf. Accessed June 24, 2021.

Semple, Kirk. 2019. Central American farmers head to the U.S., fleeing climate change. *The New York Times*. Available at www.nytimes.com/2019/04/13/world/americas/coffee-climate-change-migration.html. Accessed December 1, 2020.

Sen, Amartya. 1981. *Poverty and Famines: An Essay on Entitlement and Deprivation*. Oxford: Oxford University Press.

Sen, Amartya. 2000. *Development as Freedom*. Reprint edition. New York: Anchor.

Seo, S. Niggol, and Laura A. Bakkensen. 2016. Is tropical cyclone surge, not intensity, what kills so many people in South Asia? *Weather, Climate, and Society* 9 (2): 171–181. Available at https://doi.org/10.1175/WCAS-D-16-0059.1. Accessed September 15, 2019.

Seter, Hanne, Ole Magnus Theisen, and Janpeter Schilling. 2018. All about water and land? Resource-related conflicts in East and West Africa revisited. *GeoJournal* 83 (1): 169–187. Available at https://doi.org/10.1007/s10708-016-9762-7. Accessed September 5, 2018.

Shabnam, Nourin. 2014. Natural disasters and economic growth: A review. *International Journal of Disaster Risk Science* 5 (2): 157–163. Available at https://doi.org/10.1007/s13753-014-0022-5. Accessed June 24, 2021.

Shaik, Louise Van, Camilla Born, Elizabeth Sellwood, and Sophie de Bruin. 2019. Making Peace with Climate Adaptation. Clingendael. Available at www.clingendael.org/sites/default/files/2020-06/Making_peace_with_climate.pdf. Accessed June 24, 2021.

Sherman, Jake, and Florian Krampe. 2020. *The Peacebuilding Commission and Climate-Related Security Risks: A More Favourable Political Environment?* International Peace Institute. Available at www.ipinst.org/2020/09/peacebuilding-commission-climate-related-security-risks-a-more-favourable-political-environment. Accessed December 8, 2020.

Siddique, Abul, and A Eusof. 1987. Cyclone deaths in Bangladesh, May 1985: Who was at risk? *Tropical and Geographical Medicine* 39: 3–8.

Slackman, Michael. 2006. Lebanon talks collapse as Shiites vacate cabinet. *The New York Times*. Available at www.nytimes.com/2006/11/12/world/middleeast/12lebanon.html. Accessed December 16, 2020.

Slettebak, Rune T. 2012. Don't blame the weather! Climate-related natural disasters and civil conflict. *Journal of Peace Research* 49 (1): 163–176. Available at https://doi.org/10.1177/0022343311425693. Accessed June 24, 2021.

Sluyter, Andrew. 2003. Neo-environmental determinism, intellectual damage control, and nature/society science. *Antipode* 35 (4): 813–817. Available at https://doi.org/10.1046/j.1467-8330.2003.00354.x. Accessed June 24, 2021.

Sly, Liz. 2013. Syria's war redraws borders of the Middle East. *Washington Post*. Available at www.washingtonpost.com/world/middle_east/syrias-civil-war-tests-whether-borders-drawn-less-than-a-century-ago-willlast/2013/12/26/6718111 c-68e2-11e3-997b-9213b17dac97_story.html. Accessed December 5, 2019.

Smith, Alastair, and Alejandro Quiroz Flores. 2010. Disaster politics. *Foreign Affairs*. Available at www.foreignaffairs.com/articles/66494/alastair-smith-and-alejandro-quiroz-flores/disaster-politics. Accessed May 21, 2012.

Smith, Dan, Malin Mobjörk, Florian Krampe, and Karolina Eklöw. 2019. Climate Security: Making it #Doable. Clingendael and SIPRI. Available at www.sipri.org/publications/2019/other-publications/climate-security-making-it-doable. Accessed June 23, 2020.

Smith, Evan. 2020a. Climate change and conflict: The Darfur conflict and Syrian civil war. Undergraduate thesis, University of Vermont.

Smith, Jeffrey. 2020b. The Ugandan opposition leader survives an attack. The U.S. needs to take a stand. *Washington Post*. Available at www.washingtonpost.com/opinions/2020/12/02/uganda-bobi-wine-attacked-us-human-rights/. Accessed December 8, 2020.

Smith, Todd Graham. 2014. Feeding unrest: Disentangling the causal relationship between food price shocks and sociopolitical conflict in urban Africa. *Journal of Peace Research* 51 (6): 679–695. Available at https://doi.org/10.1177/0022343314543722. Accessed June 24, 2021.

Solana, Javier. 2008. *Climate Change and International Security: Paper from the High Representative and the European Commission to the European Council.* Available at www.consilium.europa.eu/ueDocs/cms_Data/docs/pressData/en/reports/99387.pdf. Accessed June 24, 2021.

Sommer, Alfred, and Wiley H. Mosley. 1972. East Bengal cyclone of November, 1970: Epidemiological approach to disaster assessment. *The Lancet* 299 (7759): 1030–1036. Available at https://doi.org/10.1016/S0140-6736(72)91218-4. Accessed June 23, 2021.

Steinberg, David I. 2012. The problem of democracy in the Republic of the Union of Myanmar: Neither nation-state nor state-nation? *Southeast Asian Affairs*: 220–237. Available at www.jstor.org/stable/41713996. Accessed June, 24 2021.

Sterling, Joe. 2012. Daraa: The spark that lit the Syrian flame. *CNN Digital*. Available at www.cnn.com/2012/03/01/world/meast/syria-crisis-beginnings/index.html. Accessed December 16, 2020.

Sternberg, Troy. 2011. Regional drought has a global impact. *Nature* 472 (7342): 169. Available at https://doi.org/10.1038/472169d. Accessed June 24, 2021.

Sternberg, Troy. 2012. Chinese drought, bread and the Arab Spring. *Applied Geography* 34: 519–524. Available at https://doi.org/10.1016/j.apgeog.2012.02.004. Accessed June 24, 2021.

Storlazzi, Curt D., Stephen B. Gingerich, Ap van Dongeren et al. 2018. Most atolls will be uninhabitable by the mid-21st century because of sea-level rise exacerbating wave-driven flooding. *Science Advances* 4 (4): eaap9741. Available at https://doi.org/10.1126/sciadv.aap9741. Accessed June 24, 2021.

Stover, Eric, and Patrick Vinck. 2008. Cyclone Nargis and the politics of relief and reconstruction aid in Burma (Myanmar). *JAMA* 300 (6): 729–731. Available at https://doi.org/10.1001/jama.300.6.729. Accessed June 24, 2021.

Sun, Qiaohong, Chiyuan Miao, Qingyun Duan et al. 2018. A review of global precipitation data sets: Data sources, estimation, and intercomparisons. *Reviews of Geophysics* 56 (1): 79–107. Available at https://doi.org/10.1002/2017RG000574. Accessed June 24, 2021.

Swain, Ashok, and Joakim Öjendal, eds. 2018. *Routledge Handbook of Environmental Conflict and Peacebuilding*. First edition. London; New York: Routledge.

Tammen, Ronald L., Jacek Kugler, Douglas Lemke et al. 2000. *Power Transitions: Strategies For the 21st Century*. First edition. New York: Seven Bridges Press, LLC/Chatham House.

Tan, C. K. 2020. China warns of "stronger flood" as Three Gorges Dam faces swell. *Nikkei Asian Review*. Available at https://asia.nikkei.com/Economy/Natural-disasters/China-warns-of-stronger-flood-as-Three-Gorges-Dam-faces-swell. Accessed July 21, 2020.

Tasnim, Khandker Masuma, Miguel Esteban, and Tomoya Shibayama. 2015. Observations and numerical simulation of storm surge due to Cyclone Sidr 2007 in Bangladesh. In *Handbook of Coastal Disaster Mitigation for Engineers and Planners*, edited by Miguel Esteban, Hiroshi Takagi, and Tomoya Shibayama, 35–53. Boston, MA: Butterworth-Heinemann.

Tawk, Salwa Tohmé, Mabelle Chedid, Ali Chalak, Sarah Karam, and Shadi Kamal Hamadeh. 2019. Challenges and sustainability of wheat production in a Levantine breadbasket. *Journal of Agriculture, Food Systems, and Community Development* 8 (4): 193–209. Available at https://doi.org/10.5304/jafscd.2019.084.011. Accessed June 24, 2021.

Taylor, Robert H. 2005. Do states make nations? The politics of identity in Myanmar revisited. *South East Asia Research* 13 (3): 261–286. Available at www.jstor.org/stable/23750110. Accessed December 23, 2020.

The Climate and Security Advisory Group. 2019. *A Climate Security Plan for America*. The Center for Climate & Security. Available at https://climateandsecurity.org/climatesecurityplanforamerica/. Accessed September 25, 2020.

The Economic Weekly. 1961. Migration from East Pakistan (1951–1961). Available at www.epw.in/system/files/pdf/1961_13/15/migration_from_east_pakistan_19511961.pdf. Accessed June 24, 2021.

The New Humanitarian. 2010. Why the water shortages? Available at www.thenewhumanitarian.org/news/2010/03/25/why-water-shortages. Accessed December 6, 2019.

The New York Times. 2008. Unable to help Myanmar relief efforts, U.S. Navy vessels sailing away. Available at www.nytimes.com/2008/06/04/world/asia/04iht-myanmar.1.13454790.html. Accessed April 3, 2016.

The New York Times. 2009. A year after storm, subtle changes in Myanmar. Available at www.nytimes.com/2009/04/30/world/asia/30myanmar.html. Accessed April 2, 2016.

The United Nations Accountability Project – Somalia. 2019. Neither inevitable nor accidental: The impact of marginalization in Somalia. In *War and Peace in Somalia: National Grievances, Local Conflict and Al-Shabaab*, edited by Michael Keating and Matt Waldman, 41–48. New York: Oxford University Press.

Theisen, Ole Magnus. 2008. Blood and soil? Resource scarcity and internal armed conflict revisited. *Journal of Peace Research* 45 (6): 801–818.

Theisen, Ole Magnus. 2012. Climate clashes? Weather variability, land pressure, and organized violence in Kenya, 1989–2004. *Journal of Peace Research* 49 (1): 81–96. Available at http://jpr.sagepub.com/content/49/1/81. Accessed December 1, 2016.

Theisen, Ole Magnus. 2017. Climate Change and Violence: Insights from Political Science. *Current Climate Change Reports* 3 (4): 210–221. Available at https://doi.org/10.1177/0022343311425842. Accessed June 24, 2021.

Theisen, Ole Magnus, Helge Holtermann, and Halvard Buhaug. 2012. Climate wars? Assessing the claim that drought breeds conflict. *International Security* 36 (3): 79–106. Available at https://doi.org/10.1162/ISEC_a_00065. Accessed June 24, 2021.

Thompson, Darcy, and Lina Eklund. 2017. Is Syria really a "climate war"? We examined the links between drought, migration and conflict. *The Conversation*. Available at http://theconversation.com/is-syria-really-a-climate-war-we-examined-the-links-between-drought-migration-and-conflict-80110. Accessed August 27, 2020.

Thow, A., L. Vernaccini, M. Marin Ferrer, and B. Doherty. 2018. *INFORM Global Risk Index Results 2018*. Luxembourg: Publications Office of the European Union.

Tierney, Michael J., Daniel L. Nielson, Darren G. Hawkins et al. 2011. More dollars than sense: Refining our knowledge of development finance using AidData. *World Development* 39 (11): 1891–1906. Available at https://doi.org/10.1016/j.worlddev.2011.07.029. Accessed June 24, 2021.

Tilly, Charles. 1985. War making and state making as organized crime. In *Bringing the State Back In*, edited by Dietrich Rueschemeyer, Peter B. Evans, and Theda Skocpol, 169–191. Cambridge, UK: Cambridge University Press.

Tir, Jaroslav, and Douglas M. Stinnett. 2012. Weathering climate change: Can institutions mitigate international water conflict? *Journal of Peace Research* 49 (1): 211–225. Available at https://doi.org/10.1177/0022343311427066. Accessed June 24, 2021.

Titley, David. 2017. Why is climate change's 2 degrees Celsius of warming limit so important? *The Conversation*. Available at http://theconversation.com/why-is-climate-changes-2-degrees-celsius-of-warming-limit-so-important-82058. Accessed November 19, 2020.

Todd, Zoe. 2019. By the numbers: Syrian refugees around the world. *Frontline PBS*. Available at www.pbs.org/wgbh/frontline/article/numbers-syrian-refugees-around-world/. Accessed December 1, 2020.

UCDP. 2019. *Uppsala Conflict Data Program*. Available at https://ucdp.uu.se. Accessed June 24, 2021.

UCDP/PRIO. 2017a. *Somalia*. Available at https://ucdp.uu.se/#country/520. Accessed June 24, 2021.

UCDP/PRIO. 2017b. *Somalia: Government*. Available at https://ucdp.uu.se/#/conflict/337. Accessed June 24, 2021.

UCDP/PRIO. 2018. *UCDP/PRIO Armed Conflict Dataset Version 18.1*. Available at https://ucdp.uu.se/downloads/replication_data/2019_ucdp-prio-acd-191.xlsx. Accessed July 5, 2021.

Ullman, Richard. 1983. Redefining security. *International Security* 8 (1): 129–153.

UN Disaster Management Team. 1999. Orissa super cyclone situation report 9 – India. *ReliefWeb*. Available at https://reliefweb.int/report/india/orissa-super-cyclone-situation-report-9. Accessed September 15, 2019.

UN Economic and Social Commission for Western Asia. 2016. *Strategic Review of Food and Nutrition Security in Lebanon*. Available at https://reliefweb.int/report/lebanon/strategic-review-food-and-nutrition-security-lebanon-enar. Accessed December 4, 2019.

UN Office for the Coordination of Humanitarian Affairs. 2009. *Syria Drought Response Plan – Syrian Arab Republic. ReliefWeb*. Available at https://reliefweb.int/report/syrian-arab-republic/syria-drought-response-plan. Accessed December 8, 2019.

UN Security Council. 2007. *Security Council Holds First-Ever Debate on Impact of Climate Change on Peace, Security, Hearing over 50 Speakers*. Available at www.un.org/press/en/2007/sc9000.doc.htm. Accessed June 24, 2021.

UNDP. 2007. *Poverty, Growth & Inequality in Lebanon*. Available at www.undp.org/content/dam/lebanon/docs/Poverty/Publications/Poverty,%20Growth%20and%20Inequality%20in%20Lebanon.pdf. Accessed June 24, 2021.

UNEP. 2012. *Renewable Resources and Conflict: Toolkit and Guidance for Preventing and Managing Land and Natural Resources Conflict*. Available at https://postconflict.unep.ch/publications/GN_Renewable_Consultation_ES.pdf. Accessed June 24, 2021.

UNFCCC. n.d. *What Is the United Nations Framework Convention on Climate Change?* Available at https://unfccc.int/process-and-meetings/the-convention/what-is-the-united-nations-framework-convention-on-climate-change. Accessed November 19, 2020.

United Nations. 2011. UN declares famine in two regions of southern Somalia. *UN News*. Available at https://news.un.org/en/story/2011/07/382072-un-declares-famine-two-regions-southern-somalia. Accessed March 26, 2019.

United Nations. 2019. Stop Tuvalu and "the world from sinking" UN chief tells island nation facing existential threat from rising seas. *UN News*. Available at https://news.un.org/en/story/2019/05/1038661. Accessed July 16, 2020.

United Nations Climate Security Mechanism. 2020. *Toolbox: Briefing Note*. Available at https://postconflict.unep.ch/CSM/Toolbox-1-Briefing_note.pdf. Accessed June 24, 2021.

United Nations Development Programme. 1994. *New Dimensions in Human Security*. New York: Oxford University Press.

United Nations Office for Disaster Risk Reduction. 2019. *Global Assessment Report on Disaster Risk Reduction 2019*. Available at www.undrr.org/publication/global-assessment-report-disaster-risk-reduction-2019. Accessed October 2, 2020.

United Nations Office for the Coordination of Humanitarian Affairs, Financial Tracking Service. n.d. Available at https://fts.unocha.org/. Accessed June 24, 2021.

United Nations–World Bank Group. 2018. *Pathways for Peace: Inclusive Approaches to Preventing Violent Conflict*. World Bank. Available at www.worldbank.org/en/topic/fragilityconflictviolence/publication/pathways-for-peace-inclusive-approaches-to-preventing-violent-conflict. Accessed December 7, 2020.

US Department of Commerce, National Oceanic and Atmospheric Administration. 2018. *How Do Hurricanes Form?* Available at https://oceanservice.noaa.gov/facts/how-hurricanes-form.html. Accessed September 13, 2019.

US Department of Defense. 2018. *Climate-Related Risk to DoD Infrastructure Initial Vulnerability Assessment Survey (SLVAS) Report*. Available at www.hsdl.org/?abstract&did=807779. Accessed June 24, 2021.

US Department of Energy. 2015. *Climate Change and the U.S. Energy Sector: Regional Vulnerabilities and Resilience Solutions*. Available at www.energy.gov/sites/prod/files/2015/10/f27/Regional_Climate_Vulnerabilities_and_Resilience_Solutions_0.pdf. Accessed July 8, 2021.

US Department of Energy. 2016. *Climate Change and the Electricity Sector: Guide for Assessing Vulnerabilities and Developing Resilience Solutions to Sea Level*

Rise. Available at www.energy.gov/sites/prod/files/2016/07/f33/Climate%20Change%20and%20the%20Electricity%20Sector%20Guide%20for%20Assessing%20Vulnerabilities%20and%20Developing%20Resilience%20Solutions%20to%20Sea%20Level%20Rise%20July%202016.pdf. Accessed July 5, 2021.
US Department of State. 2020. *United States Strategy to Prevent Conflict and Promote Stability*. Available at www.state.gov/wp-content/uploads/2020/12/us-strategy-to-prevent-conflict-and-promote-stability.pdf. Accessed June 25, 2021.
US Embassy in Syria. 2008a. Response: Impact of rising food/commodity prices – Syria. *Wikileaks Public Library of US Diplomacy, Syria Damascus*. Available at https://wikileaks.org/plusd/cables/08DAMASCUS311_a.html. Accessed 8 December 8, 2019.
US Embassy in Syria. 2008b. 2008 UN drought appeal for Syria. *Wikileaks Public Library of US Diplomacy, Syria Damascus*. Available at https://wikileaks.org/plusd/cables/08DAMASCUS847_a.html. Accessed December 8, 2019.
US Embassy in Syria. 2009. UN preps 2009 Syria drought appeal. *Wikileaks Public Library of US Diplomacy, Syria Damascus*. Available at https://wikileaks.org/plusd/cables/09DAMASCUS432_a.html. Accessed December 8, 2019.
US Embassy in Syria. 2010a. Drought update: WFP renews Syria aid appeal. *Wikileaks Public Library of US Diplomacy, Syria Damascus*. Available at https://wikileaks.org/plusd/cables/10DAMASCUS70_a.html. Accessed December 8, 2019.
US Embassy in Syria. 2010b. SARG sheds light on its drought concerns. *Wikileaks Public Library of US Diplomacy, Syria Damascus*. Available at https://wikileaks.org/plusd/cables/10DAMASCUS97_a.html. Accessed December 8, 2019.
USAID. n.d. The Bangladesh Cyclone of 1991. Available at https://pdf.usaid.gov/pdf_docs/Pnadg744.pdf. Accessed June 25, 2021.
USAID. 1999. India – Cyclone fact sheet #1, fiscal year (FY) 2000. *ReliefWeb*. Available at https://reliefweb.int/report/india/india-cyclone-fact-sheet-1-fiscal-year-fy-2000. Accessed September 18, 2019.
USAID. 2014. *Water & Conflict: A Toolkit for Programming*. Available at www.usaid.gov/sites/default/files/documents/1866/WaterConflictToolkit.pdf. Accessed June 25, 2021.
USAID. 2016. India. Available at www.usaid.gov/crisis/india. Accessed September 18, 2019.
USDA. 2008a. Syria: Wheat production in 2008/09 declines owing to season-long drought. Available at https://ipad.fas.usda.gov/highlights/2008/05/Syria_may2008.htm. Accessed December 6, 2019.
USDA. 2008b. Ethiopia 2008 crop assessment travel report. Available at https://ipad.fas.usda.gov/highlights/2008/11/eth_25nov2008/. Accessed June 24, 2021.
USGS and USAID. 2018. CHIRPS seasonal rainfall accumulation – Oct–Dec. Available at https://earlywarning.usgs.gov/fews/product/599. Accessed June 25, 2021.
USGS and USAID. 2019. EWX next generation viewer. Available at https://earlywarning.usgs.gov/fews/ewx/index.html?region=af. Accessed June 25, 2021.

V-Dem. 2019. *Varieties of Democracy V-9 Dataset*. Available at www.v-dem.net/en/news/v-dem-dataset-v9-released/. Accessed July 5, 2021.

V-Dem Project. 2019. *V-Dem Codebook V9*. Available at https://papers.ssrn.com/sol3/papers.cfm?abstract_id=3441060. Accessed June 25, 2021.

Venkataraman, Ayesha, Suhasini Raj, and Maria Abi-Habib. 2018. After worst Kerala floods in a century, India rejects foreign aid. *The New York Times*. Available at www.nytimes.com/2018/08/23/world/asia/india-kerala-floods-aid-united-arab-emirates.html. Accessed September 18, 2019.

Verner, Dorte, Maximillian Shen Ashwill, Jen Christensen et al. 2018. *Droughts and Agriculture in Lebanon: Causes, Consequences, and Risk Management*. The World Bank. Available at http://documents.worldbank.org/curated/en/892381538415122088/Droughts-and-Agriculture-in-Lebanon-Causes-Consequences-and-Risk-Management. Accessed December 4, 2019.

Vivekananda, Janani, Martin Wall, Florence Sylvestre, and Chitra Nagarajan. 2019. *Shoring Up Stability*. adelphi. Available at www.adelphi.de/en/publication/shoring-stability. Accessed July 23, 2020.

von Uexkull, Nina, and Halvard Buhaug. 2021. Security implications of climate change: A decade of scientific progress. *Journal Of Peace Research* 58 (1): 3–17. Available at https://doi.org/10.1177/0022343320984210. Accessed June 24, 2021.

von Uexkull, Nina, Mihai Croicu, Hanne Fjelde, and Halvard Buhaug. 2016. Civil conflict sensitivity to growing-season drought. *Proceedings of the National Academy of Sciences* 113 (44): 12391–12396. Available at https://doi.org/10.1073/pnas.1607542113. Accessed June 24, 2021.

Wagner, Gernot, and Martin L. Weitzman. 2015. *Climate Shock: The Economic Consequences of a Hotter Planet*. Princeton, NJ: Princeton University Press.

Walker, Amanda M., David W. Titley, Michael E. Mann, Raymond G. Najjar, and Sonya K. Miller. 2018. A fiscally based scale for tropical cyclone storm surge. *Weather and Forecasting* 33 (6): 1709–1723. Available at https://doi.org/10.1175/WAF-D-17-0174.1. Accessed June 24, 2021.

Wallace-Wells, David. 2017. The uninhabitable earth. Famine, economic collapse, a sun that cooks us: What climate change could wreak – sooner than you think. *New York Magazine*. Available at https://nymag.com/intelligencer/2017/07/climate-change-earth-too-hot-for-humans.html. Accessed July 5, 2021.

Wallace-Wells, David. 2019. *The Uninhabitable Earth: Life After Warming*. First edition. New York: Tim Duggan Books.

Waltz, Kenneth N. 1979. *Theory of International Politics*. Reading, MA: Addison-Wesley.

Wantchekon, Leonard. 2000. Credible power-sharing agreements: Theory with evidence from South Africa and Lebanon. *Constitutional Political Economy* 11 (4): 339–352. Available at https://doi.org/10.1023/A:1026519406394. Accessed December 25, 2020.

Watts, Nick, Markus Amann, Nigel Arnell et al. 2021. The 2020 report of The Lancet Countdown on health and climate change: Responding to converging crises. *The Lancet* 397 (10269): 129–170.

WBGU. 2007. *Climate Change as a Security Risk: Summary for Policymakers*. Available at www.qualenergia.it/sites/default/files/articolo-doc/Climate%20

Change%20as%20a%20Security%20Risk_mag2007.pdf. Accessed June 25, 2021.
Weather Underground. n.d. The 35 deadliest tropical cyclones in world history. Available at https://www.wunderground.com/hurricane/articles/deadliest-tropical-cyclones. Accessed July 5, 2021.
Webersik, Christian. 2010. *Climate Change and Security: A Gathering Storm of Global Challenges*. Santa Barbara, CA: Praeger.
Webster, Peter J. 2008. Myanmar's deadly daffodil. *Nature Geoscience*: 1–3.
Welsh, Teresa. 2020a. US State Department comes up short of a global fragility strategy. *Devex*. Available at www.devex.com/news/sponsored/us-state-department-comes-up-short-of-a-global-fragility-strategy-98116. Accessed December 8, 2020.
Welsh, Teresa. 2020b. US State Department releases global fragility strategy. *Devex*. Available at www.devex.com/news/sponsored/us-state-department-releases-global-fragility-strategy-98823. Accessed December 22, 2020.
Wendt, Alexander. 1995. Constructing international politics. *International Security* 20 (1): 71–81. Available at www.jstor.org/stable/2539217. Accessed November 15, 2020.
Werrell, Caitlin E., and Francesco Femia, eds. 2013. The Arab Spring and Climate Change. Center for American Progress, Stimson Center, and the Center for Climate and Security, Washington, DC. Available at www.americanprogress.org/issues/security/report/2013/02/28/54579/the-arab-spring-and-climate-change/. Accessed June 25, 2021.
Werrell, Caitlin, and Francesco Femia. 2017. The climate factor in Syrian instability: A conversation worth continuing. The Center for Climate & Security. Available at https://climateandsecurity.org/2017/09/08/the-climate-factor-in-syrian-instability-a-conversation-worth-continuing/. Accessed November 26, 2018.
Werrell, Caitlin E., and Francesco Femia. 2019. *The Responsibility to Prepare and Prevent: A Climate Security Governance Framework for the 21st Century*. The Center for Climate & Security. Available at https://climateandsecurity.org/the-responsibility-to-prepare-and-prevent-a-climate-security-governance-framework-for-the-21st-century/. Accessed December 1, 2020.
Werrell, Caitlin E., Francesco Femia, and Troy Sternberg. 2015. Did we see it coming?: State fragility, climate vulnerability, and the uprisings in Syria and Egypt. *SAIS Review of International Affairs* 35 (1): 29–46. Available at https://doi.org/10.1353/sais.2015.0002. Accessed June 25, 2021.
Wodon, Quentin T., Andrea Liverani, George Joseph, and Nathalie Bougnoux. 2014. *Climate Change and Migration: Evidence from the Middle East and North Africa*. The World Bank. Available at http://documents.worldbank.org/curated/en/748271468278938347/Climate-change-and-migration-evidence-from-the-Middle-East-and-North-Africa. Accessed December 8, 2019.
Woldemariam, Yohannes. 2020. Ethiopia, led by a Nobel peace winner, is looking down the barrel of civil war. *The Guardian*. Available at www.theguardian.com/commentisfree/2020/nov/19/ethiopia-conflict-civil-war-president. Accessed November 20, 2020.
Wolf, Aaron T. 1998. Conflict and cooperation along international waterways. *Water Policy* 1 (2): 251–265.

Woodward, Aylin. 2019. The film "The Day After Tomorrow" foretold a real and troubling trend: The ocean's water-circulation system is weakening. *Business Insider.* Available at www.businessinsider.com/day-after-tomorrow-was-right-and-wrong-about-climate-shifts-2019-3. Accessed July 5, 2021.

World Bank. n.d. *Climate Change Knowledge Portal.* Available at https://climateknowledgeportal.worldbank.org/. Accessed September 11, 2019.

World Bank. 2008. *Agriculture in Syria: Towards the Social Market.* Available at https://openknowledge.worldbank.org/handle/10986/16099. Accessed June 25, 2021.

World Bank. 2010. *Lebanon Agriculture Sector Note: Aligning Public Expenditures with Comparative Advantage.* Available at http://documents.worldbank.org/curated/en/685551468057242124/pdf/687920ESW0P098020120oLebanon0AgoPER.pdf. Accessed June 25, 2021.

World Bank. 2018. *Improving Lead Time for Tropical Cyclone Forecasting: Review of Operational Practices and Implications for Bangladesh.* Available at www.preventionweb.net/publications/view/58663. Accessed September 15, 2019.

World Bank. 2019a. *Agricultural Irrigated Land (% of Total Agricultural Land).* Available at https://data.worldbank.org/indicator/AG.LND.IRIG.AG.ZS. Accessed September 10, 2019.

World Bank. 2019b. *Agriculture, Forestry, and Fishing, Value Added (% of GDP) – Syrian Arab Republic, Lebanon, Jordan.* Available at https://data.worldbank.org/indicator/NV.AGR.TOTL.ZS?locations=SY-LB-JO. Accessed December 4, 2019.

World Bank. 2019c. *Employment in Agriculture (% of Total Employment).* Available at https://data.worldbank.org/indicator/sl.agr.empl.zs. Accessed June 24, 2021.

World Bank. 2019d. *GDP per Capita (Current US$) – Syrian Arab Republic, Jordan, Lebanon.* Available at https://data.worldbank.org/indicator/NY.GDP.PCAP.CD?locations=SY-JO-LB. Accessed December 4, 2019.

World Bank. 2019e. *Net ODA Received (% of Imports of Goods, Services and Primary Income).* Available at https://data.worldbank.org/indicator/DT.ODA.ODAT.MP.ZS. Accessed June 25, 2021.

World Bank. 2019f. *Population, Total – Bangladesh.* Available at https://data.worldbank.org/indicator/SP.POP.TOTL?locations=BD. Accessed September 13, 2019.

World Bank. 2019g. *Worldwide Governance Indicators.* Available at http://info.worldbank.org/governance/wgi/#faq-2. Accessed June 25, 2021.

World Bank. 2020. *Fragility, Conflict, and Violence Job Fair.* Available at http://documents1.worldbank.org/curated/en/449761580483716716/pdf/Fragility-Conflict-and-Violence-Job-Fair.pdf. Accessed June 25, 2021.

World Health Organization and the Secretariat of and the Convention on Biological Diversity. 2020. *Biodiversity and Infectious Diseases: Questions and Answers.* Convention on Biological Diversity. Available at www.cbd.int/health/infectiousdiseases. Accessed January 20, 2021.

World Meteorological Organization. 2015. *Tropical Cyclone Programme Report No. TCP-21.* Available at www.rsmcnewdelhi.imd.gov.in/uploads/report/28/28_fd9ebc_tcp-21-2015.pdf. Accessed June 25, 2021.

Wucherpfennig, Julian, Nils B. Weidmann, Lars-Erik Cederman, Andreas Wimmer, and Luc Girardin. 2011. Politically relevant ethnic groups across space and time: Introducing the GeoEPR dataset. *Conflict Management and Peace Science* 28(5): 423–437. Available at https://doi.org/10.1177/0738894210393217. Accessed July 8, 2021.

Wunsch, James S. 2000. Refounding the African state and local self-governance: The neglected foundation. *The Journal of Modern African Studies* 38 (3): 487–509. Available at www.jstor.org/stable/161708. Accessed June 25, 2021.

Wyns, Arthur. 2020. Climate change and infectious diseases. *Scientific American.* Available at https://blogs.scientificamerican.com/observations/climate-change-and-infectious-diseases/. Accessed January 20, 2021.

Xie, Michael, Neal Jean, Marshall Burke, David Lobell, and Stefano Ermon. 2016. Transfer learning from deep features for remote sensing and poverty mapping. *Proceedings of the Thirtieth AAAI Conference on Artificial Intelligence*: 3929–3935. Available at http://dl.acm.org/citation.cfm?id=3016387.3016457. Accessed March 26, 2017.

Xinhua. 2008. Myanmar assures international community best doing of its cyclone relief work: Statement – Myanmar. *ReliefWeb.* Available at https://reliefweb.int/report/myanmar/myanmar-assures-international-community-best-doing-its-cyclone-relief-work-statement. Accessed July 1, 2021.

Xu, Chi, Timothy A. Kohler, Timothy M. Lenton, Jens-Christian Svenning, and Marten Scheffer. 2020. Future of the human climate niche. *Proceedings of the National Academy of Sciences* 117 (21): 11350–11355. Available at https://doi.org/10.1073/pnas.1910114117. Accessed July 21, 2020.

Yadav, Devendra K., and Akhilesh Barve. 2017. Analysis of socioeconomic vulnerability for cyclone-affected communities in coastal Odisha, India. *International Journal of Disaster Risk Reduction* 22: 387–396. Available at https://doi.org/10.1016/j.ijdrr.2017.02.003. Accessed September 18, 2019.

Yap, D. J. 2013. Days after Yolanda's wrath, looting erupts in Tacloban City. *Inquirer.net.* Available at http://newsinfo.inquirer.net/524731/days-after-yolandas-wrath-looting-erupts-in-tacloban-city. Accessed April 12, 2017.

Yee, Vivian, and Hwaida Saad. 2019. To make sense of Lebanon's protests, follow the garbage. *The New York Times.* Available at www.nytimes.com/2019/12/03/world/middleeast/lebanon-protests-corruption.html. Accessed December 4, 2019.

Yonson, Rio, J. C. Gaillard, and Ilan Noy. 2016. *The Measurement of Disaster Risk: An Example from Tropical Cyclones in the Philippines.* Working Paper Series, School of Economics and Finance, Victoria University of Wellington,. Available at http://econpapers.repec.org/paper/vuwvuwecf/4979.htm. Accessed March 14, 2017.

Zahar, Marie-Joëlle. 2005. Power sharing in Lebanon: Foreign protectors, domestic peace, and democratic failure. In *Sustainable Peace: Power and Democracy after Civil Wars*, edited by Philip G. Roeder and Donald Rothchild, 219–240. First edition. Ithaca, NY: Cornell University Press.

Zarni, Maung, and Trisha Taneja. 2015. Burma's struggle for democracy: A critical appraisal. In *Advocacy in Conflict: Critical Perspectives on Transnational Activism*, edited by Alex de Waal, 45–67. London: Zed Books.

Zhang, David D., C. Y. Jim, George C-S Lin et al. 2006. Climatic change, wars and dynastic cycles in China over the last millennium. *Climatic Change* 76 (3): 459–477. Available at https://doi.org/10.1007/s10584-005-9024-z. Accessed October 15, 2020.

Index

Note: Locators in **bold** indicate figures, tables and maps.

1984 Ethiopian famine, 116
1985 Live Aid concert, 9
1994 Rwandan genocide, 238
2 °C threshold, of climate, 258

Ababsa, Myriam, 166
Acemoglu, Daron, 7, 46, 48
   Why Nations Fail, 48
adelphi, German-based, 232
administrative capacity, 41, 44
   state capacity, 6, 95
   weak, 75
advanced industrialized countries, 14
   climate and security discussions among, 233
Africa
   conflicts in, 29, 30, 223
   disaster risk reduction preparations in, 98
African Development Bank, 109
   projects in Somalia, 106, **107**
Age of Consequences, The (documentary), 123
agroecological zones, 73
Ahmed, Abiy, 103, 118, 238
AidData project, 56
Al Hasakah, 137, 163, 172
Alawis/Alawites, 10, 152–155, 165–166, 169
   favoritism toward, 155
Al-Shabaab militia group, 62, 94, 101, 102, 109–111, 112
   impact on international aid, 116, 241

insurgency, 228
refusal of aid, 268
Somalia invasion by Ethiopia, 268
United States and diversion of aid to, 268
Amin, Idi, 238
Amnesty International, 173
anarchy, 17, 261
Annual Agricultural Production Plan, 148
anocracies, 131, 174, 245, 256
Arab nationalism, 155
Arab Palestinians, 154
Arab Spring, 33, 120, 131, 168, 228, 229, 251
Armed Conflict Location and Event Dataset (ACLED), 75
Armenian Catholics, 154
Arria-formula, 233
Ash, Konstantin, 167, 168
Assad family, 10, 155
   Assad regime, 122, 241
   Assad, Bashar al-, 146, 153, 156
   Assad, Hafez al-, 153, 155, 156, 174–175
   Assad-Makhlouf family clan, 156
Aung San Suu Kyi, 181, 206
Australia, 1
Awami League, 214
Ayeyarwady Delta (Myanmar), 188, 219
   cyclone damage to, 12, 176

backdraft, 62, 230
Bamar, 205
Ban Ki-moon, 213, 255

323

Band Aid, 9
Bangladesh, 183
  Bangladeshi government and cyclone forecasting, 201
  cyclones in, 178–183
  discriminated minority groups in, 206
  foreign aid to, 208
  population, 185
  state capacity of, 197–205
Bangladesh Meteorological Department (BMD), 200
Bank, Andre, 167, 168
Bantu groups, 104, 111–112
  Bantu Somalis, 104
Barre, Siad, 75, 93
Barve, Akhilesh, 217
baselines
  identifying, 253–255
  shifting, 253
  use for identifying normal climatic conditions, 245
Bashir, Omar al-, 238
Bay of Bengal, 180, 182, 225
Beirut explosion, 159, 172
Bekaa Valley/region, 128, 138–140, 150, 158
  movement of farmers in, 171
Belt and Road Initiative, 232
Bengalis, 206
Benjaminsen, Tor, 30
Bennett, Andrew, 126
Bhola cyclone, 197, 204, 207
Biden, Joseph, 259
  administration, 234
  China's pledge to reduce emissions, 260
Brahmaputra River, 180
Bretthauer, Judith, 51, 102, 153
Buhaug, Halvard, 31, 253
bureaucratic capacity, 41
bureaucratic quality, 197
  countries with the worst, 198
  India's, 197
  indicator of, 198
Burke, Marshall, 31
Burundi, 50
Bush, George W., 24, 52
Bussell, Jennifer, 8, 54, 98–100
Butler, Christopher, 30

Cammett, Melani, 157
Camp Fire, California, 227
carbon dioxide
  concentration of, 1
  emissions of, 260
case selection, 72–76
  cyclones, 177
causal chains, 57
causal pathways, 32, 36, 57, 60, 125, 266
  between climate shocks and security outcomes, 67
Cederman, Lars-Erik, 50
Chamoun, Camille, 157
Chhotray, Vasudha, 215, 218
China, 233
  as a rising power, 264
  Belt and Road Initiative, 232
  emissions of carbon dioxide, 260
  sea-level rise, 225
  seasonal flooding in, 227
CHIRPS project, 84
Christians, 153
civil conflicts, 2, 256
  and climate change, 37
  in Africa, 29
  in Darfur, 3, 255
  in Somalia, 33
  indicators of, 59
civil wars. *See* civil conflicts
clean energy systems/transition, 230, 259, 263–264
climate, security link between, 232–234
climate change
  and future security outcomes, 17–18
  and general security outcomes, 37–38
  and security debate, 28–36
  international relations and, 261–264
  negative security outcomes of, 2
  securitization of, 17
climate change literature
  usefulness of, 245–249
climate disasters, 35, 49
climate hazards, 5, 6–9, 23, 25, 46–47, 51, 54, 58–60, 74, 102, 221, 267
  associated risks for, 58
  conflict and, 28, 126
  droughts, 89, *See also* drought
  effects of, 54
  extreme weather events, 58
  human security and, 227, 264
  negative security outcomes of, 4, 37
  risks to displaced populations, 228
  security outcomes, 57–60, 247
  societal effects of, 64–66

Index

state capacity and, 39
swift-onset, 2
climate migrants, 34–35, 123, 227
Climate Research Unit (CRU), 80
Climate Security Expert Network, 232n48, 233n54
climate security literature/scholarship, 246
  focus on, 37, 38, 89, 246
Climate Security Mechanism, 233
climate security/challenges, 16, 175, 223, 233–234, 235, 236, 237, 242
  and state development, 236–242
  United States and, 259
climate–conflict literature, 24, 31
Clinton, Bill, administration of, 26
CNA Corporation, 29
  CNA Military Advisory Board (MAB) report, 232, 247
Cochrane, Logan, 100
Collier, Paul, 50, 126
"Coming Anarchy, The" (Kaplan), 25
communal conflicts, 22, 30, 37, 58, 59
  causal path to, 63–70
Complex Crisis Fund, 241
complex emergencies, 63
conflicts
  climate change and, 2, 3, 58
  environmental scarcity, 47
  internal, 222–223
  measures/indicators of civil war, 59
  rainfall and, 29–32
  types of, 27
coronavirus. See COVID-19
Council on Foreign Relations, 29
countries
  and climate change impact on, 2
  state capacity and political inclusion, 162
COVID-19 crisis, 229, 239, 259, 270
  Ethiopia's elections affected by, 119
cyclones, 190
  case selection of, 177
  Cyclone Bhola, 177, 181–182, 184, 185, 213
  Cyclone Fani, 177, 182, 196
  cyclone forecasting capabilities, 200
  Cyclone Hudhud, 217
  Cyclone Nargis, 57, 62, 178, 181, 184, 188, 199, 200, 208, 209, 212, 213, 220, 225, 268
  Cyclone Phailin, 184, 195, 217

Cyclone Sidr, 13, 182, 191, 198, 202, 204, 209, 215
  in Bangladesh, India, and Myanmar, 15, 179
  in South East Asia, 176–177

Daoudy, Marwa, 166
Darfur, 255
  civil conflicts in, 117, 255
  de Waal and Homer-Dixon over, 124
Darood clan, 104
Day After Tomorrow, The (movie), 252
Dayr-al-Zur, 170
de Châtel, Francesca, 123, 141, 147, 149, 161, 164
De Juan, Alexander, 167, 168
de Waal, Alex, 112, 114, 118, 241
  anti-famine social contract, 214
  on climate change and civil war in Sudan, 121
  debate over the Darfur civil war, 124, 255
  on Ethiopia's drought, 71, 100
  Ethiopia's villagization program, 100
  on fragile states, 237
  record of major food crisis years, 82, 83
  view on Ethiopia exclusion from aid, 113
  views on EPRDF, 115
Dellmuth, Lisa, 235
democracies, 256
  food protests in, 131
Democratic Republic of Congo, 238
demographic and environmental stress (DES), 47
Denmark
  government effectiveness in, 199
  political and social exclusion/inclusion of, 105, 206
  state capacity of, 197
  tax revenues of, 152
  V-Dem dataset and, 155
Department of Agriculture and Co-operation, 217
Derg government, 10, 94, 109, 161
  donors' hostility toward, 106
  drought response of, 100
  Ethiopia after, 103
  government effectiveness under, 95
  international aid, 113, 241
  loss of power, 76
despotic power, 41, 44, 100

Deudney, Dan, 22
Development as Freedom (Sen), 52, 71
development assistance
　government expenses and, 160
　in Myanmar, Bangladesh, and India, 208
Diamond, Jared, 40, 254
disaster events, tracking, 218
Disaster Management Bureau, 201
Disaster Preparedness and Prevention Commission, 99
Disaster Risk Management and Food Security Sector (DRMFSS), 99
disaster risk reduction, 14, 44, 54–55, 198
　and emergency response in India, 203
　in Lebanon, 158
　preparation in Africa, 98
　studies of, 8
　USAID, 217
displaced populations, 160, 165–167, *See also* migration
　climate hazards risk of, 228
District Management Authority, 202
Doctors Without Borders, 116
drought, 58
　agricultural production and, 162
　in Somalia and Ethiopia, 11, 76–78
　in Syria and Lebanon, 12, 129, 132, 137–143
　migration due to, 162–166
　operationalization of, 78–79
　severe, 1
Druze populations, 137, 154
Dukhan, Haian, 166

East Africa, rainfall deviations in, **86, 90**
East Pakistan, 184, 190, 197, 204, 214
　Bengalis in, 206
　Cyclone Bhola, 181
　political exclusion, 210
Ecker, Olivier, 33, 74
El Niño, 19, 73, 89, 118
　drought in Ethiopia, 114
EM-DAT International Disaster Database, 59, 218
environment, the
　militarization of, 17
　securitization of, 17, 22
environmental change, 40
　and public health outbreaks, 39
　and security, 4
　relationship with violent conflict, 21

Environmental Change and Acute Conflicts Project (ECACP), 26
environmental migrants' movements, 34
environmental peacebuilding approaches, 235
environmental scarcity, 26–27, 47
environmental security, 20
　literature, 3, 25–28, 121, 250
　scholarship on, 5, 60
environmental threats, 25, 224
EPR. *See* Ethnic Power Relations (EPR) dataset
EPRDF. *See* Ethiopian People's Revolutionary Democratic Front (EPRDF)
Eritrea
　Ethiopia's relationship with, 94, 118, 238
　rebellion, 94
Ethiopia
　after Derg government, 103
　and Somalia cases compared, 76–78
　and Somalia evidence of international assistance to, 105–109
　Belg rainfall deviations, 82
　de Waal on the 1984 Ethiopian famine, 116
　disaster management institutions in, 99
　drought in, 71
　drought-related deaths, 9, 10
　El Niño impact on, 114
　EPRDF in, 238
　ethnic violence in, 118
　evidence of hazard exposure in, 78–93
　evidence of state capacity of, 94–102
　famine in (1984), 116
　foreign aid to, 76, 106
　governance in, 93–94
　government effectiveness, 95
　improvement of, 112
　Kiremt rainfall deviations, 83
　rainfall and temperature deviations in, 80–93
　Somalia as paired cases, 9
　state capacity of, 10
　weak administrative capacity in, 75
　weather patterns, 78–93
　within-case variation, 4, 72, 74
　World Bank projects in, 109
Ethiopian People's Revolutionary Democratic Front (EPRDF), 10, 76, 94, 99, 109, 115, 119, 238

ethnic cleansing conflicts, fleeing from, 228
ethnic minorities. *See also* marginalized groups
ethnic minorities, in Myanmar, 205
Ethnic Power Relations (EPR) dataset, 50, 51, 102–103, 153, 205, 206
ethnic-based grievances, 50
Euphrates River, 126, 134, 142
Europe
    gridded tree ring dataset of, 137
    immigration pressures in, 230
    Syrian immigrants to, 230
European Union, 14
exclusionary political institutions, 246
exclusive political institutions, 47
    Assad family and, 10
extractive institutions, 48

Failed States Index, 236
famine
    deaths from in Somalia, 71
    in Somalia, 101
    in Somalia and Ethiopia, 76–78
Famine Early Warning Systems Network (FEWSNET), USAID, 97, 99
fatter tail effects, 19
Federal Government of Somalia, 104
Feitelson, Eran, 126, 164
Femia, Francisco, 163–164, 244
Fertile Crescent, 136
Fifth Assessment Report, 20, 246, 252
    Intergovernmental Panel on Climate Change (IPCC), 20
    on human security, 224
finance
    dip in, 109
    emergency appeals for, 159–160
    external barriers to providing, 268
    overseas development assistance (ODA), 56
    Patriot Act and, 110
Financial Tracking Service (FTS), 209
Five Year Plan for 2006–2010, Syria, 148
Food and Agricultural Organization (FAO), 71, 151
foreign aid, 69, *See also* international assistance
    anticipatory, 55
    Bangladesh and, 178
    budgets, 55
    Ethiopia, 10, 76, 106

India, 216, 221
international disaster response, 13
Lebanon, 150
Myanmar and, 176, 268
negative security outcomes, 57
per capita in Myanmar, Bangladesh, and India, **209**
restrictions on, 268
Somalia, 106–109
South Asian countries and, 208
strong capacity state, 68
forensic analysis, and climate security, 244
fragile states, 236, 237
    Fragile States Index, 236
    international bodies on, 242
Fragility, Conflict, and Violence program, 241
Friedman, Tom, *Years of Living Dangerously* (documentary), 123, 169
Fritz, Hermann, 219
Fröhlich, Christine, 123, 130, 142, 147, 162, 169
Fukuyama, Francis, 6, 41, 43

Gallup, John, 254
Ganges River, 180, 226
garbage crisis, 158
Gates, Scott, 30
General Directorate of Wheat and Sugar Beet Subsidy (GDCS), 143
George, Alexander, 126
Germany, 14
    Syrians and resettlement in, 230
Glaser, Sarah, 30
Gleditsch, Nils Petter, 28, 34, 250
Gleick, Peter, 123, 134, 163, 164
    on climate change and uprisings, 123
Global Carbon Project, 260
Global Facility for Disaster Risk Reduction, 101
Global Fragility Act, 241
Global Risk Data Platform, UNEP, 58
global temperatures, 258
    efforts to prevent rising, 259
Goldsmith, Leon, 165
Gore, Al, 225
governance, 41
    corruption and, 199
    governance issues, 38
    in Ethiopia and Somalia, 93–94, **96**

328 Index

governance (cont.)
  World Bank data on, 197
Government Effectiveness, 35, 38, 94–98, 199, *See also* state capacity
  index of, 44
  World Bank's measure of, 44, 198
Grand Ethiopian Renaissance Dam, 234
Greek Orthodox, 154
Green Revolution, 17
greenhouse gas, 1
Gridded Population of the World, 188
gridded tree ring dataset of winter/spring surface moisture, 137
Guernica, 219
Guha-Sapir, Deborati, 210
Gulf Stream, 252
Gulledge, Jay, 250
Guterres, António, 222

Haggard, Stephan, 33, 131, 256
Hanson, Jonathan, 39, 44
Hariri, Rafic, 122, 158
Hawiye clans, 104, 112
hazard exposure
  evidence of in Somalia and Ethiopia, 78–93
  evidence of in Syria and Lebanon, 132–143
Health of Nations, The (Price-Smith), 39
heat wave events, 80, 259
  in India and Pakistan, 226
Hendrix, Cullen, 7, 29, 33, 41, 47, 125, 126, 131, 253, 254, 256
  case selection, 127
Hezbollah, 12, 122, 140, 151, 157–158, 240, 269
  communications network, 172
  Israel's military operations against, 160
Hinnebusch, Raymond, 156, 164, 175
Holocene, the, 250
Homer-Dixon, Thomas, 21, 26–28, 39, 60
  on climate change and civil war in Sudan, 121
  conflict types and, 27
  debate over the Darfur civil war, 124, 255
  elements of the state and, 40
  environment studies, 26
  environmentally-driven scarcity, 26
  work on origins of conflicts, 47
hoop tests, 60, 131, 171
Horn of Africa, 118, 248, 249, 255

drought in, 71
Hossain, Naomi, 204, 214–215
Hsiang, Solomon, 31
human security, 21–22
  and state security relationship, 21–22
  climate change security, 223
  concept of, 20
  concerns in Bangladesh, 228
  outcomes, 57
Humanitarian Coordination Officers, 102
humanitarian emergencies, 57, 242, 266
  resulting from climate change, 2
Huntington, Ellsworth, 254
hurricanes
  Hurricane Harvey, 1
  Hurricane Katrina, 23–24, 52, 183–184
  Hurricane Maria, 1, 24, 52
Hyogo Framework, 45, 98, 198, 201–203

Ide, Tobias, 164
inclusive political institutions, 47, 48
India, 183
  disaster risk reduction, 203
  discriminated minority groups in, 206
  Pakistan relationship, 234
  state capacity of, 197–205
Indian Meteorological Department (IMD), 183, 200, 211, 219
indirect effects, long-run climatic conditions, 253
Indus River Treaty, 234
infectious disease outbreaks, 229
INFORM index, 97, 198
infrastructural power, 6, 41, 95, 117
  service delivery and, 100, 117
instrumental rationality, 40
inter-ethnic conflict, 34
Intergovernmental Panel on Climate Change (IPCC), 28, 224, 258
  2014 report, 21
  Fifth Assessment Report, 20
internal conflicts, 21, 222–223
  climate-related, 223
international assistance, 54–57
  evidence of, 105–109, 207–210
  evidence of in Lebanon and Syria, 159–161
International Best Track Record for Climate Stewardship database, 219
International Country Risk Guide (ICRG), 42, 197

Index

International Decade for Natural Disaster Reduction, 201
International Military Council on Climate and Security, 232n48
international relations, 16, 244, 264, 271
  climate change as a parameter of, 261–264
  rivalry for reducing greenhouse gas, 261
inter-state war, 21
Irrawaddy Delta, 225
Issar, Sukriti, 157
Issofous, story of, 254

Jabhat al-Nusra, 170
Janjaweed militia, 255
Jordan River, 126

Kagame, Paul, 238
Kahl, Colin, 7, 46, 47, 60
Kaplan, Robert
  environmental studies, 26
  "The Coming Anarchy," 25
Keating, Michael, 104
Kelley, Colin, 123, 133, 163, 164
  on climate change and uprisings, 123
  drought in Syria, 136
Kerry, John, 233
Khaddour, Kheder, 155, 170
Kilcullen, David, 166
Kiribati, 23
Kniveton, Dominic, 30
Kostner, Markus, 219
Koubi, Vally, 34
Kurds, 153, 154

La Niña, 19, 73, 118
Lake Chad Basin, 242
Larkin, Emma, 211–213
Lawson, Fred, 165
Lebanon
  disaster risk reduction in, 158
  drought in, 138–141
  evidence of international assistance in, 159–161
  power-sharing agreement, 12, 143, 153, 156–158
  subsidization and, 149–150
  Syria's differences with, 171, 268
  wheat production in, 12, 141
Lemma, Melisew Dejene, 100
Levy, Marc, 4, 30, 121

Linke, Andrew, 36
Live Aid concert, 9, 76
Logic of Political Survival, 53
long-run climatic conditions, indirect effects of, 253
loss of life, 2, 6
  causal path to, 60–63
  complex emergency pathway to, **63**
  cyclone-related, 182
  drought-related in Ethiopia, 9
  in Ethiopia, 10
  in Pakistan, 181
  prevention of in India and Bangladesh, 178
  security outcomes linked to, 2, 3, 9
lower middle-income countries, 198
low-lying island nations, 34, 263
  climate change threats to, 23, 222, 224, 261, 267

Mach, Katharine, 248, 256
Mahoney, James, 126
Majid, Nisar, 109, 111–112
Malthus, Thomas, 26
Mann, Michael, 41
marginalized groups, 7, 48, 104, 111, 205
  in Somalia society, 104
Maronite Christians, 153, 154, 157
Mathews, Jessica, 25
Maxwell, Daniel, 109, 111–112
Maystadt, Jean-Francois, 33, 74, 79, 80
Mazur, Kevin, 155, 168, 169, 170
Meghna River, 180
Meierding, Emily, 32
Menkhaus, Ken, 101, 106, 116, 240
Middle East, 1, 230, 246, 267
  gridded tree ring dataset, 137
  protest movement in, 120–121
  rising temperatures in, 226
migration
  drought, 162–166
  migrants and climate hazards, 228
  reverse, 163
  through conflicts, **67**
Miguel, Edward, 29, 31
minority groups. *See* marginalized groups
Mirante, Edith, 211–213, 215, 219
Mitchell, Tom, 204, 215
mitigation, 257–261
Moran, Ashley, 44, 53
multi-ethnic states, 174

Museveni, Yoweri, 238
Myanmar, 183
  cyclones in, 178–183
  discriminated minority groups in, 205
  regime's refusal of aid, 176
  state capacity of, 197–205

National Disaster Management Council, 201
National Disaster Management Institute (NDMI), 202
National Intelligence Assessment for National Intelligence Council, 223n4
National League for Democracy, 206
National Meteorological Agency, 99
national security
  discussed by Kaplan, 26
  importance of, 22
  meaning of, 20
National Security Council, 233
national security threats, 5, 23
natural hazards, 16, 59
  Ethiopia, 100
  populations and, 2
negative security outcomes, 2, 20, 54, 57, 59–60, 69, 73, 221, 247
  and conflicts, 57
  causal pathways to, 32
  causes of, 266, 270
  climate change and, 5, 164
  climate hazards and, 4, 37
  climate shocks and, 245
  direct path to, 60
  extreme weather events and, 249
  governance issues and, 38
  historic evidence of, 3
  humanitarian emergencies as, 242
  prevention of, 239, 244
  reducing, 56
  strong capacity states, 68–70
Niazi, Amir Abdullah Khan, 182
Nordhaus, William, 258
Normalized Difference Vegetation Index (NVDI), 255n43
North Africa
  civil conflicts in, 28
  gridded tree ring dataset of, 137
  migration problems to Europe, 230
  protest movement in, 120–121
North, Douglass, 7, 46, 101
  climate shocks and, 49
  limited access orders and, 49

Obama administration, 161
Obote, Milton, 238
Obradovich, Nick, 167, 168
Odisha, 217
Odisha Relief Code, 216
OECD Development Assistance Committee, 56
Olson, Mancur, 237
open-access orders, 50
Operation Sea Angel, 182, 215
Oromo, 118
Ottoman Empire, 153
overseas development assistance (ODA), 9, 55, 56, 208–210

paired cases, 39, 73
  benefit of the regions chosen, 267
  choosing, 59
  cyclones-hit countries, 12
  different causes of negative security outcomes, 270
  droughts, 10
  Ethiopia and Somalia, 9, 76–78, 77
  exploration of, 3
  good choices for, 73
  lack of in research, 27
  learning from, 266–268
  set of, 10
  Syria and Lebanon, 10, 127, 129
  used for identification of causal claims, 126
Pakistan, 190
  cyclone-related loss of life, 181
  India relationship, 234
Pal, Indrajit, 217
Palmer Drought Severity Index, 58, 78
Paris Agreement, 258
  Trump administration withdrawal from, 260
Paris, Roland, 21
Pathways for Peace (UN-WB), 239
patrimonial marketplace, 237
Patriot Act, 110–111, 268
patronage, 237
Peace Research Institute of Oslo (PRIO), 30
peacebuilding approaches, 246
  and peacekeeping missions, UN, 233
  in weak states, 241

peacebuilding resolution of the, United
  Nations and World Bank, 240
 projects, 241
 United States, 242
People's Revolutionary Democratic Front.
  See Ethiopian People's
  Revolutionary Democratic Front
  (EPRDF)
Planetary Security Initiative (PSI), 232n48
political exclusion, 31, 105
 and agriculturally dependent countries, 33
 and conflict acceleration, 50–51
 and conflicts, 38
 and state vulnerability, 61
 and violence, 75
 East Pakistan, 210
 violence and, 262
political inclusion, 7, 46–54
 evidence of, 102–105, 152–159
 evidence of in India, Pakistan and Myanmar, 205–207
politically exclusive regimes, 153
power transitions, 260
power, types of, 41
power-sharing agreement
 Lebanon, 12, 143, 153, 156–158
 Somalia, 104
Prevention and Stabilization Fund, 241
Price-Smith, Andrew, 38, 39, 40–41
 Health of Nations, The, 39
Productive Safety Net Programme (PSNP), 99, 114
Prosperity Party, 119
protest movement
 in Lebanon, 173
protest movements, 128, 172, 175
 Arab Spring, 228
 in the Middle East and North Africa, 120–121
Protestants, 154
PRS Group, 42, 95
Puerto Rico, 225
Puntland, 101

Quiroz Flores, Alejandro, 36

Rahanweyn clan, 104–105, 111
rainfall, 249
 changes in, 19
 correlation to conflict, 29–32

Ethiopia's, 78–93, **91, 92, 93**
 patterns, 58
 scarcity of, 29, 79
 Somalia's, **87, 88**
 Syria's, **135**
 and temperature deviations in Somalia and Ethiopia, 80–93
Raleigh, Clionadh, 30, 34
Randall, Doug, 252
realism, structural theory of, 261
Red Cross-Red Crescent, 203
REDD+ forest conservation initiatives, 230
regional specialized meteorological center (RSMC), 200
Reilly, Benjamin, 214
Relief and Rehabilitation Commission, 99
Representative Concentration Pathway (RCP) 8.5, 252
Reuveny, Rafael, 34
reverse migration, 163
Robinson, James, 7, 46, 48
 Why Nations Fail, 48
Rohingya Muslims, 206, 228
Rosenblatt, Nate, 166
roving bandits, 237
Russia, 233
Rwanda, 26, 50, 238
 benevolent dictatorship government of, 239
 genocide in, 238

Sachs, Jeffrey, 254
Saffir-Simpson scale, 183–184, **186**
Salehyan, Idean, 34
Save the Children, 99
scarcity
 environmentally-driven, 26–27, 47
 types of, 26
 water, 29, 79, 128, 262
Schwartz, Peter, 252
security
 effect of climate disasters on, 35
 human security
  and the relationship with state security, 21–22
 meaning of, 20
Security Council, 233
security outcomes, 28, 37–38, 60, 67, 256, 267
 climate hazards, 57–60
 future climate change and, 17–18

security threats, 1, 5, 23, 57, 224, 228, 245, 257, 264
  climate change and, 2
  damage from other sources of, 22
Selassie, Haile, 94, 112
Selby, Jan, 118, 123, 126, 127, 163–164, 170
  on drought in Syria, 124
  protests in Syria, 168
  uses of causal analysis, 124
selectorate theory, 49, 53, 237
Sen, Amartya, 8
  Development as Freedom, 52, 71
Shia Muslims, 154
Shias, 154
shifting baselines, 253
Shiite Muslim Hezbollah groups. *See also* Hezbollah
short-term shocks, 249, 254
Sigman, Rachel, 44
single cases, risks of using, 266
Slettebak, Rune, 35
Smith, Alastair, 36
smoking gun test, 60, 130, 171
social exclusion
  in Ethiopia, Somalia, and Denmark, 105
  in Myanmar, Bangladesh, India, and Denmark, 207
social groups, measure of, 105
social market economy, 156
Somalia
  Dayr average temperature change, 84
  Dayr rainfall deviations, 81
  deaths from famine in, 76
  Ethiopia cases compared with, 9, 76–78
  Ethiopia evidence of international assistance to, 105–109
  Ethiopia explanations in outcomes in, 115–117
  evidence of hazard exposure in, 78–93
  famine-related deaths in, 71
  foreign aid to, 106–109, 108
  governance in, 93–94
  Gu average temperature change, 83
  Gu rainfall deviations, 81
  humanitarian aid to, 109–115
  overseas development assistance to, 106
  rainfall and temperature deviations in, 80–93
  security outcomes in, 39
  United States relations and, 105

World Bank and African Development Bank projects in, 106
Somalia Disaster Management Agency (SODMA), 101
Somalia Red Crescent Society, 110
Somaliland, 101
South Africa, 26
South/Southeast Asia, 1, 12, 267
  cyclones in, 176–177
  foreign aid to, 208
  rising temperatures in, 226
Standardized Precipitation Index, 58, 79
state capacity, 6–17, 246
  ability to respond to climate changes, 38–46
  Bangladesh, 178
  definition of, 38
  dimensions of, 44
  India, 178, 197–205
  low capacity countries, 198
  Myanmar, 178, 197–205
  Pakistan, 197–205
  service delivery, 38, 105
  strong, **69**
  weak, 68
state development, 223, 242, 269
  climate security, 236–242
state-centric focus, 17, 40, 151, 240, 269–270
stationarity, end of, 245, 249–253
stationary bandit, autocratic states as, 237
Steinberg, David, 111, 205
Stockholm International Peace Research Institute (SIPRI), 232
Storm Warning Center, 200
straw-in-the-wind test, 131
streetlight effect, 246
strong capacity states, 68
structural theory of realism, 261
sub-Saharan Africa, 8, 54, 267
  climate security scholarship and, 246
  effect of malaria on economic growth, 254
subsidy program, 147–149
Sudan, 29, 233, 234, 238
  Al-Shabaab militia, 102
  civil conflicts in, 3, 28, 255
  Sudanese government, support of the Janjaweed militia, 255
Sunnis, 153, 154, 165, 166, 169–170, 173, 175

Assad government support to, 148
  drought-affected, 137
  favoritism toward, 155
  Sunni Arabs, 153, 167–168
  Sunni Muslim Future Movement, 157
Swe, Chit, 211
Syria
  Arab Spring, 120
  average temperature in, 135
  civil conflicts in, 28
  drought in, 137–138
  drought-related migration from, 162–166
  evidence of international assistance in, 159–161
  exclusive political institutions in, 10
  Five Year Plan for 2006–2010, 148
  Lebanon commonality between, 174–175
  Lebanon's differences with, 171–174
  wheat production in, 141
  withdrawal from Lebanon, 268
Syrian civil war, 3, 35, 117–118, 121, 244
  climate and, 122, 125
  effect of drought on, 173
  humanitarian crisis of, 230
  migration and onset of, 123
  onset of, 128

Taif Agreement, 157
Task Team on Tropical Cyclones, World Meteorological Organization's, 180
Tasnim, Khandker Masuma, 204, 218
tax collection data, 43, 152
temperature increases, 1, 31, 258
TFG. *See* Transitional Federal Government (TFG)
Theisen, Ole Magnus, 30
threat multiplier, 29, 264
  climate as a, 247
Three Gorges Dam, 227
Tigrayan insurrection, 238
Tigrayans, 94
Tilly, Charles, 237
Transitional Federal Government (TFG), 101, 104
  Somalia's, 106
Transparency International, 198
Tripartite Core Group, 213
tropical cyclones. *See* cyclones
Trump administration, 228, 232
  climate security community, 234
  Donald Trump and Hurricane Maria, 24

Hurricane Katrina and, 52
  Paris Agreement withdrawal under, 260
Tubi, Amit, 126, 164
Tuvalu, 222
Typhoon Yolanda, 24

United States Agency for International Development (USAID). *See* USAID
Uganda, 238
Ullman, Richard, 21
Uninhabitable Earth, The (Wallace-Wells), 226
United Kingdom (UK), 14
United Nations, 106, 239
  2 °C threshold recognition by, 258
  Department of Political and Peacebuilding Affairs, 233
  insights on state fragility, 242
  UN Framework Convention on Climate Change, 258
  UN International Strategy for Disaster Reduction, 142
  UN mandate system, 153
  UN Office for the Coordination of Humanitarian Affairs (OCHA), 56–57, 159, 208
  UN Security Council, 14
  United Nations Development Programme (UNDP), 233
  United Nations Environmental Programme (UNEP), 233
  United Nations Security Council, 29, 223
  World Bank peacebuilding resolutions, 239
United States, 231
  emissions of carbon dioxide, 260
  government, insights on state fragility, 242
  important countries to USA in conflict, 14
  national security assets, 231
  security consequences of climate change, 14
  US Department of Defense, 242, 252
  US State Department, 242
  US Supreme Court, effect of ruling on aid to Somalia, 110
Uppsala Conflict Data Program, 59
USAID, 53, 102, 114, 242
  and aid to Ethiopia, 113
  asking for aid for Syria, 149
  description of drought in Ethiopia, 90–93

USAID (cont.)
    disaster risk reduction, 217
    drought preparedness initiative, 151
    India and, 182
    Office of Foreign Disaster Assistance, 57
    on Myanmar's disaster preparedness, 203
    support for CHIRPS project, 84
    USAID Famine Early Warning Systems Network (FEWSNET), 97, 99

Vanuatu, 23
V-Dem dataset, 43, 51, 105, 154–155, 206
villagization, 100
Vogt, Florian, 210
von Uexkull, Nina, 32, 36, 50, 118, 247, 253

Waldman, Matt, 104
Wall Street of Development, The, 215
Wallace-Wells, David, 19, 226
Waltz, Kenneth, 261
water mismanagement, 12, 121, 123
water scarcity, 29, 79, 128, 262
Werrell, Caitlin, 163–164, 244
West Pakistan, 214
West, Kanye, 52
wheat production, 12
    effect of drought on in Lebanon and Syria, 137–143, 171
Why Nations Fail (Acemoglu and Robinson), 48
Wikileaks, 142, 149, 163
Wilson Center, The, 229

Wine, Bobi, 238
within-case variation, 73, 182, 267
World Bank, 135, 185, 241
    assessment of subsidy program, 147–149
    insights on state fragility, 242
    measure of Government Effectiveness, 44, 198
    Myanmar 2008 cyclone, 219
    on famine in Somalia, 111
    Pathways for Peace, 239
    projects, 106, 107
    projects in Ethiopia, 109
    projects in Somalia, 106
    Worldwide Governance Indicators, 42–43, 45, 94
World Bank Climate Knowledge Portal, 80
World Food Programme (WFP), 99
    in Burma, 212
    in Syria, 161
    Somalia and, 109
World Meteorological Organization, 200
    Task Team on Tropical Cyclones, 180
World War I, 153
Worldwide Governance Data. *See* World Bank, Worldwide Governance Indicators

Yadav, Devendra, 217
Years of Living Dangerously (Friedman) (documentary), 123, 169

Zelawi, Menes, 238